The MATLAB® 5 Handbook

Springer
New York
Berlin
Heidelberg
Barcelona
Budapest
Hong Kong
London
Milan
Paris
Singapore
Tokyo

Darren Redfern
Colin Campbell

The MATLAB® 5 Handbook

Springer

Darren Redfern
Practical Approach Corporation
151 Frobisher Drive, Suite C-210
Waterloo, ON N2V 2C9
Canada

Colin Campbell
Information Systems and Technology
University of Waterloo
Waterloo, ON N2L 3G1
Canada

Library of Congress Cataloging-in-Publication Data
Redfern, Darren.
   The MATLAB 5 Handbook / [Darren Redfern, Colin Campbell].
     p.   cm.
   Includes bibliographical references (p. –) and index.
   ISBN 0-387-94200-9 (softcover : alk. paper)
   1. MATLAB.   2. Numerical analysis—Data processing.   I. Campbell,
Colin, 1959– . II. Title.
   QA297.R395   1997
   519.4´0285´53–dc20   96-10769

Printed on acid-free paper.

MATLAB is a registered trademark of the Mathworks, Inc.

©1998 Springer-Verlag New York, Inc.
All rights reserved. This work may not be translated or copied in whole or in part without the written permission of the publisher (Springer-Verlag New York, Inc., 175 Fifth Avenue, New York, NY 10010, USA), except for brief excerpts in connection with reviews or scholarly analysis. Use in connection with any form of information storage and retrieval, electronic adaptation, computer software, or by similar or dissimilar methodology now known or hereafter developed is forbidden.
The use of general descriptive names, trade names, trademarks, etc., in this publication, even if the former are not especially identified, is not to be taken as a sign that such names, as understood by the Trade Marks and Merchandise Marks Act, may accordingly be used freely by anyone.

Production managed by Anthony K. Guardiola; manufacturing supervised by Joe Quatela.
Photocomposed copy prepared from the authors' LaTeX files.
Printed and bound by Hamilton Printing Co., Rensselaer, NY.
Printed in the United States of America.

9 8 7 6 5 4 3 2 1

ISBN 0-387-94200-9 Springer-Verlag New York Berlin Heidelberg   SPIN 10424248

Contents

List of Tables	vii
List of Figures	viii
Introduction	1
MATLAB Quick Start	5
Linear Equations	21
Sparse Matrices	42
Non-Linear Equations	76
Optimization	85
Integration and Differentiation	100
Ordinary Differential Equations	105
Mathematical Functions and Operators	122
Symbolic Computations	149
Graphing Points and Curves	163
Graphing Surfaces and Volumes	197
Animation	233
Graphics Properties	246
Graphical User-Interface Functions	328

Programming in MATLAB	368
File Input/Output	408
Debugging MATLAB **Programs**	426
Advanced Data Structures	437
Object-Oriented Programming	446
Miscellaneous	458
MATLAB **Resources**	476
Index	479

List of Tables

1	Properties of Graphic Elements in Example	248
2	Properties of Graphic Elements in Example, *continued*	249
3	Properties of Graphic Elements in Example, *continued*	250
4	Properties of Graphic Elements in Example, *continued*	251
5	Properties of Graphic Elements in Example, *continued*	252
6	Properties of Graphic Elements in Example, *continued*	253
7	Properties of Graphic Elements in Example, *continued*	254
8	Properties of Graphic Elements in Example, *continued*	255
9	Alternate Commands for Changing Property Values	257
10	Default units for objects	263
11	'PaperType's and their sizes (in Portrait orientation)	283

List of Figures

1	Plot of quadratic	13
2	Plot of **para.dat** - column 2 vs. column 1	16
3	Plot of **para.dat** with best fit quadratic	19
4	Plot of baseball supply and demand model	22
5	Plot of house rotated by $60°$ four times	27
6	Comparison of CPU time to solve $Ax = b$ (for tridiagonal A) in full versus sparse storage modes, for various sizes of n.	44
7	Plot of a simple polynomial	77
8	Plot of lacrosse ball supply and demand model	78
9	Plot of lacrosse ball surplus	79
10	Surface plots of spring-mass-damper system	81
11	Contour plot of spring-mass-damper system	82
12	Plot of a simple polynomial	86
13	Plot of a paraboloid	87
14	Plot of a paraboloid, with one constraint	89
15	Plot of lacrosse ball Transportation	90
16	Solution to ODE computed by **ode23**	108
17	Solution to ODE computed by **ode45**	109
18	Spring-mass-damper system	112
19	Analytic solution to 2nd-order spring-mass-damper problem	113
20	Numeric solution to 2nd-order spring-mass-damper problem	115
21	Sample starting points for BVP form of spring-mass-damper problem	116
22	Numeric solution to BVP form of spring-mass-damper problem	118
23	Plot of $\sin(x)$	164
24	Plot of $\sin(x)$ and $\cos(x)$	165
25	Plot of $\sin(x)$ and $\cos(x)$, with axis labels, title, legend and grid lines	166

26	Plot of $\sin(x)$ with dashed line between points, and $\cos(x)$ with circle marker at points	167
27	Available line styles and markers	167
28	Plot of $\cos(x)$ and $\sin(x)$ with both markers *and* line styles.	168
29	Plot of $\sin(x)$ and $\cos(x)$, varying line width and marker size	169
30	Sample plot using 2 shades of gray	171
31	Four typical viewpoints	172
32	Plotting a function using fplot	173
33	Four combinations of linear and logarithmic axes	174
34	3-D plot with logarithmic z-axis	175
35	Polar plot of $r = \theta$ and $r = 5\cos(\theta) + 5$	176
36	Polar plot using regular linear axes	178
37	Spherical plot produced using plot3 on cartesian axes	179
38	Sample data and histogram	181
39	Histogram with bins of **non-uniform** width	183
40	"Rose plot" - polar histogram	184
41	Using fill to shade a histogram	185
42	Plotting complex matrix elements using compass, feather, and quiver	186
43	Plotting a complex 3×2 matrix using quiver	188
44	Plotting points on surface $z(x,y)$ using plot3	199
45	Plotting mesh surface through $z(x,y)$ using mesh	199
46	Plotting contours below mesh surfaces using meshc, and plotting vertical lines around surfaces using meshz.	201
47	Omitting mesh lines in one direction using waterfall	201
48	Plotting matrices using 'YDir', 'normal/reverse'	202
49	Visualizing the complex numbers in the matrix we wish to pass to mesh	204
50	Mesh plot of matrix after "zeroing-out" complex elements	205
51	Mesh plot of matrix after replacing complex elements with NaN	206
52	Mesh plot of matrix defined over *polar* grid	207
53	Plot of (x, y, z) points, where (x, y) points are irregularly distributed	208
54	Mesh plot of interpolated (XI, YI, ZI) points on grid, above original irregularly distributed (x, y) points.	210
55	Surface plots of (x, y, z) points defined on: (1) a coarse grid, and (2) a finer grid.	211

56	Surface plots of (x, y, z) points, varying the shading to be proportional to: (1) X + Y, and (2) abs(Z).	211
57	Plotting contours below shaded surface using surfc	212
58	Plotting contours through surface defined by X, Y, and Z	215
59	Contours labels	216
60	Contours lines at specific z-values (labelled)	217
61	Contours lines at specific z-values, plotted in 3-D (unlabelled)	217
62	Plotting surfaces using pcolor	218
63	Plotting surfaces using image	219
64	Plotting volumetric slices using slice	220
65	Mesh and surface plots from triangularized data (top) and interpolated data (bottom).	222
66	Voronoi diagram	223
67	Hierarchy of graphics objects, using quads.m as an example. (Same as Figure 73.)	247
68	Light gray axes area with medium-gray figure in the background	258
69	Positioning figure in the middle of the screen using pixels and normalized units	265
70	First figure with three sets of axes created using the subplot command	267
71	Second figure with one set of axes created (and positioned) using the axes command	267
72	Four plots in one figure. axes was called directly to specify the position, rather than using subplot.	270
73	Hierarchy of graphics objects, using quads.m as an example. (Handles in parentheses are rounded to one decimal place.)	271
74	Plot customized using only non-set commands	273
75	Plot customized using *only* set commands	275
76	Four most popular fonts (Courier, Times, Helvetica, Symbol)	277
77	Plot showing results of changing 'LineStyle' of individual *negative* contour lines to "dashed"	279
78	Default 'PaperPosition' is an 8 × 6 inch print area centered on page	281
79	The value of a figure's 'PaperSize' property is influenced by changes to one or more of its 'PaperUnits', 'PaperOrientation' and 'PaperType' properties	282
80	The value of a figure's 'PaperPosition' property is *not* affected by changing between portrait and landscape mode	283
81	Hierarchy of graphics objects, using quads.m as an example. (Same as Figure 73.)	288

82	Fahrenheit-to-Celsius conversion	329
83	Fahrenheit-to-Celsius conversion, using a "slider"	332
84	Fahrenheit-to-various conversion, using a "popup" menu	334
85	Fahrenheit-to-various conversion, using three radiobuttons	336
86	Fahrenheit-to-various conversion, using three "check" boxes	337
87	Fahrenheit-to-various conversion, using three pushbuttons	339
88	Question "dialog box" for quitting conversion program, created using questdlg	341
89	Menu added along top of figure window	342
90	Entries in `File` submenu	343
91	Dialog box displayed by uigetfile	344
92	Entries in "Format" submenu	346
93	Entries in "Digits" subsubmenu	347
94	Context plot	349
95	Guide Control Panel	352
96	Guide Property Editor	353
97	Guide Menu Editor	355
98	Planetary motion of 0.5 Earth years	369
99	Planetary motion of 0.5 Earth years	369
100	Tree diagram showing scripts and functions used by `solar.m`	371
101	Effect of increasing eccentricity, *e*, on the shape of an ellipse	373
102	Relationship between eccentricity (e) and minor axis length (b)	374
103	MATLAB *Editor/Debugger* window.	433

Introduction

How to Use This Handbook

The MATLAB Handbook is a complete reference tool for the MATLAB computation language, and is written for all MATLAB users, regardless of their discipline or field(s) of interest. All the built-in mathematical, graphic, and system-based commands available in MATLAB 5[1] are detailed herein.

Overall Organization

One of the main premises of The MATLAB Handbook is that most MATLAB users approach the system to solve a particular problem (or set of problems) in a specific subject area. Therefore, all commands are organized in logical subsets that reflect these different categories (e.g., linear equations, ordinary differential equations, surface and volume graphics, etc.) and the commands within a subset are explained in a similar language, creating a tool that allows you quick and confident access to the information necessary to complete the problem you have brought to the system.

In addition, because there is much information about MATLAB that is very difficult to express on a purely command-by-command basis, each subject is prefaced with an introductory section. Here, detailed examples are given, profiling some of the most common applications of the commands in that particular section. There is also an introductory session, titled MATLAB Quick Start, which is intended to get those readers not already familiar with MATLAB off to a flying start.

Cross Referencing

One of the most important goals of The MATLAB Handbook is to provide pointers to appropriate information so that you are able to solve your problems quickly and efficiently. The MATLAB Handbook is rife with valuable references, presented so as to be accessible yet not clutter the information to which they are attached.

There are two types of references in The MATLAB Handbook. First, each command listing has a *See also* section that points you to commands, within that section or

---

[1] Actually, we have updated this handbook for MATLAB 5.1. If you are using a later version of MATLAB, please refer to the *Release Notes* for further information.

elsewhere, that contain related information. Second, if you know the name of a command but are unsure to which subject area it belongs, there is a complete alphabetical index of all MATLAB commands at the end of the handbook.

Individual Command Entries

While the information contained in the entries for each individual command is unique, the format in which the information is presented is identical across all entries. The following "dummy" example illustrates the various elements that are used throughout this book.

M_{out} = **acommand**(M, num)
Performs a command on matrix M at level num.
Output: If M contains only real values, a single real value is returned. Otherwise, if M contains *any* complex values, a 2-element vector is returned.
Argument options: (M) to limit the computation to the first level. ♦ (V) to perform the computation on vector V. Only first level computation is available for vectors.
Additional information: If performing the operation acommand at level num is mathematically inconsistent, an error message is returned.
See also: bcommand, *anothercommand*

The *command call*, **acommand**(M, num), gives the command name as well as its most common type of parameter sequence. The command name itself (by which the command entries are alphabetically sorted) is in a special typeface (e.g., **acommand**). The command call's parameter sequence (as well as the parameter sequences found in other elements of a command entry) normally represents placeholders for the actual input you use when calling the command. For example, a placeholder of num could be replaced with the numeric values 2, 75.4, or −0.11. When a parameter sequence contains an element that appears in italics, for example *nobalance*, it means that the word *nobalance* is to be entered as is in the command, not replaced with some other value. Such elements most frequently occur with predefined options and input values.

Whenever possible, the expected data type of a parameter is specified with one of the following abbreviations:

angle	an angle in radians
boolean	*true* or *false*
complex	a complex value or expression
expr	an expression
exprseq	an expression sequence
filename	a file name
fnc	a function
ineq	an inequation
int	an integer
n, m, i, j, posint	a positive integer
name	a name to be assigned a value by the command
num, a, b	a numeric value

M	a two-dimensional matrix
option	one of a set of predefined options
subexpr	a subexpression
V	a one-dimensional matrix, vector
var	an unassigned variable

The above abbreviations deal with parameter types that are encountered across all disciplines covered by *The* MATLAB *Handbook*. There are also dozens of other data types that are specific to individual areas; these are detailed in the introductions to each chapter. [Note: when a sequence of parameters is represented with ..., for example, $M_1, ..., M_n$, do not confuse this with MATLAB's line continuation characters ...—one is a short-hand representation for n values, while the other is a language structure.]

Following the command call is a short description of how the command works on the given parameter sequence. This is meant to give you enough information to use the command in most instances. If more information is needed, the *Argument options* and *Additional information* sections should be read.

The *Output* listing gives some idea of what type of output (i.e., what data types) to expect from the most common calling structure. This is extremely helpful when you either want to dissect the answer for further use or correctly include the command within another command or a procedure. This information can also be found, in part, in the command call line to the left of the = sign.

The *Argument options* listing provides valid variations to the parameter sequence and brief explanations of their functioning. If any alternate parameter sequence is of paramount importance, there is an individual command listing to discuss it. ♣ characters appear in this section to separate multiple parameter sequences.

The *Additional information* section lists just that—additional information about the command. This could include, among other things, special pointers to other command entries, brief descriptions of algorithms used to compute the command, or warnings about dangerous combinations of parameters. ♣ characters appear in this section to distinguish separate items.

The *See also* section gives pointers to other command listings within that section (in normal typeface) or in other sections (in *italic* typeface) which contain related information or work in conjunction with the initial command. When searching for a command in another section, it is best to consult the index at the back of this book for an exact page location.

MATLAB's On-line Help System

All versions of MATLAB come with an on-line help facility containing many pages of brief descriptions. *The* MATLAB *Handbook* adds to this facility.

To view the on-line help for any MATLAB command, simply enter help name, where name is the command for which you want information. For example, the command help plot displays the on-line help page for the plot command. Another useful tool is the lookfor command, which scans *all* the on-line help files for the occurrence of a given keyword.

Another useful tool is the MATLAB Help Desk, which can be accessed by entering the helpdesk command. This opens a browser that contains HTML and PDF versions of much of the printed documentation.

Where to Go for More Information

The range of MATLAB books, courseware, and third-party applications is constantly growing. Apart from the standard MATLAB product documentation (available from The MathWorks) there are books on using MATLAB in subject areas from beginning engineering to control theory (available from various publishers), and there are many more volumes currently being written by authors in the academic and commercial fields.

In addition, there have been many scholarly papers, reports, and theses written around the MATLAB system. For more information on these or other MATLAB materials, contact the vendor who sold you MATLAB.

Many other sources of information are available on the World Wide Web, including the official company web site at **http://www.mathworks.com** and our own MATLAB web site at **http://www.pracapp.com/matlab/**, which contains many of the coding examples used in this handbook as well as any updates to the basic material.

In the meantime, there is no teacher like practical use for learning the intricacies of MATLAB. Take your copy of *The* MATLAB *Handbook* and go at it!

Acknowledgements

We would like to thank all the people at The MathWorks, the University of Waterloo, and Springer-Verlag (New York) for their consistent support. Thanks go out to those who helped directly with the creation of this book, including Bruce Barber, David Doherty, Peter Trogos, Dave Wakstein, Jim Tung, Naomi Bullock, Cristina Palumbo, and Cleve Moler.

MATLAB Quick Start

This chapter covers all you need to get started *right away* with MATLAB; subsequent chapters assume that you have read this introductory chapter.

- What is MATLAB?
- Entering and exiting MATLAB
- Interactive use of MATLAB
- Non-interactive use of MATLAB
- Prompting for values of variables
- Creating your own functions
- Using existing MATLAB functions
- Plotting functions
- Reading in data files
- Working with matrices
- "Free" format output using save

There is more detailed information on these topics in the remaining chapters of this handbook.

What is MATLAB?

In simplest terms, MATLAB is a computer environment for performing calculations.

MATLAB is a contraction of "Matrix Laboratory," which hints at the fact that originally MATLAB was designed as a more convenient tool (than C or FORTRAN) for the manipulation of matrices. It has since added more functionality, and it still remains a better tool in general for scientific computation.

Numerical computations and graphics can be done with MATLAB. MATLAB is a procedural language, combining an efficient programming structure with a bevy of predefined mathematical commands. While simple problems can be solved interactively with MATLAB, its real power shows when given calculations that are extremely cumbersome or tediously repetitive to do by hand.

Entering and Exiting MATLAB

The method of invoking the MATLAB program varies from system to system, but is generally intuitive. On Unix systems, type: matlab at the operating system prompt. On Microsoft Windows and Macintosh systems, double-click on the MATLAB icon. Once in MATLAB, something like the following should appear on your screen:

```
            < M A T L A B (R) >
    (c) Copyright 1984-97 The MathWorks, Inc.
              All Rights Reserved
                  Version 5.1

  To get started, type one of these commands: helpwin, helpdesk, or demo

>>
```

The last thing to appear on the screen is the MATLAB prompt. This is where you enter commands, e.g., quit to exit MATLAB.

Interactive Use of MATLAB

MATLAB commands can be issued interactively or run from command files (known as *scripts*). Typically, you can use a combination of both methods. Let's do some simple calculations interactively. For example, consider the quadratic equation:

$$ax^2 + bx + c = 0.$$

Calculate the two roots of this quadratic equation using the well-known formula

$$x = \frac{-b \pm \sqrt{b^2 - 4ac}}{2a}.$$

Because MATLAB is a *numeric* computation program, you must define particular values of the parameters a, b, and c before you can use them in another function. The equals sign (=) is used to denote assignment of a value to a variable.

Type the following, pressing **Enter** or **Return** at the end of each line.

```
a = 1
b = 0
c = -2
```

As you can see, there is no need to end each line with a standard *terminator* (e.g., semicolon (;) or colon (:)), such as in other languages. MATLAB knows a command has come to an end when it reaches a carriage return. The semicolon terminator, which we will get into shortly, does have meaning in MATLAB, but it is not necessary in most cases.

Instruct MATLAB to calculate *one* root.

```
x1 = (-b + sqrt(b^2 - 4*a*c) ) / (2*a)
```

MATLAB displays the result:

```
x1 =
    1.414
```

The basic mathematical operators for creating mathematical expressions are:

+	addition
-	subtraction (negation)
*	multiplication
/	division
^	exponentiation

Now, tell MATLAB to calculate the second root. To save typing, edit your previous command by doing the following.

1. Press the ↑ cursor key on your keyboard to recall the previous command(s).

2. Press the ← cursor key to move the cursor left, press Backspace to delete the +, and type in a - in its place.

3. Similarly, change x1 to x2.

4. Press Enter .

MATLAB executes the resulting command:

```
x2 = (-b - sqrt(b^2 - 4*a*c) ) / (2*a)
```

and the result is the other root:

```
x1 =
   -1.414
```

Notes

- If you make a mistake entering an expression or command, MATLAB gives you an error message. For example:

    ```
    a = 1+
    ```

    produces

    ```
    ??? a = 1+
            |
    Missing operator, comma, or semicolon.
    ```

- For the most part, blank spaces can be added at will and are automatically removed by MATLAB's parser if redundant. Sometimes, however, blank spaces have special meaning, such as when they are used to separate the elements of a matrix. Any other special considerations for blank spaces will be detailed in appropriate places throughout this handbook.

Non-interactive Use of MATLAB

MATLAB commands can be put into a file, and you can then instruct MATLAB to execute those commands.

Let's put all the above commands into a file called **abc.m**. The name can be any alphanumeric name you like so long as

- the first character is a letter, and
- the file has a **.m** extension.

These files are frequently created when using MATLAB, and are called *M-files*. To create one, first you must invoke a text editor in your operating system, which allows you to enter *text* (words, numbers) and save it in a file. The following are the text editors most commonly used.

Windows or Mac: We recommend the editor that comes with MATLAB for Windows. The most convenient way to invoke this editor is to simply use the **New** and **M-File** options from the **File** menu in MATLAB.

Unix: vi and **emacs** are the most popular. However, we *strongly* recommend that those new to Unix use **pico** or some other "user-friendly" editor, if available.

You can invoke Unix operating system commands from within MATLAB. Simply put an ! in front of the desired Unix command. For example:

```
!vi abc.m
```

or

```
!pico abc.m
```

Using ! saves you the bother of quitting MATLAB (quit), invoking the operating system command (**vi abc.m**), and then re-invoking MATLAB (**matlab**). As well, any changes made to an M-file in this manner are immediately recognized by MATLAB.

Once in the editor, type the following MATLAB commands:

```
a = 1
b = 0
c = -2
x1 = (-b + sqrt(b^2 - 4*a*c)) / (2*a)
x2 = (-b - sqrt(b^2 - 4*a*c)) / (2*a)
```

After you have finished, save the file as **abc.m** and quit from the text editor. Once back in MATLAB, you can check the contents of abc.m by issuing

```
type abc.m
```

To *run* the commands in **abc.m** through MATLAB, issue

```
abc
```

MATLAB automatically searches your directory structure for the file abc.m and executes its contents as if you had typed them in from the keyboard.

Notes

- It is not necessary for you to enter the type command before you run the contents of the file.

Prompting for Values of Variables

To change the values of a, b, and c you must edit your **abc.m** file, change the value(s), and then run **abc.m** again. One way around this is to modify your program in **abc.m** to prompt you for the values of a, b, and c each time it runs.

```
a = input('Enter a: ');
b = input('Enter b: ');
c = input('Enter c: ');

x1 = (-b + sqrt(b^2 - 4*a*c)) / (2*a)
x2 = (-b - sqrt(b^2 - 4*a*c)) / (2*a)
```

When abc.m is re-run, the following appears on the screen.

```
Enter a:
```

Type in the values following the prompts, for example:

```
Enter a: 1
Enter b: -2
Enter c: 5
```

The semicolon (;) at the end of the three input commands tells MATLAB not to display the value read in. *Without* the semicolons, the screen would show the following.

```
Enter a: 1
a = 1.0000

Enter b: -2
a = -2.0000

Enter c: 5
a = 5.0000
```

The above values for a, b, and c produce complex numbers as the resulting roots of the quadratic equation.

```
x1 = 1.0000 + 2.0000i
x2 = 1.0000 - 2.0000i
```

where i represents $\sqrt{-1}$.

Creating Your Own Functions

It would be convenient to have a *function* which returned the roots of a quadratic equation. If such a function existed and were called quadroot, then your MATLAB program (**abc.m**) could be rewritten as:

```
% -------------------------------
% Revised "abc.m" (program file):
% -------------------------------

a = input('Enter a: ');
b = input('Enter b: ');
c = input('Enter c: ');

[x1, x2] = quadroot(a,b,c);
x1
x2
```

MATLAB treats anything that appears after the % character on a line as comments and does not attempt to compute it.

You can write your own functions, such as quadroot; in this case you must, since no such function exists in MATLAB. The MATLAB code for the quadroot function must go into a separate file with the same name as the function (**quadroot**) and an extension of **.m**.

```
% -------------------------------
% "quadroot.m" (function file):
% -------------------------------

function [x1, x2] = quadroot(a, b, c)

    radical = sqrt(b^2 - 4*a*c);

    x1 = (-b + radical) / (2 * a);
    x2 = (-b - radical) / (2 * a);

end
```

When writing functions, it is common practice to put semicolons after each statement to be executed. This prevents any intermediate results from being displayed (which may be plentiful and cumbersome). The lines in the body of quadroot are indented here strictly to improve readability.

Once you have modified **abc.m** and created **quadroot.m** (as shown above), re-run **abc.m** to check that you get the same results.

Notes

- The form of all functions is

```
function [out1, ..., outn] = name(in1, ..., inm)

    calculations

end
```

where the name of the function is name. If there is only one *output* argument, then you can omit the [ ]:

```
function out = name(in1, ..., inm)

    calculations

end
```

If there are no *input* arguments, then you can omit the ( ):

```
function [out1, ..., outn] = name
    calculations
end
```

- You can display the first block of comment lines in a .m file by issuing the help command. For example:

```
help quadroot

-----------------------------
"quadroot.m" (function file):
-----------------------------
```

The first block of comments is meant to tell the user about the function, *what* it does, and how to *use* it. Subsequent comments in the body of a file are to help other programmers understand *how* the function works.

Using Existing MATLAB Functions

MATLAB already possesses a function for calculating roots of polynomials, appropriately enough called roots.

For a brief description of how to use it, issue

```
help roots

ROOTS  Find polynomial roots.
    ROOTS(C) computes the roots of the polynomial whose coefficients
    are the elements of the vector C. If C has N+1 components,
    the polynomial is C(1)*X^N + ... + C(N)*X + C(N+1).

    See also POLY, RESIDUE, FZERO.
```

roots takes a vector as a parameter. To define a vector, use one of the following methods.

```
a = 1
b = 0
c = -2
coeffs = [a b c]
```

or

```
coeffs = [1 0 -2]
```

To pass the vector coeffs to roots, you can either pass the name of that vector or pass the vector itself directly. For example:

```
x = roots(coeffs)
```
or:
```
x = roots([1 0 -2])
```

The results are returned in a vector (x) as well. You can extract the individual roots from x.

```
x1 = x(1)
x2 = x(2)
```

This returns outputs resembling:

```
x1 =

    1.4142

x2 =

    -1.4142
```

By using this built-in roots function, the quadroot function can be greatly simplified.

```
function [x1, x2] = quadroot (a, b, c)

    x = roots([a b c]);

    x1 = x(1);
    x2 = x(2);

end
```

Plotting Functions

To plot a quadratic (or any function, for that matter), you must define it first. For starters, define it in an M-file. Use a text editor to create **myquad.m** containing:

```
function y = myquad(x)
    a = 1;  b = 0;  c = -2;

    y = a*x^2 + b*x + c;
end
```

Then plot this function using:

```
fplot('myquad', [-3 3])
```

The forward quotes around myquad are necessary to prevent MATLAB from evaluating myquad until inside fplot. The second argument to fplot, [-3 3], is a vector which defines the beginning and end of the range of *x*-values over which to plot.

In this version of the function, if you wish to change a, b, and c, then you have to edit the function file, **myquad.m**. This is not convenient, especially if you want to plot many different examples of the function. Fortunately, however, you can define a, b, and c as *global* variables so that their values can be set from *outside* of **myquad.m**. Change **myquad.m** to look like the following.

```
function y = myquad(x)
    global a b c

    y = a*x^2 + b*x + c;
end
```

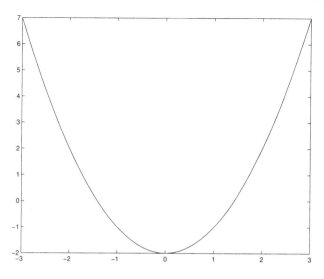

FIGURE 1. Plot of quadratic

Now, to plot different quadratics, invoke fplot as before, but first declare a, b, and c as global and assign them values.

```
global a b c
a = 1;  b = 0;  c = -2;

fplot('myquad', [-3 3])
```

You may have noticed that you need a global statement both in the function myquad and in the on-line session that calls it. Global variables must be declared global in *all* contexts where they are used, not just in one or another.

The resulting plot is shown in Figure 1.

Notes

- MATLAB has a function for evaluating polynomials, called polyval, which simplifies myquad.m to

    ```
    function y = myquad(x)
        global a b c

        y = polyval([a b c], x);
    end
    ```

    For more information, see the command listing for polyval.

To print a figure, click on its **File** menu and **Print** menu item.

Reading in Data Files

Use a text editor to create a data file called **para.dat** containing the following data which we will subsequently analyse.

```
10    1.1
20    3.9
30    9.1
40   15.9
50   25
```

The data do not have to be in any particular columns, as long as there are an equal number of values on each line (as in a matrix) and there is a space or comma between each number on a line.

Read **para.dat** into MATLAB by issuing

```
load para.dat
```

This creates a matrix in MATLAB called para (the same name as the file, but without the **.dat** extension).[2] You can display the variable para using

```
para
```

and MATLAB displays

```
para =
   10.0000    1.1000
   20.0000    3.9000
   30.0000    9.1000
   40.0000   15.9000
   50.0000   25.0000
```

This is the standard way of displaying a MATLAB matrix. As you can see, the values 10, 20, ..., 50 have been converted into floating-point values.

Another way to define this matrix is to enter within the session

```
para = [10    1.1
        20    3.9
        30    9.1
        40   15.9
        50   25. ]
```

or, more compactly, using a semicolon to separate rows;

```
para = [10 1.1;  20 3.9;  30 9.1;  40 15.9;  50 25.]
```

or, one element at a time,

```
para(1,1) = 10
para(1,2) = 1.1
 ...
para(5,1) = 50
para(5,2) = 25.
```

---

[2] You can use any extension you like for your data files. However, we recommend you avoid **.m** (used for scripts and functions), and **.mat** (used for *binary* data files).

(The two values in round brackets, ( ), refer to the row and column positions, respectively.)

However, it is often most convenient to enter data into a file and *load* it into MATLAB. Also, much of the data you will be dealing with may have been created by another application and saved in just such a data file.

Notes

- Floating-point numbers can be entered in two ways. For example, the value 123.45 can be written as:

    ```
    123.45
    ```

    or in exponential (scientific) notation as:

    ```
    1.2345e2
    ```

Working with Matrices

Matrices are the most common variable type in MATLAB. In most cases, you will want to store your input and output values in matrices of one dimension (vectors) or two dimensions (matrices).

There are several matrix operators that you will need to use. Among the more basic ones are:

M.'	transpose of M
M'	complex conjugate transpose of M
+	matrix addition
-	matrix subtraction
*	matrix multiplication
.*	element-by-element multiplication
^	matrix exponentiation
.^	element-by-element exponentiation

Note that most of these operators are identical to the ones used for creating non-matrix expressions. This is one of the great simplicities of MATLAB.

The size function in MATLAB is handy for determining the size of a matrix that has been read in from a file.

```
[m, n] = size(para)
```

and MATLAB displays

```
m =
     5

n =
     2
```

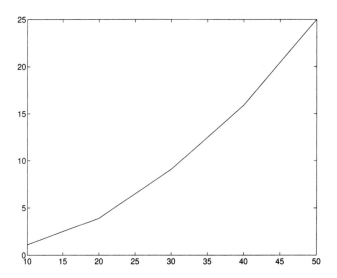

FIGURE 2. Plot of **para.dat** - column 2 vs. column 1

Extract the first column from para and store it in a column vector called x. Similarly, store the second column in y.

```
x = para(:, 1)
y = para(:, 2)
```

The first line can be read as: "vector x is assigned the values from matrix para in all the rows of column 1." The colon standing alone is a short form for 1:m, when m is the number of rows in the matrix.

The result is:

```
x =
    10.0000
    20.0000
    30.0000
    40.0000
    50.0000

y =
     1.1000
     3.9000
     9.1000
    15.9000
    25.0000
```

Plot these two vectors against one another.

```
plot(x,y)
```

The result is the plot shown in Figure 2.

Now, perform a linear regression of y against x and x^2 and see how well a quadratic model fits our data. The function in question looks familiar.

$$y = ax^2 + bx + c$$

Quick Start

MATLAB has a built-in function to calculate the best fit polynomial to a set of data. Call help polyfit for information on the statistical assumptions that polyfit makes about your data.

```
abc = polyfit(x,y,2)   % degree 2

abc =
    0.0101   -0.0106    0.1600
```

Now we can compute the polynomial at our x values using the polyval function

```
ypred = polyval(beta, x)
              abc
```

which results in

```
ypred =
    1.0686
    4.0057
    8.9714
   15.9657
   24.9886
```

Let's compare these results with those obtained using the *matrix* technique for calculating a, b, and c. If you are unfamiliar with this technique, it doesn't matter as we are using it as an example to teach you MATLAB, not statistics.

The technique involves forming a matrix X containing 3 columns:

1. a column with the x^2 values,
2. a column with the x values,
3. a column with all "1"s.

```
X = [ x.^2   x   ones(m,1) ]
```

The result is:

```
X =
       100        10        1
       400        20        1
       900        30        1
      1600        40        1
      2500        50        1
```

MATLAB is case sensitive. X and x are two different variables. The function ones returns a matrix of all ones. Here we asked for an m × 1 matrix of ones.

This is your first exposure to matrix operators. The notations x^2 and x.^2 look similar but behave differently. x^2 is the same as x*x and means "multiply matrix x by matrix x in the classical matrix multiplication sense." x.^2 is the same as x.*x and means "multiply each element of x by each element of x".

Then the coefficients a, b, and c are calculated as follows.

$$\beta = (X^T X)^{-1}(X^T y)$$

$$a = \beta_1$$

$$b = \beta_2$$
$$c = \beta_3$$

Now you can calculate *a*, *b*, and *c* using the matrix formula given above.[3]

```
beta = inv(X' * X) * (X' * y)

a = beta(1)
b = beta(2)
c = beta(3)
```

The result is

```
[a b c]

ans =
    0.0101   -0.0106    0.1600
```

To check the accuracy of the model, you can evaluate the quadratic equation at the data's *x*-values. One way is as follows:

```
for i=1:m
    ypred(i) = a*x(i)^2 + b*x(i) + c
end
```

Or you can avoid the *explicit* for loop by using

```
ypred = a*x.^2 + b*x + c
```

Whichever way you choose to calculate ypred, the result is the same.

```
ypred =
    1.0686
    4.0057
    8.9714
   15.9657
   24.9886
```

You can plot the observed *y*-values and the predicted quadratic function on the same graph.

```
plot(x,y,'+')      % data values

global a b c;
a = beta(1);   b = beta(2);   c = beta(3);

hold on
fplot('myquad', [x(1) x(m)])
hold off

print -deps beta.eps
```

The result is the plot shown in Figure 3.

---

[3]Computing the inverse of $X^T X$ to solve for β is not efficient. See the *Linear Equations* chapter under *L-U decomposition* for a more efficient technique.

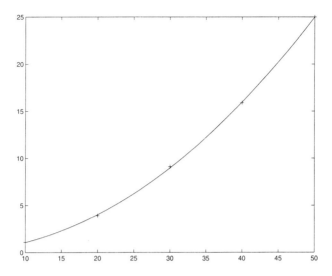

FIGURE 3. Plot of **para.dat** with best fit quadratic

Notes

- The statement hold on tells MATLAB not to erase the plot created by plot(x, y, '+') when the next plot, created by fplot('myquad', [x(1) x(m)]), is done. Instead, the two plots are combined into one.

"Free" format output using save

Results can be written to a file using the save command. For example:

```
save para.out x y ypred -ascii -double
```

or, equivalently,

```
out = [x y ypred]
save para.out out -ascii -double
```

The filename (para.out) is listed first, then the variables (x y ypred) are listed, followed by any options (e.g., -ascii -double).

The -ascii option specifies the file is written out in *ASCII* or "human-readable" format. The default is to write the file in a more compact *binary* format, which computer programs (but not humans) can read much more efficiently than ASCII.

The -double option specifies the results are written using full *double precision*. MATLAB does its calculations internally to 16-digit precision on most computers, but by default only writes ASCII data in 7-digit precision.

You can view the resulting file using

```
type para.out
```

The file should look similar to the following. (We have truncated each number here to save space.)

```
1.00e+01    1.10e+00    1.06e+00
2.00e+01    3.89e+00    4.00e+00
3.00e+01    9.09e+00    8.97e+00
4.00e+01    1.59e+01    1.59e+01
5.00e+01    2.50e+01    2.49e+01
```

para.out can be re-read into a new MATLAB variable para which contains these columns using

```
load para.out
```

which is similar to how you read **para.dat** earlier.

Linear Equations

This chapter examines mathematical operations on linear equations in matrix form. Many of the functions discussed here are built into MATLAB (in the *compiled* C language), and form building blocks for many other functions implemented in the *interpreted* MATLAB language.

If scientific computing is at the "heart" of MATLAB, then surely linear equations are at its "soul"!

In this chapter we will look at MATLAB functions for:

- Linear equations solution
- Eigenvalues and eigenvectors

Linear Equations Solution

Consider the graph in Figure 4 which represents supply and demand for some product, say baseballs. [4]

In a competitive marketplace, the price gravitates to an *equilibrium point* where the two lines intersect. [5] To find this (*Price*, *Quantity*)-point, you need to solve the linear equations

$$(Supply) \quad Quantity = 10 + 20 Price$$
$$(Demand) \quad Quantity = 100 - 10 Price$$

Re-expressing these in the more conventional form

$$(Supply) \quad -20 Price + Quantity = 10$$
$$(Demand) \quad 10 Price + Quantity = 100$$

---

[4] Tempted as we are to honor our national sport and use hockey pucks, we'd like instead to pay tribute to the Toronto Blue Jays, the 1992 and 1993 World Series Champions.

[5] In the next chapter we will look at a *non-competitive* marketplace.

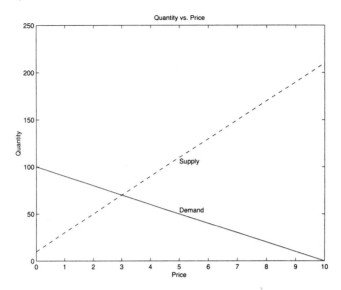

FIGURE 4. Plot of baseball supply and demand model

you can easily convert these equations to the equivalent matrix form:

$$\begin{bmatrix} -20 & 1 \\ 10 & 1 \end{bmatrix} \begin{bmatrix} Price \\ Quantity \end{bmatrix} = \begin{bmatrix} 10 \\ 100 \end{bmatrix}$$

The problem is now in the form $Ax = b$, which can be readily solved. There are two fundamental ways of doing this in MATLAB.

Method 1: Calculate $x = A^{-1}b$

One way to solve for x in $Ax = b$ mathematically is to multiply both sides of this equation on the *left* by A^{-1}.[6]

$$A^{-1}Ax = A^{-1}b$$

which simplifies to

$$x = A^{-1}b$$

This is easily calculated in MATLAB using the inv function.

```
A = [-20 1
      10 1]

b = [ 10
     100]

x = inv(A)*b
```

---

[6]Remember that matrix multiplication is not commutative: *AB* does not equal *BA*, in general.

The solution is

```
x =
    3.0000
   70.0000
```

Therefore, the Supply and Demand lines cross at the point where *Price* is equal to $3.00, and *Quantity* is 70 baseballs. The resulting sales revenue is $210.00.

Method 2: Calculate $x = A\backslash b$

If b has only 1 column, as in this case, then it is far more efficient to use MATLAB's *left division* operator (\).

To solve for x in $Ax = b$, divide through on the *left* by A to obtain

$$A\backslash Ax = A\backslash b$$

which simplifies to

$$x = A\backslash b$$

This is easily calculated in MATLAB using

```
x = A \ b
```

Again, the result is

```
x =
    3.0000
   70.0000
```

The \ operator is more easily understood if demonstrated with numbers.

```
half = 1 / 2         % Both assignments
                     % produce the same
half = 2 \ 1         % result:  half = 0.5
```

(You may find it easier to remember the meaning of \ by exaggerating the backslash in your mind's eye, so that it creates a "lop-sided" fraction.)

```
          \ 1
half =     \
          2 \
```

When performing *left division* by A, MATLAB does *not* actually compute the inverse of A. Instead it performs what is known as an *L-U decomposition* of A, from which x can be quickly calculated. Contrast this to the first method, where to calculate the inverse of A MATLAB performs not 1 but n of these left divisions (where n is the number of columns of A). So, in our 2×2 example, computing x using inv(A)

```
x = inv(A) * b
```

is actually equivalent to

```
x = (A \ [1 0
         0 1]) * b
```

Multiple Right-Hand Sides

Consider the following matrix A and right-hand sides b1, b2, and b3: [7]

```
degrees = pi / 180;
a = sin(45*degrees);

%-----1--2--3--4---5--6--7--8---9-10-11-12--13

A = [-a  0  0  1   a  0  0  0   0  0  0  0   0
     -a  0 -1  0  -a  0  0  0   0  0  0  0   0
      0 -1  0  0   0  1  0  0   0  0  0  0   0
      0  0  1  0   0  0  0  0   0  0  0  0   0

      0  0  0 -1   0  0  0  1   0  0  0  0   0
      0  0  0  0   0  0 -1  0   0  0  0  0   0
      0  0  0  0  -a -1  0  0   a  1  0  0   0
      0  0  0  0   a  0  1  0   a  0  0  0   0

      0  0  0  0   0  0  0 -1  -a  0  0  a   0
      0  0  0  0   0  0  0  0  -a  0 -1 -a   0
      0  0  0  0   0  0  0  0   0 -1  0  0   1
      0  0  0  0   0  0  0  0   0  0  1  0   0

      0  0  0  0   0  0  0  0   0  0  0 -a  -1];

b1 = [0  5  0  0   0  0  0  0   0  1  0  0   0]';
b2 = [0  4  0  0   0  0  0  0   0  2  0  0   0]';
b3 = [0  3  0  0   0  0  0  0   0  3  0  0   0]';
```

We could solve this for each right-hand side separately, or solve all three at once:

```
b = [b1 b2 b3];

f=flops;      x1 = A\b1;    flops-f
f=flops;      x2 = A\b2;    flops-f
f=flops;      x3 = A\b3;    flops-f

f=flops;      x  = A\b;     flops-f
```

The number of *flops* (floating-point operations) in each case is as follows:

Solution	Flops	Total
A\b1	2,559	
A\b2	2,559	
A\b3	2,559	7,677
A\b	3,209	3,209

Doing the solution all in one step is the most efficient, because MATLAB only has to do the LU decomposition of A once (rather than three times) before performing the triangular solves.

---

[7] This example is taken from the *Sparse Matrices* chapter, and is described in detail there. In that chapter we also see how MATLAB can take advantage of the sparsity of this matrix (about 5% nonzeros). Using sparse matrix techniques, MATLAB can solve these three systems of equations much faster than here — 381 flops compared to 3209.

However, sometimes you do not know *beforehand* how many right-hand sides you are going to have. In those cases, you can first compute the LU factors and then save them (L and U) for subsequent re-use:

```
f=flops;      [L, U] = lu(A);      flops-f

f=flops;      x1 = U \ (L \ b1);   flops-f
f=flops;      x2 = U \ (L \ b2);   flops-f
f=flops;      x3 = U \ (L \ b3);   flops-f

f=flops;      x  = U \ (L \ b);    flops-f
```

L and U are lower and upper triangular matrices, respectively, so performing the "\" operation with them is much quicker than with A. The results are as follows:

Solution	Flops	Total
lu(A)	1,378	
U \ (L \ b1)	390	
U \ (L \ b2)	390	
U \ (L \ b3)	390	2,548
lu(A)	1,378	
U \ (L \ b)	1,170	2,548

Clearly, having computed the L and U factors previously, there is now no performance penalty for individually solving with the three righthand sides.

Eigenvalues and Eigenvectors

MATLAB can compute the eigenvalues (and eigenvectors) of a matrix, which allow a matrix to be decomposed into the form

$$A = X\Lambda X^{-1},$$

where X is a matrix of eigenvectors, and Λ (lambda) is a diagonal matrix containing the corresponding eigenvalues.

Matrices are often decomposed into this form to speed up mathematical operations, such as repetitive multiplication.

$$\begin{aligned} A^3 &= AAA \\ &= (X\Lambda X^{-1})(X\Lambda X^{-1})(X\Lambda X^{-1}) \\ &= X\Lambda(X^{-1}X)\Lambda(X^{-1}X)\Lambda X^{-1} \\ &= X\Lambda\Lambda\Lambda X^{-1} \\ & X\Lambda^3 X^{-1} \end{aligned}$$

Because Λ is diagonal, raising it to a power is much quicker than with the original matrix, A.

$$\begin{bmatrix} \lambda_1 & 0 & 0 \\ 0 & \lambda_2 & 0 \\ 0 & 0 & \lambda_3 \end{bmatrix}^3 = \begin{bmatrix} \lambda_1^3 & 0 & 0 \\ 0 & \lambda_2^3 & 0 \\ 0 & 0 & \lambda_3^3 \end{bmatrix}$$

For example, consider the following *rotation matrix*.

$$A = \begin{bmatrix} \cos(\frac{\pi}{3}) & -\sin(\frac{\pi}{3}) \\ \sin(\frac{\pi}{3}) & \cos(\frac{\pi}{3}) \end{bmatrix}$$

It can be used to rotate the points in an object by $60°$ ($\frac{\pi}{3}$ radians), simply by multiplying A by the (x,y)-coordinates of the points. The following MATLAB program rotates a house (represented by a 2×9 matrix of 9 *xy*-points) four times, and plots the result at each step (see Figure 5). It also pauses for one second between each rotation for an animated effect.

```
A = [cos(pi/3) -sin(pi/3)
     sin(pi/3)  cos(pi/3)]

house0 = [0 1 3 3 2 1 1 1 3
          0 0 0 1 2 1 0 1 1]

house1 = A*house0
house2 = A*A*house0
house3 = A^3*house0
house4 = A^4*house0

hold on
axis([-pi, pi,  -pi, pi]);
axis('equal');

plot(house0(1,:), house0(2,:));   pause(1);
plot(house1(1,:), house1(2,:));   pause(1);
plot(house2(1,:), house2(2,:));   pause(1);
plot(house3(1,:), house3(2,:));   pause(1);
plot(house4(1,:), house4(2,:));   pause(1);

print -deps houses.eps
hold off
```

Continuing with the example, use the eig function to compute the eigenvalues of our square matrix, A.

```
A = [cos(pi/3) -sin(pi/3)
     sin(pi/3)  cos(pi/3)]

vals = eig(A)
```

which produces a vector with the two eigenvalues of A. [8]

```
vals =
   0.5 + 0.8660i
   0.5 - 0.8660i
```

---

[8] The MATLAB output in this section has been condensed slightly.

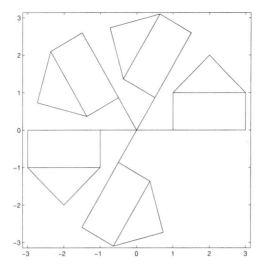

FIGURE 5. Plot of house rotated by $60°$ four times

If you specify two output parameters, then MATLAB interprets them as matrices and calculates eigenvectors as well. For example:

```
[X,L] = eig(A)
```

produces eigenvectors in the columns of X

```
X =
  -0.7071         -0.7071
   0.7071i        -0.7071i
```

and the corresponding eigenvalues are stored on the main diagonal of L.

```
L =
   0.5 + 0.8660i         0
        0           0.5 - 0.8660i
```

Now check the solution. Either of the following should result in a matrix equal to our original A matrix.

```
X * L * inv(X)
```

or

```
X * L / X
```

And, as you can see, it does.

```
ans =
   0.5000   -0.8660
   0.8660    0.5000
```

Working with Eigenvectors and Eigenvalues

Let's extract the individual eigenvectors, x_1 and x_2, from X:

```
x1 = X(:,1)          % column 1
x2 = X(:,2)          % column 2
```

and the eigenvalues, λ_1 and λ_2, from L:

```
lambda1 = L(1,1)
lambda2 = L(2,2)
```

Each λ_i and x_i value must satisfy the basic property of eigenvalues and eigenvectors. That is, there exist n values of λ_i (where n is the number of rows/columns in square matrix A), and corresponding vectors x_i, such that Ax_i and $\lambda_i x_i$ are equal; i.e.,

$$Ax_i = \lambda_i x_i.$$

Verify that for both our λ and x values, $Ax - \lambda x = 0$.

```
A*x1 - lambda1*x1
A*x2 - lambda2*x2
```

In both cases, we get:

```
ans =

     0
     0
```

The determinant of A (written $|A|$) can be calculated using the det function in MATLAB.

```
det(A)
```

which results in

```
1
```

However, if you have already calculated the eigenvalues, then you might decide to save some execution time by directly computing the determinant from the product of the eigenvalues. For example:

```
L(1,1) * L(2,2)
```

also results in

```
1
```

For larger matrices you *could* use a for loop. With a 10×10 matrix the syntax would be

```
d = 1;
for i=1:10
    d = d * L(i,i);
end
d
```

However, in terms of efficiency, coding effort, and clarity, this is not the "best" technique. A better way is to use the diag function in MATLAB to extract the diagonal elements of L and place them in a one-dimensional vector.

```
vals = diag(L);
```

Then, use the prod function to calculate the product of the elements in that vector.

```
d = prod(vals);
```

This can also be done in one step.

```
d = prod( diag(L) )
```

Notes

- diag is dual-purpose, and can be used in reverse. That is, diag will take a vector and form a diagonal matrix from it. For example, using the vector vals computed earlier:

    ```
    L = diag(vals)
    ```

    results in

    ```
    L =
       0.5 + 0.8660i        0
            0           0.5 - 0.8660i
    ```

Calculating Eigenvalues from First Principles

(Note: You would seldom do the following in a *practical* application, but it is a good review exercise in Linear Algebra and MATLAB fundamentals.)

Eigenvalues are values λ_i which satisfy the equation

$$Ax_i = \lambda_i x_i$$

This can be re-written as

$$(A - \lambda_i I)x = 0$$

or

$$Bx = 0,$$

where B is equal to $A - \lambda_i I$. For this equation to be satisfied for non-zero values of x, it is required that B be a "singular" matrix, i.e., $|B| = 0$.

$|A - \lambda_i I|$ actually expands out into a polynomial expression in terms of λ_i, which is referred to as the *characteristic polynomial*. You can ask MATLAB to calculate the coefficients of this polynomial.

```
p = poly(A)

p =
    1.0000   -1.0000    1.0000
```

Recall that MATLAB uses vectors to store the coefficients of polynomials. For example, $x^2 + 2x + 3$ is represented by the vector [1 2 3].

Hence, using lambda as the variable, the characteristic polynomial for A is

$$p(\lambda_i) = \lambda_i^2 - \lambda_i + 1.$$

The *roots* of the polynomial are values λ_i, where $p(\lambda_i) = 0$. These values satisfy the equation $|A - \lambda_i I| = 0$ and hence are the *eigenvalues* of A. You can calculate the roots of p using

```
vals = roots(p)
```

which results in

```
vals =
   0.5 + 0.8660i
   0.5 - 0.8660i
```

which are the same as the eigenvalues computed previously with eig.

Matrix Multiplication

As mentioned above, a common application of eigenvalues and eigenvectors is to speed up successive matrix multiplications. For example, to compute A^{120} you could either use the following:

```
A120 = A^120
```

or

```
[X,L]  = eig(A)
A120   = X * L.^120 / X
```

Then, the rotation could be applied to the house matrix.

```
house120 = A120 * house
```

Either way, the result will be the same. The house completes 20 full rotations, then is back to its original position. Notice that the result is the identity transformation, which corresponds to the rotation matrix for 0, 2π, 4π, ... radians (0, 360, 720, ... degrees):

$$A^{120} = \begin{bmatrix} cos(0) & -sin(0) \\ sin(0) & cos(0) \end{bmatrix}$$

Matrix Exponential

It can be proven (through Taylor Series expansion) that the exponential of a matrix, e^A, is equal to

$$e^A = Xe^\Lambda X^{-1},$$

and that, since Λ is diagonal, e^Λ is equal to

$$e^\Lambda = \begin{bmatrix} e^{\lambda_1} & 0 & 0 \\ 0 & e^{\lambda_2} & 0 \\ 0 & 0 & \dots \end{bmatrix}.$$

MATLAB has a function called expm which calculates e^A directly.[9] Let's try it with our matrix A.

```
expm(A)

ans =
     1.0681   -1.2559
     1.2559    1.0681
```

You get the same answer when calculating the matrix exponential the long way.

```
[X,L] = eig(A)

X * diag(exp(diag(L))) / X

ans =
     1.0681   -1.2559
     1.2559    1.0681
```

Notice that because $e^0 \neq 0$ we were careful to only exponentiate the *diagonal* elements of L.

MATLAB also has routines for calculating the square root of a matrix (sqrtm) and the log of a matrix (logm).

Command Listing

M_b = **balance**(M)
Balances square matrix M to improve the condition of its eigenvalues.
Output: A square matrix is assigned to M_b.
Argument options: [M_d, M_b] = balance(M) to return both the balanced matrix M_b and the diagonal matrix M_d
Additional information: Balancing alleviates ill-conditioning of the resulting eigenvector matrix by focussing the effects to a scaling factor in the diagonal. ✤ For a further discussion of balancing, see the on-line help file. ✤ Unless told otherwise, the eig command automatically performs balancing of the input matrix. ✤ You should understand the theory and application of balancing before using balance, so you know when and why to apply it.
See also: eig, cond

---

[9]Contrast this with the exp function, which calculates the exponential of each element of a matrix.

[M$_{r1}$, M$_{r2}$] = cdf2rdf(M$_{eig,cmplx}$, M$_{d,cmplx}$)
Converts the results of the eig command from *complex diagonal form* to *real block diagonal form*.
Output: Real-valued matrices are assigned to M$_{r1}$ and M$_{r2}$.
Additional information: The individual columns of V$_{r1}$ are no longer eigenvectors, but each block in the matrix, associated with a 2 × 2 block in M$_{r2}$, spans the corresponding invariant vectors.
See also: eig, rsf2csf

M$_C$ = chol(M)
Computes the Cholesky factorization of matrix M.
Output: If M is positive definitive, an upper triangular matrix is assigned to M$_C$. If M is not positive definitive, an error message is printed and no value is assigned to M$_C$.
Argument options: [M$_C$, int] = chol(M) to return both an upper triangular matrix M$_C$ and an integer. If M is positive definitive, int is 0. Otherwise, int is a positive integer and M$_C$ is an upper triangular matrix of order int−1 such that M$_C$' * M$_C$ = M(1:int-1, 1:int-1). Either way, no error message is printed.
Additional information: A fully successful result of this command, M$_C$, satisfies the equation M$_C$' * M$_C$ = M. ♣ If M is positive definitive, then reordering the columns of M with colperm can increase the efficiency of chol.
See also: cholinc, symbfact, colperm

M$_C$ = cholinc(M, num)
Computes the incomplete Cholesky factorization of matrix M with a *drop tolerance* of num.
Output: If M is positive definitive, an upper triangular matrix is assigned to M$_C$. If M is not positive definitive, an error message is printed and no value is assigned to M$_C$.
Argument options: [M$_C$, int] = chol(M) to return both an upper triangular matrix M$_C$ and an integer. If M is positive definitive, int is 0. Otherwise, int is a positive integer and M$_C$ is an upper triangular matrix of order int−1 such that M$_C$' * M$_C$ = M(1:int-1, 1:int-1). Either way, no error message is printed.
Additional information: See the on-line help file for additional options to cholinc.
See also: chol, luinc

M$_c$ = compan(V)
Computes the companion matrix of V, a vector of polynomial coefficients.
Output: A square matrix is assigned to M$_c$.
Additional information: The eigenvalues of the companion matrix equal the roots of the polynomial represented by V.
See also: roots, eig, poly

num = cond(M)
Computes the 2-norm condition number of matrix M.
Output: A numeric value is assigned to num.
Additional information: The 2-norm condition number of M is a measure of sensitivity to errors in M and equals the ratio of the largest singular value to the smallest singular value. ♣ In practice, rcond is more often used.

See also: svd, condeig, condest, rcond, norm, rank

num = **condeig**(M)
Computes the condition number, with respect to eigenvalues, of matrix M.
Output: A numeric value is assigned to num.
Additional information: These numbers are the related to the angles between eigenvectors.
See also: eig, cond, condest, rcond

[num, V] = **condest**(M)
Computes the 1-norm condition number estimate of matrix M.
Output: A numeric value is assigned to num and a vector v such that absA v = absA absv / num is assigned to V.
Argument options: [num, V] = condest(M, 1) to print information about each iteration as it is performed. ♣ [num, V] = condest(M, -1) to print ratios of the results to the computed condition number and the result of rcond.
Additional information: num is a *lower* bound for the 1-norm condition number.
♣ In practice, rcond is more often used.
See also: cond, condeig, rcond, normest

num = **det**(M)
Computes the determinant of square matrix M.
Output: A numerical value is assigned to num.
Additional information: It is not recommended that you use det to check for singularity of large matrices; rcond is much more efficient.
See also: rcond, inv, lu, rref

V = **eig**(M)
Computes the eigenvalues of square matrix M.
Output: A vector of numerical values is assigned to V.
Argument options: $[M_{eig}, M_d]$ = eig(M) to assign a diagonal matrix of eigenvalues to M_d and a full matrix whose columns are the corresponding eigenvectors to M_{eig}.
♣ $[M_{eig}, M_d]$ = eig(M, 'nobalance') to specify that an initial balancing not be performed.
♣ V = eig(M_a, M_b) or $[M_{eig}, M_d]$ = eig(M_a, M_b) to solve for the *generalized* eigenvalues and eigenvectors.
Additional information: All eigenvectors computed are *right eigenvectors*. To compute the *left eigenvectors* of M, pass M' to eig. ♣ Initial balancing of M is automatically done with the balance command. ♣ Each eigenvector is scaled (normalized) so that its norm is 1.0.
See also: balance, norm, qz, schur, hess, rosser, symrcm, wilkinson

M = **gallery**(name, ...)
Returns a Higham test matrix, corresponding to name.
Output: A matrix is assigned to M.

Additional information: For a list of available names, see the on-line help file.
♣ Most names require additional parameters, and many return more than one output.
See also: hilb, hadamard, wilkinson

M = gallery(3)

Returns a poorly conditioned 3 × 3 matrix.
Output: A 3 × 3 matrix is assigned to M.
Additional information: To find the condition number of a matrix, use cond.
See also: hilb, rcond, cond

M = gallery(5)

Returns a 5 × 5 matrix with interesting eigenvalue properties.
Output: A 5 × 5 matrix is assigned to M.
Argument options: M = gallery(3) to return a badly conditioned 3 × 3 matrix.
Additional information: To find the eigenvalues of a matrix, use eig.
See also: hilb, hadamard, eig, cond

M = hadamard(n)

Creates the n × n Hadamard matrix.
Output: A square matrix is assigned to M.
Additional information: The elements of Hadamard matrices are all either 1 or −1. All columns are orthogonal to one another.
See also: gallery, hankel, compan, vander, toeplitz

M = hankel(V)

Creates the square Hankel matrix whose first column equals vector V.
Output: A square matrix with dimensions equal to the number of elements of V is assigned to M.
Argument options: M = hankel(V_c, V_r) to create the Hankel matrix whose first column and last row equal V_c and V_r, respectively. The last element of V_c overrides the first element of V_r.
Additional information: Each antidiagonal (i.e., diagonal running from lower left to upper right) of a Hankel matrix has the same element throughout all positions.
♣ For the one parameter case, all elements below the main antidiagonal are set to 0.
See also: hadamard, compan, vander, toeplitz

M_H = hess(M)

Computes the Hessenberg form of square matrix M.
Output: A matrix of equal dimensions to M is assigned to M_H.
Argument options: [M_u, M_H] = hess(M) to assign the Hessenberg form to M_H and a unitary matrix to M_u, such that M = M_u*M_H*M_u'.
Additional information: Entries below the first subdiagonal of Hessenberg matrices are all zeroes.
See also: eig, qz, schur

M = **hilb**(n)
Computes the n × n Hilbert matrix.
Output: A square matrix is assigned to M.
Additional information: The element in column c and row r of a Hilbert matrix has a value of $\frac{1}{c+r+1}$. ♣ Hilbert matrices are notoriously poorly conditioned.
See also: invhilb, cond, gallery(3)

M_{inv} = **inv**(M)
Computes the inverse of square matrix M.
Output: A square matrix is assigned to M_{inv}.
Additional information: If M is poorly scaled or sufficiently close to being singular, an error message is displayed. On some types of computers a result is produced even if these conditions are encountered. ♣ Avoid over-dependence on inverting matrices. In many cases, the matrix division operator is more efficient. ♣ For more information, see the on-line help file for inv.
See also: det, lu, rcond, rref, pinv, invhilb

M = **invhilb**(n)
Computes the inverse of the n × n Hilbert matrix.
Output: A square matrix is assigned to M.
Additional information: For values of n less than approximately 12, the exact values of the inverse hilbert matrix can be calculated. ♣ Performing this calculation with inv(hilb(n)) introduces several different types of round-off errors.
See also: hilb

V_x = **lscov**(M_A, V_b, M_{cov})
Computes the least square solution of M_A*x = V_b given covariance matrix M_{cov}.
Output: A vector is assigned to V_x.
Additional information: For more information on the computations performed, see the on-line help file.
See also: \, nnls

[M_{lp}, M_u] = **lu, num**(M)
Performs incomplete factorization of matrix M, with *drop tolerance* of num.
Output: A permutation of a lower triangular matrix is assigned to M_{lp} and a purely upper triangular matrix is assigned to M_u.
Additional information: See the on-line help file for more information.
See also: lu, cholinc

[M_{lp}, M_u] = **luinc**(M)
Factors matrix M into lower and upper triangular matrices.
Output: A permutation of a lower triangular matrix is assigned to M_{lp} and a purely upper triangular matrix is assigned to M_u.
Argument options: [M_l, M_u, M_p] = lu(M) to store a purely lower triangular matrix in M_l and the corresponding permutation matrix in M_p.
Additional information: The results are such that M = M_{lp} * M_u or M_p * M = M_l * M_u.
See also: qr, inv, det, rcond, rref

V_{nn} = **nnls**(M, V)

Solves the system of equations $Mx = V$ in the non-negative least squares sense.
Output: A vector of non-negative value is assigned to V_{nn}.
Argument options: $[V_{nn}, V_{dual}]$ = **nnls**(M, V) to return the *dual vector* in V_{dual}. ♣ V_{nn} = **nnls**(M, V, tol) to specify a tolerance of tol.
Additional information: The solution from nnls may not fit as well as that from using the \ operator, but the result contains no negative values.
See also: \

num = **norm**(M)

Computes the 2-norm of matrix M, its largest *singular* value.
Output: A numeric value is assigned to num.
Argument options: num = norm(M, 1) to compute the 1-norm, the largest *column* sum of M. ♣ num = norm(M, inf) to compute the infinity norm of M, its largest *row* sum. ♣ num = norm(M, 'fro') to compute the F-norm, or sqrt(sum(diag(M'*M))). ♣ num = norm(V, int), where V is a vector, to compute sum(abs(V).^int)^(1/int). ♣ num = norm(V, inf) to compute max(abs(V)). ♣ num = norm(V, -inf) to compute min(abs(V)). ♣ For more information on values of norm over vectors, see the on-line help page.
See also: svd, cond, condest, normest, rcond, min, max, abs

num = **normest**(M)

Estimates the z-norm of matrix M.
Output: A numeric value is assigned to num.
Argument options: num = normest(M, tol) to use the tolerance tol. Default tolerance is 1.e-6. ♣ [num, num$_{cnt}$] = normest(M, tol) to return the number of iterations performed in num$_{cnt}$.
Additional information: This command works best with sparse matrices. normest works with full matrices but takes significantly more time. ♣ For more information on the algorithm used to find the estimate, see the on-line help page.
See also: cond, condeig, norm, condest, rcond

M_n = **null**(M)

Computes the null space of matrix M.
Output: A matrix whose columns represent an orthonormal basis is assigned to M_{inv}.
Additional information: The number of columns of M_n represents the nullity of M.
See also: orth, svd

M_n = **orth**(M)

Computes the range space of matrix M.
Output: A matrix whose columns represent an orthonormal basis is assigned to M_{inv}.
Additional information: The number of columns of M_n represents the rank of M.
See also: null, svd

M = **pascal**(n)

Computes the n × n Pascal matrix.
Output: A square matrix is assigned to M.

Argument options: M = pascal(n, 1), to compute the lower triangular Cholesky Factor of the n × n Pascal matrix. This matrix is always its own inverse. ✦ M = pascal(n, 2), to compute a variation of the Pascal matrix that is the cube root of the n × n identity matrix.

Additional information: The elements in a Pascal matrix are taken from a Pascal's triangle laid on its side such that the top of the triangle is the upper-left element of the matrix. ✦ The inverse of a Pascal matrix has integer elements.

See also: inv, chol

M_{ps} = **pinv**(M)

Computes the Moore-Penrose pseudoinverse of matrix M.
Output: A matrix of the same dimensions as M' is assigned to M_{psinv}.
Argument options: M_{ps} = pinv(M, num) to treat singular values less then num as zero. The default tolerance is max(size(A)) * norm(A) * eps.
Additional information: the pseudoinverse has most of the qualities of the true inverse, even for matrices that are not square. ✦ For a listing of the criteria for a pseudoinverse, see the on-line help file.
See also: inv, norm, svd, rank, qr

[M, V] = **polyeig**(M_0, M_1, ..., M_p)

Solves the degree p polynomial eigenvalue problem given matrices M_0 through M_p.
Output: A matrix containing the eigenvectors is assigned to M and a vector containing the eigenvalues is assigned to V.
Additional information: Each input matrix must be of identical size, n. ✦ The output matrix M will be of size n × n*p, and the output vector V will be of size n*p.
See also: eig, qz, size

[M_{un}, M_{ut}] = **qr**(M)

Performs orthogonal-triangular decomposition on matrix M.
Output: A unitary matrix is assigned to M_{un} and an upper triangular matrix is assigned to M_{ut}.
Argument options: [M_{un}, M_{ut}, M_p] = qr(M) to store a permutation matrix in M_p such that M * M_p = M_{un} * M_{ut}. ✦ qr(M) to return output similar to that from the LINPACK subroutine *ZQRDC*.
Additional information: In the most common calling sequence, the result is such that M = M_{un} * M_{ut}. ✦ This command is often used to solve linear systems that have more equations than unknowns. For more information, see the on-line help file.
See also: qrdelete, qrinsert, orth, lu, null

[$M_{un,new}$, $M_{ut,new}$] = **qrdelete**(M_{un}, M_{ut}, n)

Removes the nth column from the result of a call to qr.
Output: A new unitary matrix is assigned to $M_{un,new}$ and a new upper triangular matrix is assigned to $M_{ut,new}$.
Additional information: See the entry for qr for more details.
See also: qr, qrinsert

[M$_{un,new}$, M$_{ut,new}$] = qrinsert(M$_{un}$, M$_{ut}$, V, n)
Add a column with elements defined by vector V before the nth column of the result of a call to qr.
Output: A new unitary matrix is assigned to M$_{un,new}$ and a new upper triangular matrix is assigned to M$_{ut,new}$.
Additional information: To insert a column after the last column, set n equal to the total number of columns plus one. ♣ See the entry for qr for more details.
See also: qr, qrdelete

[M$_{ut1}$, M$_{ut2}$, M$_{lt}$, M$_{rt}$, M$_{eig}$] = qz(M$_1$, M$_2$)
Computes the QZ factorization of square matrices M$_1$ and M$_2$.
Output: Upper triangular matrices are assigned to M$_{ut1}$ and M$_{ut2}$. Left and right transformation matrices are assigned to M$_{lt}$ and M$_{rt}$. The general eigenvector matrix is assigned to M$_{eig}$.
Additional information: The results of qz are such that M$_{lt}$ * M$_1$ * M$_{rt}$ = M$_{ut1}$, M$_{lt}$ * M$_2$ * M$_{rt}$ = M$_{ut2}$, and M$_1$ * M$_{eig}$ * diag(M$_{ut2}$) = M$_2$ * M$_{eig}$ * diag(M$_{ut1}$). ♣ For more information, see the on-line help file.
See also: eig

n = rank(M)
Computes the rank of matrix M.
Output: A non-negative integer value is assigned to n.
Argument options: n = rank(M, num) to include any singular values from M that are larger than tolerance num. The default tolerance is max(size(M)) * norm(M) * eps.
Additional information: The rank of a matrix is a reflection of its number of singular values of a certain magnitude.
See also: svd, norm

num = rcond(M)
Computes an estimate of the reciprocal of the condition number of matrix M.
Output: A numeric value between 0 and 1 is assigned to num.
Additional information: The 1-norm of M is used in this calculation. ♣ rcond is more efficient but less accurate than cond.
See also: svd, condeig, condest, cond, norm, rank

M = rosser
Represents a matrix whose eigenvalues are difficult to compute.
Output: An 8 × 8 matrix with integer elements is assigned to M.
Additional information: While many programs have difficulty with this matrix, MATLAB's QR algorithm (implemented in eig) handles it easily. ♣ To see the elements or the eigenvalues, simply try it out for yourself.
See also: eig, wilkinson

M$_{new}$ = rref(M)
Computes the row-reduced echelon form of matrix M.
Output: A matrix with the same dimensions as M is assigned to M$_{new}$.

Argument options: M_{new} = rref(M, num$_{tol}$) to set the tolerance used for determining negligible column elements to num$_{tol}$. The default value is max(size(M)) * eps * norm(M, inf).
Additional information: Gauss-Jordan elimination with partial pivoting is used to compute the row-reduced echelon form. ♣ To see a movie of the reduction process, use rrefmovie.
See also: rank, inv, lu, rrefmovie

M_{new} = **rrefmovie(M)**

Computes the row-reduced echelon form of matrix M and displays a movie detailing the computations involved.
Output: A matrix with the same dimensions as M is assigned to M_{new}.
Argument options: M_{new} = rrefmovie(M, num$_{tol}$) to set the tolerance used for determining negligible column elements to num$_{tol}$. The default value is max(size(M)) * eps * norm(M, inf).
Additional information: Initially, the matrix M is printed to the screen and you are prompted to press a key. After each key press the current matrix is overlaid with the next step of the computation.
See also: rref, rank, inv, lu

[M_{Sc}, M_{uc}] = **rsf2csf(M_{Sr}, M_{ur})**

Converts a matrix in *real Schur form* to the equivalent matrix in *complex Schur form*.
Output: Two matrices with the same dimensions as the input matrices are assigned to M_{Sc} and M_{uc}.
Additional information: For more information on *real Schur form* and *complex Schur form*, see the listing for schur.
See also: schur, cdf2rdf

M_S = **schur(M)**

Computes the Schur decomposition of matrix M.
Output: If M contains only real values, then the *real Schur form* matrix with the same dimensions as M is assigned to M_S. Otherwise, the *complex Schur form* matrix with the same dimensions as M is assigned to M_S.
Argument options: [M_S, M_u] = schur(M) to assign a unitary matrix M_u such that M = M_u * M_S * M_u' and M_u * M_u' = eye(size(M)).
Additional information: The *complex Schur form* matrix has the eigenvalues of M on the main diagonal. ♣ The *real Schur form* matrix has the real eigenvalues of M on the main diagonal and the complex eigenvalues in 2 × 2 blocks on the diagonal. ♣ To convert from the *real* to the *complex* form, use the rsf2csf command.
See also: eig, hess, qz, rsf2csf

num = **subspace(M_1, M_2)**

Computes the angle between the two subspaces represented by the columns of matrices M_1 and M_2.
Output: An angle value (in radians) is assigned to num.
Argument options: subspace(V_1, V_2) to compute the angle between two vectors.

Additional information: The input matrices must be of equal dimensions.
See also: rank

V = svd(M)
Computes the singular values of matrix M.
Output: A vector is assigned to V.
Argument options: $[M_{u1}, M_d, M_{u2}]$ = svd(M) to compute diagonal matrix M_d and unitary matrices M_{u1} and M_{u2} such that $M = M_{u1} * M_d * M_{u2}$. ✦ $[M_{u1}, M_d, M_{u2}]$ = svd(M, 0) computes smaller versions of matrices M_d and M_{u1}.
Additional information: For more information on the matrices produced, see the on-line help file. ✦ If a singular value is too difficult to find a warning message is displayed.
See also: lu, qr

M = toeplitz(V_c, V_r)
Creates the Toeplitz matrix defined columnwise by vector V_c and rowwise by V_r.
Output: A matrix is assigned to M.
Argument options: M = toeplitz(V) to create the symmetric Toeplitz matrix defined by vector V. This is also known as a *Hermitian* Toeplitz matrix.
Additional information: If the first elements of the two input vectors disagree, the element of V_c takes precedence. ✦ The vectors V_c and V_r do not need to be of equal lengths. ✦ For more information about how a Toeplitz matrix is formed, see the on-line help file or create a simple one yourself.
See also: hankel, vander

num = trace(M)
Computes the trace of matrix M.
Output: A numerical value is assigned to num.
Additional information: The trace of a matrix is the sum of the elements along its main diagonal, i.e., sum(diag(M)).
See also: sum, diag, det, eig

M_{new} = tril(M)
Extracts the lower triangular part of matrix M.
Output: A lower triangular matrix is assigned to M_{new}.
Argument options: M_{new} = tril(M, n) to extract the lower matrix elements starting with the nth superdiagonal of M. ✦ M_{new} = tril(M, -n) to extract the lower matrix elements starting with the nth subdiagonal of M.
Additional information: All matrices returned are of equal dimensions to M.
See also: triu, diag

M_{new} = triu(M)
Extracts the upper triangular part of matrix M.
Output: An upper triangular matrix is assigned to M_{new}.
Argument options: M_{new} = triu(M, n) to extract the upper matrix elements starting with the nth superdiagonal of M. ✦ M_{new} = triu(M, -n) to extract the upper matrix elements starting with the nth subdiagonal of M.

Additional information: All matrices returned are of equal dimensions to M.
See also: tril, diag

M = **vander(V)**
Creates the Vandermonde matrix whose columns are defined by the elements of vector V.
Output: A square matrix with dimension equal to the length of V is assigned to M.
Additional information: The columns of a Vandermonde matrix contain powers of the various elements in V. ♣ Vandermonde matrices are typically used in polynomial fitting routines. ♣ For more information about how a Vandermonde matrix is formed, see the on-line help file or create a simple one yourself.
See also: toeplitz, vander, polyfit

M = **wilkinson(n)**
Creates an n × n matrix whose eigenvalues are difficult to compute.
Output: An n × n matrix with integer elements is assigned to M.
Additional information: To see the elements or the eigenvalues, simply try it out for yourself.
See also: gallery, eig, rosser

Sparse Matrices

The functions covered in this chapter include:

Defining: sparse, speye, spdiags, spconvert, spalloc

Utility: full, find, nonzeros, nnz, nzmax, spfun

Visualizing: spones, spy, gplot

Other: See *Command Listing* section

Introduction

Sparse matrices are matrices that consist mostly of zeros. The non-zero elements typically account for 5% or fewer of the elements.

If you identify a matrix as sparse, MATLAB will store only those non-zero elements, together with information on their (i, j) location in the matrix.

For example, consider the following 1000×1000 matrix:

```
-2  1  0  0  0  0  . . .   0  0  0  0  0  0
 1 -2  1  0  0  0  . . .   0  0  0  0  0  0
 0  1 -2  1  0  0  . . .   0  0  0  0  0  0
 0  0  1 -2  1  0  . . .   0  0  0  0  0  0
 0  0  0  1 -2  1  . . .   0  0  0  0  0  0
 .  .  .  .  .  .           .  .  .  .  .  .
 .  .  .  .  .  .           .  .  .  .  .  .
 .  .  .  .  .  .           .  .  .  .  .  .
 0  0  0  0  0  0  . . .  -2  1  0  0  0  0
 0  0  0  0  0  0  . . .   1 -2  1  0  0  0
 0  0  0  0  0  0  . . .   0  1 -2  1  0  0
 0  0  0  0  0  0  . . .   0  0  1 -2  1  0
 0  0  0  0  0  0  . . .   0  0  0  1 -2  1
```

Notice that the matrix consists of all zeros except along the main diagonal (-2s) and the first subdiagonal and superdiagonal (1s).[10] In fact with only 2,998 non-zero elements out of 1,000,000 the sparsity of this matrix is just 0.3%.

---

[10] This matrix, in various sizes, is the so-called *second difference operator* and arises in "finite difference" techniques for approximating solutions to partial differential equations.

Storing this as a *full* matrix in MATLAB would typically require 1,000,000 8-byte storage locations: **8 megabytes of RAM**.

But storing this as a *sparse* matrix typically requires only **38 kilobytes of RAM**:

- 2,998 8-byte storage locations for the *non-zero elements* (24 kilobytes of RAM), plus
- 2,998 4-byte storage locations for the *row indices* of each non-zero element (12 kilobytes of RAM), plus
- 1,000 4-byte storage locations as *pointers to the start of each column* (4 kilobytes of RAM).

[Typically, a real $m \times n$ matrix containing *nnz* non-zero entries occupies: $8 \times nnz + 4 \times (nnz + n)$ bytes, in sparse storage mode. A complex matrix requires about twice as much memory to store the elements; the total needed is: $16 \times nnz + 4 \times (nnz + n)$ bytes.]

Furthermore, the execution time to perform operations on the sparse matrix is significantly reduced as well. In Figure 6 we plot the execution time to solve $Ax = b$ (using $x = A\backslash b$) against various sizes, n, of this matrix. The trials are summarized below: [11]

Size	Sparse	Full
	0.03	0.05
n = 100	0.03	0.07
	0.05	0.05
	0.67	1.63
n = 500	0.67	1.50
	0.65	1.50
	2.77	7.07
n = 1000	2.73	6.55
	2.73	6.30
	11.47	29.37
n = 2000	11.30	25.88
	11.08	26.03

As it says in the abstract to [1]:

> The matrix computation language and environment MATLAB is extended to include sparse matrix storage and operations. The only change to the outward appearance of the MATLAB language is a pair of commands to create full or sparse matrices. Nearly all the operations of MATLAB now apply equally well to full or sparse matrices, without any explicit action by the user.

---

[11] These times are in seconds of CPU time, and were obtained on a Sun SPARCstation 10.

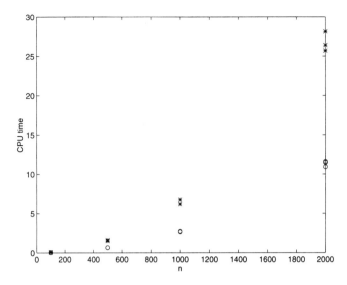

FIGURE 6. Comparison of CPU time to solve $Ax = b$ (for tridiagonal A) in full versus sparse storage modes, for various sizes of n.

The sparse data structure represents a matrix in space proportional to the number of nonzero entries, and most of the operations compute sparse results in time proportional to the number of arithmetic operations on nonzeros.

Defining Sparse Matrices

Consider a smaller (5 × 5) version of the above tridiagonal matrix. We can enter it as a full matrix in the usual way:

```
A = [-2   1   0   0   0
      1  -2   1   0   0
      0   1  -2   1   0
      0   0   1  -2   1
      0   0   0   1  -2];
```

and then convert it to a sparse matrix using the sparse function:

```
S = sparse(A)
```

MATLAB displays the (i,j) locations of the non-zero elements, and the elements themselves:

```
S =
    (1,1)        -2
    (2,1)         1
    (1,2)         1
    (2,2)        -2
    (3,2)         1
    (2,3)         1
```

```
(3,3)        -2
(4,3)         1
(3,4)         1
(4,4)        -2
(5,4)         1
(4,5)         1
(5,5)        -2
```

If you now issue the whos command, it reports:

```
Name    Size    Elements  Bytes  Density  Complex

 A      5 by 5     25      200    Full      No
 S      5 by 5     13      176    0.5200    No

Grand total is 38 elements using 376 bytes
```

Notice that S occupies 176 bytes: $8 \times 13 + 4 \times (13 + 5)$.

Rather than create a full matrix and convert it to sparse (we may not have enough RAM), we can create a sparse matrix directly with one of:[12]

- sparse(i, j, s, m, n, nzmax)
- spconvert([i j s])
- spdiags(A, d, m, n)
- speye(n)
- spalloc(m, n, nzmax)

sparse(i, j, s, m, n, nzmax): With the sparse function you provide the row (i) and column (j) indices, and the non-zero values (s) in individual vectors. For example, to construct the above 5×5 tridiagonal matrix, use:

```
i = [ 1  2   1  2  3   2  3  4   3  4  5   4  5];
j = [ 1  1   2  2  2   3  3  3   4  4  4   5  5];

s = [-2  1   1 -2  1   1 -2  1   1 -2  1   1 -2];

A = sparse(i, j, s, 5, 5)
```

If i contains m (the largest row number), and j contains n (the largest column number) you can omit m and n from the call, and sparse will deduce them:

```
A = sparse(i, j, s)
```

nzmax can be omitted unless you expect to add more elements to the matrix, in which case it should be set to the total number of non-zero elements you expect to eventually have.

spconvert(A): With spconvert you provide a matrix containing three columns of values similar to the *i*, *j* and *s* values in the sparse description above.

---

[12] We only cover the most common form of the following functions. Refer to the *Command Listing* section for other useful forms.

```
A =     [1      1      -2
         2      1       1
         1      2       1
         2      2      -2
         3      2       1
         2      3       1
         3      3      -2
         4      3       1
         3      4       1
         4      4      -2
         5      4       1
         4      5       1
         5      5      -2]

S = spconvert(A);
```

If the last row or column of the actual *sparse* matrix is all zeros, then spconvert cannot correctly deduce the size, $m \times n$, from the elements in A.

In this case, one of the rows of A can be used to specify the size of the matrix by entering the values of: m, n, and 0 on it. For example, we can add the size to the bottom using:

```
A = A ; ...
    [5 5 0];            % optional in this case

S = spconvert(A);
```

Note: spconvert is usually used with data read from a file into a matrix, e.g., using load.

spdiags(A, d, m, n): In cases like this where the data is defined along the main diagonal and subdiagonals, spdiags is very convenient to use. For example:

```
A = [1 -2  1
     1 -2  1
     1 -2  1
     1 -2  1
     1 -2  1];

S = spdiags(A, [-1 0 1], 5, 5);
```

full(S) converts S to full storage mode so that we can display it in traditional format:

```
full(S)

ans =

    -2     1     0     0     0
     1    -2     1     0     0
     0     1    -2     1     0
     0     0     1    -2     1
     0     0     0     1    -2
```

The second argument of spdiags: [-1, 0, 1], specifies the diagonals. -1 refers to the first subdiagonal (immediately below the main diagonal), 0 refers to the main diagonal, and 1 refers to the first superdiagonal.

For much larger matrices we need a more compact way of defining A, the matrix with the diagonal elements, when they are all the same:

```
n = 5;
b = ones(n, 1);

A = [b -2*b b];
```

Before we leave spdiags, it is important to be aware of how it handles the extra elements in the columns of A. Consider the following A matrix containing 3 diagonals:

```
A = [1   11   111
     2   22   222
     3   33   333
     4   44   444
     5   55   555];

S = spdiags(A, [-1 0 1], 5, 5);

full(S)
```

The result is:

```
11   222     0     0     0
 1    22   333     0     0
 0     2    33   444     0
 0     0     3    44   555
 0     0     0     4    55
```

Can you detect which elements from A are missing?

speye(n): speye is the sparse equivalent of the eye function for producing identity matrices. For example:

```
A = eye(5);

S = speye(5)
```

results in:

```
S =
   (1,1)       1
   (2,2)       1
   (3,3)       1
   (4,4)       1
   (5,5)       1
```

whos reports:

```
Name        Size      Elements    Bytes    Density    Complex

  A        5 by 5        25        200      Full        No
  S        5 by 5         5         80      0.2000      No

Grand total is 30 elements using 280 bytes
```

spalloc(m, n, nzmax): You can change or add elements to a sparse matrix using assignment statements. You can even delete elements from storage by assigning them to 0.

Each sparse matrix has a predetermined storage area for nzmax non-zero elements, where nzmax varies from matrix to matrix. (In the above examples nzmax was the same as the number of elements being initially-defined and no more.)

When you store more than nzmax elements into a sparse matrix, MATLAB must allocate a larger storage area and copy *all* the old elements, plus the new elements into it. Hence it is a good idea to increase the size of a matrix before adding a lot of elements to it.

One way is to use sparse as mentioned above, i.e.:

```
S = sparse([], [], [], m, n, nzmax)
```

The other is using spalloc. For example, we can allocate space for (and define) the 13 non-zero elements in our 5 × 5 *second difference operator* using:

```
S = spalloc(5, 5, 13)
for i=1:4,   S(i  ,i+1) =  1;   end
for i=1:5,   S(i  ,i  ) = -2;   end
for i=1:4,   S(i+1,i  ) =  1;   end
```

To find out what nzmax is for a matrix use the nzmax function. For example: nzmax(S) returns 13 in this case.

Operations on Sparse Matrices

As stated in [1], MATLAB is designed so that:

> The **value** of the result of an operation does not depend on the storage class [full or sparse] of the operands, although the **storage class** of the result may.

Full matrix operands produce full results (except with functions like sparse, etc. which create sparse matrices).

In the case of sparse or mixed (sparse and full) operands, the following rules apply:

Operation:	Result:	S or F:	Comment:
size(S)	[m, n]	FULL	Functions which produce a FIXED length
nnz(S)	n	FULL	result (no matter the size of the inputs)
...			produce FULL results.
zeros(m,n)	A	FULL	
ones(m,n)	A	FULL	Functions of FIXED length vectors (or
randn(m,n)	A	FULL	scalars) that produce matrices produce
eye(n)	A	FULL	FULL results ...
...			
sparse(m,n)	A	sparse	
sprandn(m,n,d)	A	sparse	... unless you use their sparse equivalents!
speye(n)	A	sparse	
-S, ~S	A	sparse	
chol(S)	A	sparse	The remaining functions of one matrix that
diag(S)	A	sparse	produce matrices (or vectors) generally
max(S)	v	sparse	return sparse results from a sparse operand
sum(S)	v	sparse	(and FULL results from a FULL operand).
...			
S + F	A	FULL	
S * F	A	FULL	Binary operators produce sparse results
S \ F	A	FULL	from sparse operands, and FULL results
F \ S	A	FULL	from FULL operands.
S | F	A	FULL	
			With mixed operands, the results are FULL
S .* F		sparse	with some operators and sparse with
S & F		sparse	others.
			Matrix concatenation involving one or
[F F; F S]	A	sparse	more sparse matrices produce a sparse
			result. (Otherwise the result is FULL.)
S(a:b, c:d)	A	sparse	Submatrix indexing of a sparse matrix results in a sparse matrix.

Finally, if you perform submatrix indexing on the **left** side of an assignment statement, it leaves the storage class of the matrix on the left unaltered:

```
F(a:b, c:d) = S        <-- F remains full
S(a:b, c:d) = F        <-- S remains sparse
```

Utility Functions

We have already covered two utility function: (1) full(S) for converting a sparse matrix to full storage mode, and (2) nzmax(S) for determining the number of storage locations reserved for the non-zero elements of a sparse matrix.

Four other useful functions are:

1. nnz(S)

2. nonzeros(S)

3. [i, j, s] = find(S)

4. spfun('fun', S)

nnz(S): This function returns the number of non-zero elements in S. For example, using S from above, nnz(S) returns 13.

nonzeros(S): This function returns the actual non-zero elements in a column vector. However, it is not as useful as find which also returns the row and column indices of the elements.

[i,j, s] = find(S): This function returns the row and column indices and the values of each non-zero element. For example, in the case of S from the previous section:

```
[i,j, s] = find(S);

A = [i j s]
```

produces:

```
A =
    1    1   -2
    2    1    1
    1    2    1
    2    2   -2
    3    2    1
    2    3    1
    3    3   -2
    4    3    1
    3    4    1
    4    4   -2
    5    4    1
    4    5    1
    5    5   -2
```

spfun('fun', S): This function evaluates the specified function on all the **non-zero** elements of S.

For example:

```
A = spfun('exp', S);

full(A)
```

looks like:

```
0.1353    2.7183         0         0         0
2.7183    0.1353    2.7183         0         0
     0    2.7183    0.1353    2.7183         0
     0         0    2.7183    0.1353    2.7183
     0         0         0    2.7183    0.1353
```

Compare this to:

```
B = exp(S);

full(B)
```

which evaluates exp on **all** elements of S, be they nonzero *or* zero. This results in a *not* particularly sparse matrix:

```
0.1353    2.7183    1.0000    1.0000    1.0000
2.7183    0.1353    2.7183    1.0000    1.0000
1.0000    2.7183    0.1353    2.7183    1.0000
1.0000    1.0000    2.7183    0.1353    2.7183
1.0000    1.0000    1.0000    2.7183    0.1353
```

Visualization Functions

The three main functions for visualizing sparse matrices are:

1. spones(S)

2. spy(S)

3. gplot(S, xy)

spones(S) This function returns a matrix with the same sparsity pattern as S, but with all the non-zero elements replaced by ones.

For example, continuing with our matrix S from the last section (5 × 5 *second difference operator*):

```
U = spones(S);

full(U)
```

results in:

```
1    1    0    0    0
1    1    1    0    0
0    1    1    1    0
0    0    1    1    1
0    0    0    1    1
```

Alternatively, you may want to try the format + command which causes Matlab to display matrix elements as "+", "-", or blank, depending on whether they are positive, negative or zero. For example:

```
format +
S
```

produces:

```
S =

 - +
 + - +
   + - +
     + - +
       + -
```

To restore the default output options, issue format by itself.

spy(S) Plots the matrix with a dot representing non-zero elements. For example, using the above S matrix, spy(S) produces:

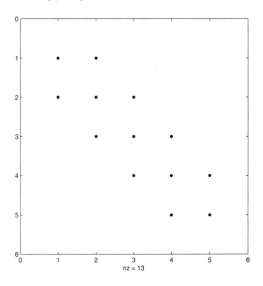

gplot(S, xy) Sometimes a sparse matrix, S, contains the connections between nodes i and j in a network rather than a set of linear equations to be solved.

Consider the following diagram:

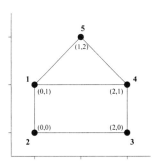

The following matrix, S, contains a nonzero value, S(i,j), wherever nodes i and j are connected. The matrix xy contains the coordinates of each node. The whole object is plotted using gplot(S, xy):

```
S = spalloc(5, 5, 6);

S(1,2) = 1;
S(2,3) = 1;
S(3,4) = 1;
S(4,5) = 1;
S(5,1) = 1;
S(1,4) = 1;

xy = zeros(5, 2);
```

```
xy(1,:) = [0.0  1.0];
xy(2,:) = [0.0  0.0];
xy(3,:) = [2.0  0.0];
xy(4,:) = [2.0  1.0];
xy(5,:) = [1.0  2.0];

subplot(2,1, 1)
    spy(S)
    title('spy(S)')

subplot(2,1, 2)
    gplot(S, xy)
    title('gplot(S, xy)')
    axis equal
    axis([-1 3   -1 3])

print -deps sxy.eps
```

The result is the following two subplots:

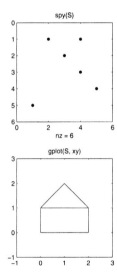

Example 1 - Truss

Consider the following "truss" which is fixed on the left side, but allows horizontal motion on the right. We wish to determine all the forces F_1 - F_{13} in this truss. We also want to know the force bearing down on the left support and on the right support:

Sparse Matrices

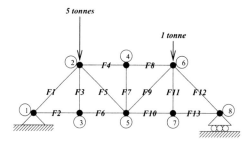

From Mechanics we know that this truss is *statically determinant* because the number of joints ($j=8$) and the number of members ($m=13$) satisfy the equation:

$$2j - 3 = m$$

The structure is in equilibrium (not moving) so we can assert that the horizontal and vertical forces at each joint sum to 0. Letting $a = \sin 45° = \cos 45°$, we have the following 13 equations involving the 13 unknown forces:

Joint 2	$\sum F_x = -aF_1 + F_4 + aF_5 = 0$
	$\sum F_y = -aF_1 - F_3 - aF_5 - 5 = 0$
Joint 3	$\sum F_x = -F_2 + F_6 = 0$
	$\sum F_y = F_3 = 0$
Joint 4	$\sum F_x = -F_4 + F_8 = 0$
	$\sum F_y = -F_7 = 0$
Joint 5	$\sum F_x = -aF_5 - F_6 + aF_9 + F_{10} = 0$
	$\sum F_y = aF_5 + F_7 + aF_9 = 0$
Joint 6	$\sum F_x = -F_8 - aF_9 + aF_{12} = 0$
	$\sum F_y = -aF_9 - F_{11} - aF_{12} - 1 = 0$
Joint 7	$\sum F_x = -F_{10} + F_{13} = 0$
	$\sum F_y = F_{11} = 0$
Joint 8	$\sum F_x = -aF_{12} - F_{13} = 0$

Here are these equations represented in matrix form, and solved in Matlab (once converted to sparse form):

```
degrees = pi / 180;
a = sin(45*degrees);

%-----1--2--3--4---5--6--7--8---9-10-11-12--13

A = [-a   0   0   1    a   0   0   0    0   0   0   0   0
     -a   0  -1   0   -a   0   0   0    0   0   0   0   0
      0  -1   0   0    0   1   0   0    0   0   0   0   0
      0   0   1   0    0   0   0   0    0   0   0   0   0

      0   0   0  -1    0   0   0   1    0   0   0   0   0
      0   0   0   0    0   0  -1   0    0   0   0   0   0
      0   0   0   0   -a  -1   0   0    a   1   0   0   0
      0   0   0   0    a   0   1   0    a   0   0   0   0

      0   0   0   0    0   0   0  -1   -a   0   0   a   0
      0   0   0   0    0   0   0   0   -a   0  -1  -a   0
```

```
                0 0 0 0   0 0 0 0   0 -1  0  0    1
                0 0 0 0   0 0 0 0   0  0  1  0    0

                0 0 0 0   0 0 0 0   0  0  0 -a   -1];
    b = [ 0 5 0 0   0 0 0 0   0  1  0  0    0]';

    S = sparse(A);
    b = sparse(b);

    F = S\b;

    Fleft  = a*F(1)
    Fright = a*F(12)
```

The solution is:

```
    Fleft =
         -4

    Fright =
         -2
```

For very large trusses we want to avoid creating a full matrix, and so specify the non-zero elements of A by their i and j location:

```
    degrees = pi / 180;
    a = sin(45*degrees);

    i = [ 1  1  1  2  2  2  3  3  4  5  5  6, ...
          7  7  7  7  8  8  8  9  9  9 10 10, ...
         10 11 11 12 13 13];

    j = [ 1  4  5  1  3  5  2  6  3  4  8  7, ...
          5  6  9 10  5  7  9  8  9 12  9 11, ...
         12 10 13 11 12 13];

    s = [-a  1  a -a -1 -a -1  1  1 -1  1 -1, ...
         -a -1  a  1  a  1  a -1 -a  a -a -1, ...
         -a -1  1  1 -a -1];

    S = sparse(i,j,s, 13,13);

    i = [2 10];
    j = [1  1];
    s = [5  1];

    b = sparse(i,j,s, 13,1);

    F = S\b;

    Fleft  = a*F(1)
    Fright = a*F(12)
```

Let's plot the solution over top of the truss. First we need to define the connectivity matrix, C, and the coordinates matrix xy:

```
    C = spalloc(8,8, 13);

    C(1,2) = 1;    % F1;
    C(1,3) = 2;    % F2;

    C(2,3) = 3;    % F3;
```

Sparse Matrices

```
C(2,4) = 4;     % F4;
C(2,5) = 5;     % F5;

C(3,5) = 6;     % F6;

C(4,5) = 7;     % F7;
C(4,6) = 8;     % F8;

C(5,6) = 9;     % F9;
C(5,7) = 10;    % F10;

C(6,7) = 11;    % F11;
C(6,8) = 12;    % F12;

C(7,8) = 13;    % F13;

spy(C + C')

print -deps conn1.eps

%-------------------------------------------

xy = [0.0  0.0
      1.0  1.0
      1.0  0.0
      2.0  1.0
      2.0  0.0
      3.0  1.0
      3.0  0.0
      4.0  0.0];

gplot(C, xy);
axis('equal')

print -deps conn2.eps
```

This program produces the following spy and gplot plots:

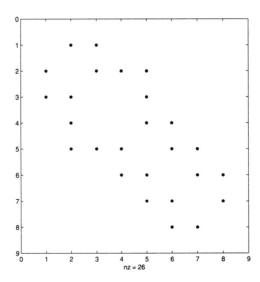

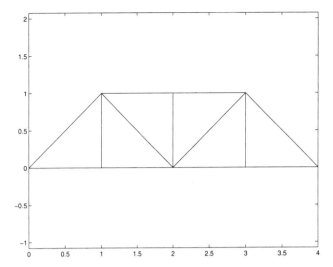

Finally, a graphical view of the solution is shown below. Where the force in a member is positive (under tension), a solid line is drawn. Where the force is negative (under compression), a dashed line is drawn. Where the force is zero, a dotted line is drawn:

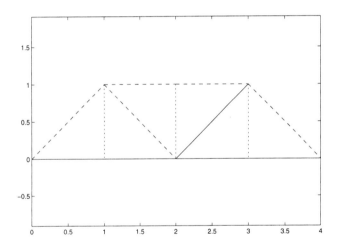

This was produced by issuing:

```
forcplot(C, xy, F)
axis equal
print -deps forc.eps
```

where `forcplot.m` contains:

```
function forcplot(C, xy, F)

[i,j,m] = find(C);

[rows,cols] = size(m);

for k=1:rows*cols
     if F(m(k))  < 0,  linestyle = '--';  end
```

```
            if F(m(k)) == 0,  linestyle = ':';   end
            if F(m(k)) >  0,  linestyle = '-';   end

            plot([xy(i(k),1) xy(j(k),1)], ...
                 [xy(i(k),2) xy(j(k),2)], ...
                 linestyle)
            hold on
    end

    hold off
```

Permutations for Sparse Factorizations

MATLAB provides several functions to permute the columns (and rows) of a sparse matrix in such a way that lu (and chol[13]) factorizations often execute faster and result in smaller factors.

The functions include:

colmmd: Column *minimum degree* ordering.

colperm: Order columns based on non-zero count.

symmmd: Symmetric *minimum degree* ordering.

symrcm: Reverse Cuthill-McKee ordering to reduce profile or bandwidth. Good for "long, skinny" problems.

The types of matrices they work on are as follows. Also shown is the function for which they produce a good pre-ordering.

	Function:	Asym-metric?	Symmetric Indefinite?	Sym. Pos. Definite?	Good pre-ordering for:
\	c = colmmd(S)	√	√		lu(S(:,c))
	p = colperm(S)	√	√	√	lu(S(:,p))
		-	-	√	chol(S(p,p))
\	s = symmmd(S)	-	√	-	lu(S(s,s))
\		-	-	√	chol(S(s,s))
	r = symrcm(S)	√	√	-	lu(S(r,r)) or
		-	-	√	chol(S(r,r)) or

The first column of the above table indicates that the preordering functions colmmd and symmmd are automatically used by MATLAB in computing sparse S\b.

Here is a flow chart showing the steps followed by S\b.

---

[13] C=chol(A), where A must be Hermitian positive definite (which includes real symmetric positive definite), produces an upper-triangular C such that C' * C = A.

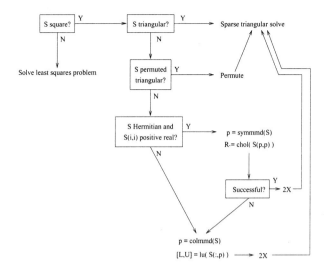

Contrast this with full F\b which simply invokes lu (if F is not triangular) and performs triangular solve(s).

You can actually monitor the workings of sparse S\b (and other sparse operations). Diagnostic output can be turned on by issuing:

```
none     = 0
some     = 1
too_much = 2

spparms('spumoni', some)
```

Now when we ask for the forces to be computed in the truss example:

```
F = S\b;
```

we get the following extra output:

```
    sp\: is A triangular? no.
    sp\: is A morally triangular? no.
    sp\: is A a candidate for Cholesky? no.
    sp\: LU factor and solve two triangular systems.
*   sp\: column minimum degree.
*   colmmd: withholding 0 dense rows.
*   mmd: 4 stages.
    sprealloc in spmult: 13 1 1 14
```

In the above output, "*Is A morally triangular?*" is short for asking: "*Is A a permutation of a triangular matrix?*"

You can suppress the automatic re-ordering done by sparse S\b by issuing:

```
spparms('autommd', 0)
```

in which case the diagnostic output for F=S\b will not contain the three lines shown above with "*".

Example 2 - Sparse Random Matrix S

Let's create a 5% sparse 100 × 100 random matrix S and right-hand side b: [14]

```
rand( 'seed', 123456789);
randn('seed', 123456789);

            n = 100;
                density = 0.05;
                            rc = 1.0;
    S = sprandn(n, n, density, rc);

    b = sprandn(n, 1, density);
```

A spy plot of S and lu(S) is shown below:

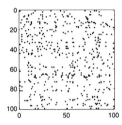

Let's compute the various permutations:

```
c = colmmd(S);
p = colperm(S);

s = symmmd(S);
r = symrcm(S);
```

Then let's determine the number of *flops* (floating-point operations) needed to solve Sx = b. (First we will disable the automatic use of colmmd (and symmmd) by sparse "\" and "/".)

```
spparms('autommd', 0)
f = flops;    F  = S\b;          f = flops-f

f = flops;    Fc = S(:,c)\b;     f = flops-f
f = flops;    Fp = S(:,p)\b;     f = flops-f

f = flops;    Fs = S(s,s)\b(s);  f = flops-f
f = flops;    Fr = S(r,r)\b(r);  f = flops-f
spparms('autommd', 1)
```

Note: In the case of symmmd and symrcm we permuted the rows of S (not just the columns). Therefore we had to permute the corresponding rows on the right-hand side, b.

---

[14] We use sprandn to produce S and b. For S we specify that we want its reciprocal condition number (rc) to be approximately 1.0. sprandn uses both the uniform (rand) and normal (randn) random number generators in Matlab, and here we set their starting seeds so that this sequence of commands always produces the same matrix and vector in S and b.

Note: To compare the actual solutions we must first unpermute them:

```
Fc(c) = Fc;
Fp(p) = Fp;

Fs(s) = Fs;
Fr(r) = Fr;
```

The results are as follows. (The timing shown with a "\" indicates the permutation that would have been chosen by S\b if "autommd" had not been disabled.) Also shown are the number of non-zeros in the lu factorization of S for each permutation:

	Permutation	flops	nnz(lu(S(...)))
	No permutation	31,194	2,208
\	c = colmmd(S)	27,485	1,714
	p = colperm(S)	11,879	1,192
	s = symmmd(S)	33,446	2,167
	r = symrcm(S)	44,953	2,421

Clearly the default permutation saves significant time and memory. However, it is not the best choice in this case: colperm has performed significantly better. As it says in [1], colperm "is sometimes a good reordering for matrices with very irregular structures, especially if there is great variation in the nonzero counts of rows or columns."

spy plots of the colmmd and colperm permutations and their lu factorizations are shown below:

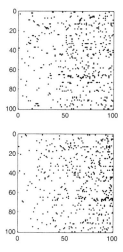

Notice in the spy plot for S(:,p) that the columns get denser (nnz increases) as your eyes move to the right.

For comparison, here are the corresponding spy plots for symmmd and symrcm:

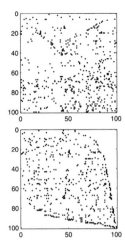

Notice that the permutation produced by symrcm does its best to reduce the bandwidth of S.

So the best choice in this case is to override the automatic colmmd permutation, and use colperm.

Example 3 - Sparse Symmetric (Positive Definite) Random Matrix S

Let's create a 5% sparse 100×100 **symmetric** (positive definite) random matrix S, and right-hand side b:[15]

```
rand( 'seed', 123456789);
randn('seed', 123456789);

            n = 100;
            density = 0.05;
                rcond = 1.0;
                    kind = 1;
S = sprandsym(n, density, rcond, kind);

b = sprandn(n, 1, density);
```

A spy plot of S and chol(S) is shown below:

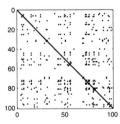

We show the Cholesky factorization results (chol) because that is what S\b will use in the case of this matrix.

---

[15] Here we specify (using kind=1) that S is to have exactly $\frac{1}{rc}$ as the condition number.

The results of computing S\b with each permutation are as follows:

	Permutation	flops	nnz(chol(S(_,_)))
	No permutation	3,590	385
	c = colmmd(S)	2,885	333
	p = colperm(S)	2,577	317
\	s = symmmd(S)	2,566	319
	r = symrcm(S)	2,659	324

Clearly the default permutation (symmmd) saves significant time and memory compared to no permutation at all. As it says in [1], "The most generally useful symmetric preordering in MATLAB is minimum degree, obtained by the function s = symmmd(S)."

However, in this case colperm and symrcm also perform comparably well.

Spy plots of the colmmd and colperm results are shown below:

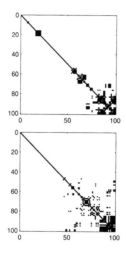

Here are the spy plots for symmmd and symrcm:

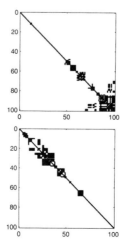

Example 1 Re-visited - Sparse "Long, Thin" Matrix S

Let's return to our truss example. Its sparsity is about 5%:

```
>> nnz(S) / prod( size(S) )

ans =
    0.0496
```

A spy plot of S and lu(S) is shown below:

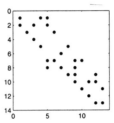

The results of computing S\b with each permutation are as follows:

	Permutation	flops	nnz(lu(S(...)))
	No permutation	206	38
\	c = colmmd(S)	210	39
	p = colperm(S)	210	38
	s = symmmd(S)	209	42
	r = symrcm(S)	185	37

Clearly the only permutation to have any significant impact on performance is symrcm, with a 10% reduction ($\frac{185-206}{206} \times 100\%$) in flops. symrcm is known for its efficacy in "long, thin" (basically 1-dimensional) problems like this. Even with such a relatively small matrix, this effect is evident.

spy plots of colmmd and colperm results are shown below:

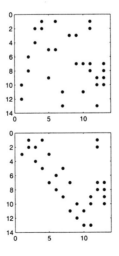

Here are the spy plots for symmmd and symrcm:

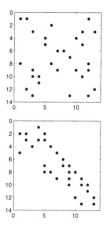

Example 1 Re-visited - Multiple right-hand sides

Consider our truss example again, and three different scenarios of external forces that we want to study:

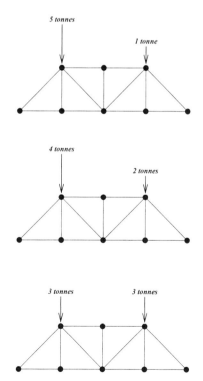

We could solve this for each right-hand side separately, or solve all three at once:

```
b1 = [ 0  5  0  0    0  0  0  0    0  1  0  0    0]';
b2 = [ 0  4  0  0    0  0  0  0    0  2  0  0    0]';
b3 = [ 0  3  0  0    0  0  0  0    0  3  0  0    0]';

b1 = sparse(b1);
b2 = sparse(b2);
b3 = sparse(b3);
b  = [b1 b2 b3];

f=flops;     F1 = S\b1;      flops-f
f=flops;     F2 = S\b2;      flops-f
f=flops;     F3 = S\b3;      flops-f
f=flops;     F  = S\b;       flops-f

Fleft  = a*F(1,:)
Fright = a*F(12,:)
```

The number of flops in each case is as follows:

Solution	Fleft	Fright	Flops	Total
S\b1	-4.0	-2.0	208	
S\b2	-3.5	-2.5	208	
S\b3	-3.0	-3.0	197	613
S\b			381	381

Doing the solution all in one step is the most efficient because MATLAB only has to do the lu decomposition of S once (rather than three times) before performing the triangular solves. [16]

However, sometimes you do not know beforehand how many right-hand sides you are going to have. In that case you can first compute the LU factors, and then save them (L and U) for subsequent re-use.

We have to first remember to apply the appropriate permutation. S\b applies one automatically, but lu does not.

Using colmmd would in general be a good choice for a non-symmetric matrix such as this (that's what S\b used above), so we will use it here for comparison with our previous results: [17]

```
f=flops;     c = colmmd(S);
             [L, U] = lu(S(:,c));    flops-f

f=flops;     F1 = U \ (L \ b1);      flops-f
f=flops;     F2 = U \ (L \ b2);      flops-f
f=flops;     F3 = U \ (L \ b3);      flops-f
f=flops;     F  = U \ (L \ b);       flops-f

F(c,:) = F;

Fleft  = a*F(1,:)
Fright = a*F(12,:)
```

---

[16] Or the Cholesky factorization, if S were Hermitian positive definite (which includes real symmetric positive definite).

[17] We are using colmmd for comparison with our previous results even though we know that symrcm performs significantly better for this particular "long, skinny" matrix and underlying basically one-dimensional problem. We will cover how to use symrcm later.

The results are as follows:

Solution	Fleft	Fright	Flops	Total
colmmd & lu			118	
U \ (L \ b1)	-4.0	-2.0	88	
U \ (L \ b2)	-3.5	-2.5	88	
U \ (L \ b3)	-3.0	-3.0	77	371
colmmd & lu			118	
U \ (L \ b)			253	371

As you can see, there was no performance improvement associated with computing all three solutions simultaneously.

Notes

- If we had chosen to use symrcm instead we would have changed the first three lines above to:

```
f=flops;     r = symrcm(S);
             [L, U] = lu(S(r,r));      flops-f

f=flops;     F1 = U \ (L \ b1(r));     flops-f
```

Eigenvalues and Eigenvectors of Sparse Matrices

The eig function in MATLAB accepts sparse matrices that are complex Hermitian (which includes real symmetric matrices).[18] It only finds the eigenvalues though, and not the eigenvectors. For example:

```
n = 5;

S = [ones(1,n);
     ones(n-1,1) speye(n-1,n-1)];

full(S)
```

produces the following real symmetric arrow-shaped matrix:

```
ans =
     1    1    1    1    1
     1    1    0    0    0
     1    0    1    0    0
     1    0    0    1    0
     1    0    0    0    1
```

and:

```
eig(S)
```

---

[18] The eigs function, on the other hand, accepts matrices that are not necessarily complex Hermitian, optionally returning only a few of the eigenvalues and eigenvectors.

produces:

```
ans =
   -1.0000
    3.0000
    1.0000
    1.0000
    1.0000
```

References

[1] John R. Gilbert, Cleve Moler, and Robert Schreiber, *Sparse Matrices in Matlab: Design and Implementation*, **SIAM Journal on Matrix Analysis and Applications**, Volume 13, Number 1, pp. 333-356, January 1992.

Command Listing

M_x = bicg(M_a, V_b)
Solves the linear system, $M_a * M_x = V_b$, using the biconjugate gradients method.
Output: A matrix is assigned to M_x.
Argument options: M_x = bicg(M_a, V_b, num_{tol}) to specify a tolerance num_{tol}. ✦ M_x = bicg(M_a, V_b, num_{max}) to specify the maximum number of iterations, num_{tol}.
Additional information: There are many other options that can be provided to bicg. See the on-line help file for information.
See also: bicgstab, cgs, pcg, gmres, qmr

M_x = bicgstab(M_a, V_b)
Solves the linear system, $M_a * M_x = V_b$, using the biconjugate gradients *stabilized* method.
Output: A matrix is assigned to M_x.
Argument options: M_x = bicgstab(M_a, V_b, num_{tol}) to specify a tolerance num_{tol}. ✦ M_x = bicgstab(M_a, V_b, num_{max}) to specify the maximum number of iterations, num_{tol}.
Additional information: There are many other options that can be provided to bicgstab. See the on-line help file for information.
See also: bicg, cgs, pcg, gmres, qmr

M_x = cgs(M_a, V_b)
Solves the linear system, $M_a * M_x = V_b$, using the conjugate gradients squared method.
Output: A matrix is assigned to M_x.
Argument options: M_x = cgs(M_a, V_b, num_{tol}) to specify a tolerance num_{tol}. ✦ M_x = cgs(M_a, V_b, num_{max}) to specify the maximum number of iterations, num_{tol}.
Additional information: There are many other options that can be provided to cgs. See the on-line help file for information.
See also: bicg, bicgstab, pcg, gmres, qmr

V = **colmmd**(M_{sp})

Computes the minimum degree ordering for sparse matrix M_{sp}.
Output: A permutation vector is assigned to V.
Additional information: To create the permuted matrix, simply enter $M_{new} = M(:, V)$.
✣ The resulting permutation is one automatically used by the \ operator. ✣ The behavior of this command can be altered with spparms. ✣ For more information and examples, see the on-line help page.
See also: symmmd, symrcm, colperm, chol, *spparms*

V = **colperm**(M_{sp})

Reorders the columns of sparse matrix M_{sp} so that the columns with the least non-zero entries are furthest to the left.
Output: A permutation vector is assigned to V.
Additional information: To create the permuted matrix, simply enter $M_{new} = M(:, V)$.
See also: symmmd, symrcm colmmd, chol

[V_1, V_2, V_3] = **dmperm**(M_{sp})

Computes permutation of sparse square matrix M_{sp} that transform it into block triangular form.
Output: Row and column vectors are assigned to V_1 and V_2, respectively. A block boundary vector is assigned to V_3.
Argument options: V = dmperm(M_{sp}) to return a maximum matching for M_{sp}. ✣ [V_1, V_2, V_3, V_4] = **dmperm**(M_{sp}) to perform operation on a non-square matrix. V_3 and V_4 are assigned row and column boundary values defining the blocks.
Additional information: For an example of usage, see [1].
See also: sprank

V = **eigs**(M)

Computes *some* of the eigenvalues of square matrix M.
Output: A vector of numerical values is assigned to V.
Argument options: V = eigs(M, int) to return int eigenvalues. ✣ [M_{eig}, M_d] = eig(M) to assign a diagonal matrix of eigenvalues to M_d and a full matrix whose columns are the corresponding eigenvectors to M_{eig}.
Additional information: See the on-line help file for more details about available options for eigs.
See also: eig, svds

M = **full**(M_{sp})

Converts sparse matrix M_{sp} into a fully stored matrix.
Output: A fully stored matrix is assigned to M.
Additional information: If the input matrix is already stored fully, no change is made. ✣ There are several advantages to both full and sparse storage. For more information, see the on-line help file.
See also: sparse

M_x = gmres(M_a, V_b, num)

Solves the linear system, $M_a * M_x = V_b$, using the generalized minimum residual method with initial iterate num.
Output: A matrix is assigned to M_x.
Argument options: M_x = gmres(M_a, V_b, num, num$_{tol}$) to specify a tolerance num$_{tol}$. ✤ M_x = gmres(M_a, V_b, num, num$_{max}$) to specify the maximum number of iterations, num$_{tol}$.
Additional information: There are many other options that can be provided to gmres. See the on-line help file for information.
See also: bicg, bicgstab, cgs, pcg, qmr

int = issparse(M)

Determines if matrix M is stored as a sparse matrix.
Output: If M is stored sparsely, a value of 1 is assigned to int. Otherwise, a value of 0 is assigned to int.
See also: sparse, full

n = nnz(M)

Calculates the number of nonzero elements in matrix M.
Output: A non-negative integer value is assigned to n.
Additional information: This command is useful when dealing with sparse matrices.
See also: nonzeros, nzmax, whos, issparse

V = nonzeros(M)

Returns the non zero elements of matrix M, ordered by column.
Output: A vector of nonzero elements is assigned to V.
Additional information: The indices of the nonzero elements are not returned, just the values.
See also: nnz, nzmax, find, whos, issparse

n = nzmax(M)

Determines the number of storage locations used for the elements of matrix M.
Output: A non-negative integer value is assigned to n.
Additional information: When used on matrices with full storage, n is simply equal to the number of elements (of any value) in M. ✤ When used on matrices with sparse storage, n is greater than or equal to the number of nonzero elements in M. ✤ For more information, see the on-line help file.
See also: nonzeros, nnz, size, whos, issparse

M_x = pcg(M_a, V_b)

Solves the linear system, $M_a * M_x = V_b$, using the preconditioned conjugate gradients method.
Output: A matrix is assigned to M_x.
Argument options: M_x = pcg(M_a, V_b, num$_{tol}$) to specify a tolerance num$_{tol}$. ✤ M_x = pcg(M_a, V_b, num$_{max}$) to specify the maximum number of iterations, num$_{tol}$.
Additional information: There are many other options that can be provided to pcg. See the on-line help file for information.
See also: bicg, bicgstab, cgs, gmres, qmr

M_x = qmr(M_a, V_b)

Solves the linear system, $M_a * M_x = V_b$, using the quasi-minimal residual method.
Output: A matrix is assigned to M_x.
Argument options: M_x = qmr(M_a, V_b, num_{tol}) to specify a tolerance num_{tol}. ✤ M_x = qmr(M_a, V_b, num_{max}) to specify the maximum number of iterations, num_{tol}.
Additional information: There are many other options that can be provided to qmr. See the on-line help file for information.
See also: bicg, bicgstab, cgs, gmres, pgs

M_{sp} = spalloc(m, n, int)

Allocates the memory for an initially zero m × n sparse matrix which will hold at most int non-zero entries.
Output: A sparse matrix full of zeroes is assigned to M_{sp}.
Additional information: This command is more efficient than simply adding non-zero entries to a matrix without preset allocation. ✤ The value int can be exceeded, but after that point efficiency is sacrificed.
See also: sparse, issparse, nnz, nonzeros

M_{sp} = sparse(M)

Converts fully stored matrix M into a sparse matrix.
Output: A sparse matrix is assigned to M_{sp}.
Argument options: M_{sp} = sparse(V_i, V_j, V_{num}, m, n), where V_i, V_j, and V_{num} are vectors of equal size representing the row and column positions of non-zero entries and the corresponding entry itself, to create an m × n sparse matrix and assign it to M_{sp}. ✤ M_{sp} = sparse(V_i, V_j, V_{num}, m, n, int) to allocate space for a total of int non-zero elements when creating the matrix. ✤ M_{sp} = sparse(V_i, V_j, V_{num}) to compute the dimensions of the resulting matrix from max(V_i) and max(V_j). ✤ M_{sp} = sparse(m, n) to create an m × n matrix with no non-zero entries.
Additional information: If any elements in V_{num} are zero, they are removed along with the corresponding elements of V_i and V_j. ✤ If any pair of corresponding elements in V_i and V_j is equal to another pair of corresponding elements, then their corresponding V_{num} elements are summed and the redundancy is removed. ✤ Use the sixth parameter (or the spalloc command) whenever possible to maximize efficiency.
See also: full, spalloc, issparse, nnz, nonzeros, spones, sprand, speye

M_{aug} = spaugment(M)

Creates the augmented matrix which is associated with the least squares method for matrix M.
Output: If M is an m × n matrix, then a sparse m + n × m + n matrix is assigned to M_{aug}.
Argument options: M_{aug} = spaugment(M, num) to multiply the identity matrix in the upper left-hand corner of the augmented matrix by the scalar value num. The default value is 1.
Additional information: For more information on the characteristics of augmented matrices and their uses, see the on-line help file.
See also: spparms

M_{sp} = **spconvert(M)**

Converts a $3 \times n$ matrix M, where the three columns represent the rows and columns of the nonzero entries and their respective values, into a sparse matrix.

Output: A sparse matrix is assigned to M_{sp}.

Argument options: M_{sp} = spconvert($M_{4 \times n}$), where $M_{4 \times n}$ is a $4 \times n$ matrix with the third and fourth rows represent the real and complex components of the nonzero values, to create sparse matrix of complex values.

Additional information: This command is typically used when reading in nonzero values from an external ASCII file. ♦ If the input matrix contains a [m, n, 0] or [m, n, 0, 0] entry, then the size of the resulting matrix is m × n. Otherwise, the size of the resulting matrix is taken from the highest row and column values.

See also: sparse, spalloc, issparse, nnz, nonzeros

M_{sp} = **spdiags(M, V_{diag}, m, n)**

Creates a sparse matrix of dimensions m × n whose diagonals specified by the elements of V_{diag} are set to the related values in the columns of M.

Output: A sparse matrix is assigned to M_{sp}.

Argument options: M_{out} = spdiags(M, V_{diag}, M_{in}) to replace the diagonals in M_{in} specified by the elements of V_{diag} with the appropriate values from M. ♦ M_{out} = spdiags(M, V) to extract the elements in the specified diagonals from M and place them in M_{out}. ♦ [M_{out}, V_{out}] = spdiags(M) to extract the elements in all the diagonals from M. The elements are placed in the columns of M_{out} and the corresponding diagonal specifiers in V_{out}.

Additional information: A positive value in V_{diags} specifies a superdiagonal in the result. A negative value in V_{diags} specifies a subdiagonal in the result. ♦ For an example of usage, see the tutorial portion of this chapter.

See also: diag

M_{sp} = **speye(n)**

Creates the sparse n × n identity matrix.

Output: A square sparse matrix is assigned to M_{sp}.

Argument options: M_{sp} = speye(m, n) to create a sparse m × n matrix with ones on the main diagonal (starting in the upper-left corner) and zeroes elsewhere.

Additional information: For large identity matrices, sparse allocation is much more memory efficient than full representation.

See also: eye, spones, sparse, zeros

M_{new} = **spfun('fnc', M_{sp})**

Applies the function fnc to each nonzero element of sparse matrix M_{sp}.

Output: A sparse matrix of the same dimensions as M_{sp} is assigned to M_{new}.

Additional information: The function fnc must take a matrix as its only parameter and act upon each individual element.

See also: sparse

M_{new} = **spones(M_{sp})**

Replaces all nonzero entries in sparse matrix M_{sp} with values equal to 1.

Output: A sparse matrix with the same dimensions as M_{sp} is assigned to M_{new}.

Additional information: Using sum in conjunction with spones allows you to count the number of nonzero entries in each row or column of M_{sp}.
See also: speye, spones, ones, sum, nnz

spparms(str, num)
Sets sparse matrix algorithm parameter str to the value num.
Output: Not applicable
Argument options: spparms(V) to set *all* the available parameters to the related values in vector V. ♣ spparms('*default*') to reset all parameters to their default values. ♣ spparms('*tight*') to set minimum degree ordering parameters to their *tight* settings. ♣ num = spparms(str) to return the current setting of parameter str. ♣ V = spparms to return the current setting of all the parameters in vector V. ♣ [M, V] = spparms to return the vector of current settings and the related parameter names in character matrix M.
Additional information: While there are 10 individual parameters that can be set, only two ('*spumoni*' and '*autommd*') are frequently used. See the on-line help file for more details.
See also: symmmd, colmmd, \

M_{sp} = sprand(m, n, num)
Creates an m × n sparse matrix, with a nonzero density of approximately num, of uniformly distributed random values.
Output: An m × n sparse matrix is assigned to M_{sp}.
Argument options: M_{sp} = sprandn(m, n, num_d, num_{rc}) to specify that the reciprocal condition number of M_{sp} be num_{rc}. ♣ M_{sp} = sprandn(M) to create a random sparse matrices with the same sparsity structure as M and assign it to M_{sp}.
Additional information: For more information on reciprocal condition numbers, see the on-line help file.
See also: sprandsym, sprandn, rand

M_{sp} = sprandn(m, n, num)
Creates an m × n sparse matrix, with a nonzero density of approximately num, of normally distributed random values.
Output: An m × n sparse matrix is assigned to M_{sp}.
Argument options: M_{sp} = sprandn(m, n, num_d, num_{rc}) to specify that the reciprocal condition number of M_{sp} be num_{rc}. ♣ M_{sp} = sprandn(M) to create a random sparse matrices with the same sparsity structure as M and assign it to M_{sp}.
Additional information: The random entries are generated with a mean of 0 and a variance of 1. ♣ For more information on reciprocal condition numbers, see the on-line help file.
See also: sprandsym, randn

M_{sp} = sprandsym(n, num)
Creates an n × n square symmetric sparse matrix, with a nonzero density of approximately num, of normally distributed random values between 0 and 1.
Output: An n × n sparse matrix is assigned to M_{sp}.

Argument options: M_{sp} = sprandsym(n, num$_d$, num$_{rc}$) to specify that the reciprocal condition number of M_{sp} be num$_{rc}$. ✤ M_{sp} = sprandn(M) to create a random symmetric sparse matrices with the same sparsity structure as M and assign it to M_{sp}. ✤ M_{sp} = sprandsym(n, num$_d$, num$_{rc}$, 1) to use a random Jacobi rotation of a positive definite diagonal matrix to compute M_{sp}. ✤ M_{sp} = sprandsym(n, num$_d$, num$_{rc}$, 2) to use a shifted sum of outer products to compute M_{sp}. ✤ M_{sp} = sprandsym(M, num$_d$, num$_{rc}$, 3) create a random symmetric sparse matrices with the same sparsity structure as M and with a condition number approximately equal to 1/num$_{rc}$.
Additional information: The random entries are generated with a mean of 0 and a variance of 1. ✤ For more information on reciprocal condition numbers, see the on-line help file for sprandn.
See also: sprandn

n = sprank(M_{sp})
Computes the structural rank of sparse matrix M_{sp}.
Output: A non-negative integer is assigned to n.
Additional information: The structural rank is also known as the *maximum traversal* or *maximum assignment* of M_{sp}. ✤ The structural rank of M_{sp} is always greater than or equal to rank(M_{sp}).
See also: rank, dmperm

spy(M_{sp})
Displays a plot of the sparsity pattern of sparse matrix M_{sp}.
Output: A two-dimensional plot with dots corresponding to the nonzero elements of M_{sp} is displayed.
Argument options: spy(M_{sp}, str) to use the color represented by str for the dots instead of the default yellow. ✤ spy(M_{sp}, num) to change the marker size to num.
Additional information: The two optional arguments to spy can be used in combination.
See also: sprandsym, sprandn

V = svds(M)
Computes *some* of the singular values of matrix M.
Output: A vector is assigned to V.
Argument options: V svds(M, int) to find the int largest singular values of M.
Additional information: For more information on available options and the matrices produced, see the on-line help file.
See also: svd, eigs

V = symmmd(M)
Computes a permutation of symmetric positive definite matrix M, such that the new matrix has a sparser Cholesky factorization.
Output: A permutation vector is assigned to V.
Additional information: To create the permuted matrix, simply enter M_{new} = M(V, V). ✤ The resulting permutation is one automatically used by the \ operator. ✤ On occasion, symmmd also works for symmetric indefinite matrices. ✤ The behavior of

this command can be altered with spparms. ♣ For more information and examples, see the on-line help page.
See also: colmmd, symrcm, colperm, chol, *spparms*

V = symrcm(M)

Computes a symmetric reverse Cuthill-McKee ordering of the matrix M.
Output: A permutation vector is assigned to V.
Additional information: To create the permuted matrix, simply enter $M_{new} = M(V, V)$. ♣ The permuted matrix tends to have its nonzero entries closer to the diagonal (i.e., it has a narrower bandwidth). ♣ For sparse real symmetric M, it is usually more efficient to compute the eigenvalues of M(V, V). ♣ For more information and examples, see the on-line help page.
See also: colmmd, symmmd, colperm, chol, eig

Non-Linear Equations

This chapter examines solving non-linear equations. The three MATLAB functions covered are:

- roots finds *all* roots of a single polynomial, $f(x) = \sum_{i=0}^{n} a_i x^i$.
- fzero finds *one* root of a single function in one variable, $f(x)$.
- fsolve finds *one* root of one or more functions in one or more variables, $f_i(x_j)$.

Both roots and fzero are part of the basic MATLAB package. fsolve, however, is part of the *Optimization Toolbox* which is sold separately.

There are also several methods for solving non-linear equations provided in the *Symbolic Toolbox*, which are demonstrated in the *Symbolic Computations* chapter.

roots - Finding All Roots of a Polynomial

Consider the following simple 2nd-degree polynomial defining a parabola

$$f(x) = x^2 - 2$$

which is shown in Figure 7.

As seen previously, this polynomial can be defined as a vector

```
p = [1 0 -2]
```

and its roots may now be found using

```
p_roots = roots(p)
```

which returns

```
p_roots =
    1.4142
   -1.4142
```

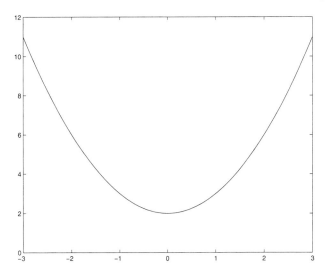

FIGURE 7. Plot of a simple polynomial

Notes

1. To create the plot in Figure 7, you must first create an M-file which evaluates it at any point.

   To be more efficient, create the following function, myval, which evaluates *any* polynomial defined by the vector my at any point x:

   ```
   function y = myval(x)
   global my
   
   y = polyval(my, x);
   end
   ```

   Then, to plot this particular polynomial, do as follows:

   ```
   global my
   
   my = p
   fplot('myval', [-3 3])
   
   % Plot y=0 for x from -3 to 3:
   
   hold on
   plot([-3 3], [0 0])
   hold off
   
   print -deps parabola.eps
   ```

fzero - Finding One Root of f(x)

fzero finds one root of a single function of one variable, f(x). For example, consider the graph in Figure 8 representing supply and demand for some product, say lacrosse balls.[19]

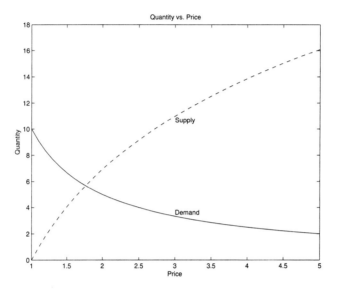

FIGURE 8. Plot of lacrosse ball supply and demand model

In a competitive marketplace, the price gravitates to an equilibrium point where the two lines intersect. To find this (Price, Quantity)-point, you need to solve the non-linear equations

$$(Supply) \quad Quantity = 10\log(Price)$$
$$(Demand) \quad Quantity = 10/Price$$

In other words, you wish to find the value of *Price* which satisfies

$$10\log(Price) = 10/Price$$

which can be re-expressed as

$$surplus(Price) = 10\log(Price) - 10/Price$$

where the *surplus(Price)* function indicates by how much supply exceeds demand for any particular value of *Price*. The plot of this function is shown in Figure 9.

---

[19] Again, no slight against hockey, but lacrosse is actually the *official* national sport of Canada.

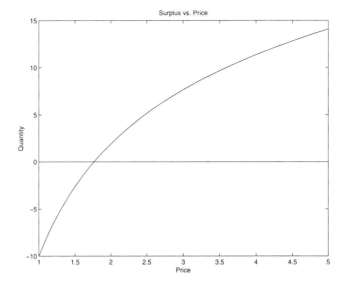

FIGURE 9. Plot of lacrosse ball surplus

You can invoke the fzero function in MATLAB to solve for the value of *Price* that makes *surplus(Price)* equal to 0.

First, create an M-file, surplus.m, which defines the function.

```
function s = surplus(Price)
s = (10.0) .* log(Price) - (10.0) ./ Price;
```

Then, invoke the fzero function to solve for the critical price.

```
          guess = 0;
            tol = eps;
              trace = 1;
price = fzero('surplus', guess, tol, trace)
```

We have indented the first three lines above simply to make the assignments line up with the variables in the invocation of fzero. While not required by MATLAB, this can save time, especially when checking long lists of arguments by hand.

guess is a guess at where the desired root is. The algorithm for fzero uses this as a starting point for its computations. (It was not necessary to set up this variable; you could have directly input 0 as the second parameter to fzero. However, using a variable clarifies the meaning of parameters.)

tol is the *relative tolerance*, which specifies how accurate the result should be. The default value for tol is eps.

trace controls whether or not fzero is to print the value of the surplus function at points where it is invoked during the solution process. The default value for trace is 0, which suppresses the extra printing; any other value prints the extra messages.

Both tol and trace can be omitted from the call to fzero, in favor of the default values. Then this example can be abbreviated to

```
price = fzero('surplus', guess)
```

The result of invoking fzero as above is

```
price =
   -0.0003 + 0.0080i
```

This complex value is not at all what was expected. Try again, but with a non-zero initial guess.

```
            guess = 1;
            tol = eps;
            trace = 1;
price = fzero('surplus', guess, tol, trace)
```

The result is much better.

```
price =
   1.7632
```

Now check the answer.

```
surplus(price)

ans =
   8.8818e-16
```

The result is a number very close to 0. (The actual value will vary between different types of computers.)

fsolve - Finding One Root of a System $f_i(x_j)$

fsolve finds one root of one or more functions of one or more variables, $f_i(x_j)$. For example, consider the following equation for the position of a mass on the end of a spring, suspended in a viscous liquid.[20]

$$x(t, \zeta) = e^{-\zeta \omega t} \left(\frac{\zeta \sin(\omega \sqrt{1-\zeta^2} t)}{\sqrt{1-\zeta^2}} + \cos(\omega \sqrt{1-\zeta^2} t) \right),$$

where ζ (zeta) is the *damping value* $(0 < \zeta < 1)$.[21]
The velocity for this system at any time is given by

$$v(t, \zeta) = -\frac{\omega e^{-\zeta \omega t} \sin(\omega \sqrt{1-\zeta^2} t)}{\sqrt{1-\zeta^2}}.$$

Set ω equal to 1, and plot $x(t, \zeta)$ and $v(t, \zeta)$ for various values of ζ. (see Figure 10.)

---

[20] These equations have been simplified by assuming an initial position of $x(0) = 1$ and an initial velocity of $v(0) = 0$. Also assume that the natural frequency, ω_n, is 1.
[21] Refer to the *Differential Equations* chapter for an explanation of ω and ζ.

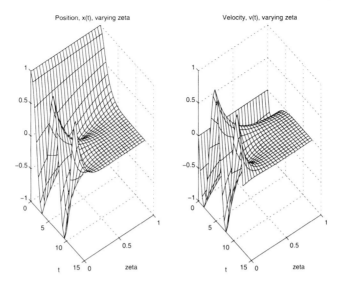

FIGURE 10. Surface plots of spring-mass-damper system

Notice that initially (at t=0), $x(0, \zeta) = 1$ and $v(0, \zeta) = 0$ for all values of ζ. You wish to find the (t, ζ)-point where the velocity is again 0, but the position is half its original value. In other words, you want to solve the system of equations

$$x(t, \zeta) - \frac{1}{2}x(0, \zeta) = 0$$
$$v(t, \zeta) - v(0, \zeta) = 0$$

It helps to create a contour plot of $x(t, \zeta) = 0.5$ and $v(t, \zeta) = 0$ to see if there are any (t, ζ)-points where both these conditions are true. Such a plot is shown in Figure 11, and you can see that one solution is near the point $(t = 6.0, \zeta = 0.1)$.

The first step to finding the *exact* point is to create an M-file containing the two functions to be solved. Here it is called smd.m.

```
function der_yv = smd(t, yv)

global m omega_n zeta;        % "t" a vector?

    y = yv(1);                % Extract y and v
    v = yv(2);                %-----------------

%--------------------------------------------------
% Compute derivative of y(t) and v(t):
%--------------------------------------------------

    der_y = v;
    der_v = 0/m - (2 * omega_n * zeta) * v ...
              - (omega_n)^2 * y;

%--------------------------------------------------

    der_yv(1) = der_y;        % Store in der_yv
```

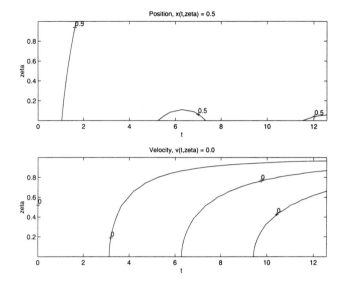

FIGURE 11. Contour plot of spring-mass-damper system

```
            der_yv(2) = der_v;       %----------------
```

Then you create a main program to call fsolve and find where smd.m returns zero values.

```
global omega x_target v_target;

omega = 1;

%-------------------------------------------------

x_target = 1/2;
v_target = 0;

                t_zeta0 = [6. .1]';

                    options = foptions;
                    options(1) = 0;
                    options(2) = 10^(-4);
                    options(3) = 10^(-4);
                    options(5) = 0;
                    options(7) = 0;

t_zeta = fsolve('smd', t_zeta0, options);

t    = t_zeta(1)
zeta = t_zeta(2)

%-------------------------------------------------

x_target = 0;
v_target = 0;

smd([t, zeta])
```

And the result of all this is

```
t =
    6.3213

zeta =
    0.1097

ans =
    0.5000   -0.0000
```

where the two values in ans represent the values of x and v at the given values of t and zeta.

In the main program, two global variables are used, x_target and v_target, allowing you to vary the target values of x and v from outside the smd.m function. Once you have found the values of t and zeta corresponding to this point (using fsolve), evaluate x(t,zeta) and v(t,zeta) at that point, simply by setting x_target and v_target to 0.

The default values of the five options that fsolve uses have actually been declared in the program. Another way of initializing the options vector is to use the built-in foptions function.

```
options = foptions;
```

options(1) controls whether or not fsolve is to print intermediate results during the solution process. options(2) is a *relative tolerance* which specifies how accurate the solution should be. options(3) is used as a tolerance in the computation of the functions being solved. options(5) specifies the algorithm to be used. The default of 0 specifies the *Gauss-Newton* method. A value of 1 specifies the *Levenberg-Marquardt* method. options(7) specifies the *line search procedure* to be used. The default of 0 specifies a mixed quadratic and cubic line search procedure. A value of 1 specifies a purely cubic line search procedure.

Command Listing

M = **fsolve**(fnc, M$_{init}$)

Solves a system of non-linear equations, returned by the function fnc, given initial guesses for the solutions in matrix M$_{init}$. (This command is in the *Optimization Toolbox*.)

Output: A matrix of numeric values is assigned to M.

Argument options: M = fsolve(fnc, M$_{init}$, options) to specify various control parameters for fsolve. For a listing of the elements of options that are actually used in fsolve, see the example in the tutorial section. ✤ M = fsolve(fnc, M$_{init}$, options, 'grad', expr$_1$, expr$_2$, ..., expr$_n$) to pass the parameters expr$_1$ through expr$_n$ directly to function fnc. This can eliminate the need for using global variables. ✤ [M, name] = fsolve(fnc, M$_{init}$) to return the parameters used in finding the solution in name. name(10) represents a count of the function evaluations made.

Additional information: The function fnc can either be found in a standard M-file (fnc.m) or specified in the parameter sequence of fsolve as a string. ✤ The default

values for the optional parameters can be found in the vector *foptions*. ✦ See the tutorial section of this chapter for an example of using this command.
See also: roots, fsolve

num_0 = **fzero(fnc, num)**

Computes a zero (root) of the one-variable function fnc near to the value num.
Output: A numeric value is assigned to num_0.
Argument options: num_0 = fzero(fnc, num, num_{tol}) to find a zero to within a tolerance of num_{tol}. Default is the value eps. ✦ num_0 = fzero(fnc, num, num_{tol}, num_{tr}) where num_{tr} is a nonzero value to trace the iterations used to find the zero
Additional information: The function fnc can either be found in a standard M-file (`fnc.m`) or specified in the parameter sequence of fzero as a string.
See also: roots, fsolve

V_c = **roots(V_r)**

Computes the roots of the polynomial whose coefficients are represented by the elements of row vector V_r.
Output: A column vector containing the roots of the polynomial is assigned to V_c.
Argument options: V_r = roots(V_c), where V_c is a column vector containing the roots of a polynomial, to assign the coefficients of that polynomial to row vector V_r.
Additional information: Coefficients are ordered in descending powers. ✦ Computed roots are in no specific order. ✦ For the most part, poly and roots are the inverse functions for each other. ✦ For more information about the algorithms used, see the on-line help file. ✦ To find a root near to a specific value, use fzero.
See also: fzero, poly, polyval, conv, residue

zerodemo

Runs a demonstration script showing how to use the fzero command.
Output: Not applicable.
Additional information: Various commands are automatically entered for you.
✦ Occasionally, you will be prompted to strike any key to continue the demonstration. ✦ For a complete list of demonstrations available on your platform, see the on-line help file for demos.
See also: fzero, quaddemo, odedemo, fftdemo, fplotdemo

Optimization

This chapter looks at minimizing or maximizing[22] functions, with or without constraints.

The four MATLAB functions covered are:

- fmin finds a minimum of a function of 1 variable, $f(x)$.
- fmins finds a minimum of a function of one or more variables, $f(x_i)$.
- constr finds a minimum of a function of one or more variables, $f(x_i)$, subject to constraints.
- lp finds the minimum of a *linear* function of one or more variables, $f(x_i)$, subject to *linear* constraints.

Both fmin and fmins come as part of MATLAB. However, constr and lp are part of the *Optimization Toolbox,* which contains numerous other functions and is sold separately.

The solution found by these functions is not guaranteed to be the *global* minimum of your function, only a *local* minimum. You may need to adjust your initial guess to improve the result.

Chapter E04 of the *NAG Toolbox* (available separately) contains several optimization functions written in FORTRAN which can be used from within MATLAB.

`fmin` - Minimize $f(x)$

Consider the following simple 2nd-degree polynomial defining a parabola:
$$f(x) = x^2 - 2$$
which is shown in Figure 12.

Since no special functions exist in MATLAB to find the minimum of a *polynomial*, treat the polynomial like any other non-linear function. First, create an M-file to define the function to be minimized, e.g., `myfun.m`:

---

[22] All of these MATLAB functions find the *minimum* of a function. To find the *maximum* of a function, simply ask MATLAB to minimize a function which returns the *negative* of your function value.

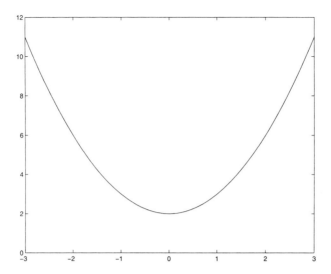

FIGURE 12. Plot of a simple polynomial

```
function y = myfun(x)

y = x.^2 - 2;
```

Then a local minimum in the range $x_1 \leq x \leq x_2$ may be found using fmin as follows:

```
x1 = -2;
x2 = 2;
    options = foptions;
    options(1)  = 0;
    options(2)  = 1e-4;
    options(14) = 500;

x = fmin('myfun', x1, x2, options)

y = myfun(x)
```

which returns results resembling:

```
x =
  -5.5511e-17

y =
  -2
```

foptions returns an 18-element vector containing the default values of optional control parameters for MATLAB's optimization routines. Not all optimization routines use all 18 options, however. The fmin routine uses only the three options specified above (1, 2, and 14), and the values listed here are actually the *default* values which foptions provides.

If options(1) is non-zero, then fmin displays intermediate results during execution. options(2) is used as a tolerance in the computation of x. options(14) is the maximum number of intermediate steps that fmin may take in computing x.

If the default values are acceptable, then the above invocation of fmin may be abbreviated to

```
x = fmin('myfun', x1, x2)

y = myfun(x)
```

Notes

- Both fmins and constr could be used in this example. Both commands are, however, more flexible than is needed in this case. See the next sections for descriptions of problems that *require* fmins and constr.

fmins - Minimize $f(x_i)$

fmins can be used to minimize functions of one or more variables. Consider the following simple function of two variables (shown in Figure 13) which defines a paraboloid:

$$f(x,y) = x^2 + y^2 - 2$$

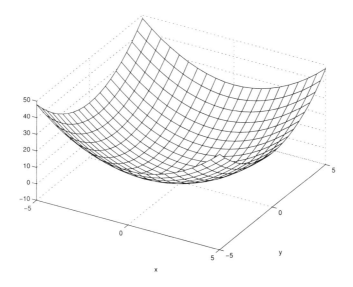

FIGURE 13. Plot of a paraboloid

First, create an M-file to define the function, e.g., myfun2.m:

```
function z = myfun2(xy)

x = xy(1);
y = xy(2);

z = x.^2 + y.^2 - 2;
```

Then a local minimum may be found using fmins. Notice that you must provide an initial guess (xy0). Here the guess is that the minimum value is at x = 1 and y = 1:

```
                          xy0     = [1., 1.];
                          options = foptions;
                          options(1)  = 0;
                          options(2)  = 1e-4;
                          options(3)  = 1e-4;
                          options(14) = 200;

xy = fmins('myfun2', xy0, options)

z = myfun2(xy)
```

The results resemble:

```
xy =
  1.0e-15 *
  -0.7772    -0.7772

z =
  -2
```

The fmins routine uses four of the optional value in options. options(1), options(2), and options(14) have the same meaning as they do with fmin. options(3) is used as a tolerance in the computation of the function being optimized.

Notes

- The fminu command, in the *Optimization Toolbox*, provides similar functionality as fmins but uses a different algorithm. Refer to the *Optimization Toolbox User's Guide* for details.

constr - Minimize $f(x_i)$, Subject to Constraints

constr finds a minimum of a function of one or more variables, $f(x_i)$, *subject to constraints*. First, consider a simple example with only one constraint (an inequality), and then look at a more typical example with several constraints (inequalities and equalities).

constr - a Simple Example

Add a single constraint to the previous problem, and solve it using constr.

$$f(x, y) = x^2 + y^2 - 2$$
$$x + 5 \leq y$$

A graphical representation of this problem is shown in Figure 14.

First create an M-file that returns the function value (f) and the constraint value (g) at a point (xy), e.g., myfun2c.m:

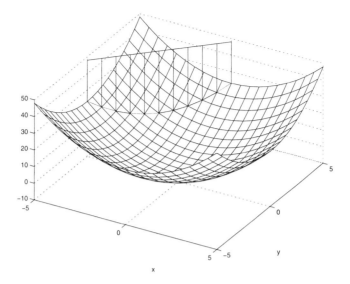

FIGURE 14. Plot of a paraboloid, with one constraint

```
function [f,g] = myfun2c(xy)

x = xy(1);
y = xy(2);

f = x.^2 + y.^2 - 2;
g = x + 5 - y;

end
```

Then call the constr function to find an xy that minimizes f, such that $g \leq 0$.[23] Notice that with constr you must provide an initial guess, xy0. While this initial guess need not satisfy all the constraints provided, two values that do, x = -4 and y = 1, are given for the initial guess.

```
xy0 = [-4., 1.];
    options = foptions;
    options(1)  = 0;
    options(2)  = 1e-4;
    options(3)  = 1e-4;
    options(13) = 0;
    options(14) = 2 * 100;

xy = constr('myfun2c', xy0, options)

[f, g] = myfun2c(xy)
```

The results of this are:

```
xy =
    -2.5000    2.5000
```

---

[23] In this example, there is just one constraint, so g returns a scalar value. However, in the next example, g returns a vector of five values.

```
f =
    10.5000

g =
    -4.4409e-16
```

Notice that the value of g is practically zero, indicating that the solution lies *along the edge* of the constraint, which is what we would expect from the graph of the function in Figure 14.

As you can see, the constr routine uses five of the optional value in options. options(1), options(2), options(3), and options(14) have the same meaning as they do with fmins. options(14) has a default value of one hundred times the number of variables in the function being optimized (2*100 in this example). options(13) specifies the number of *equality* constraints. There are none in this example; it only has one constraint and it is an *inequality* constraint.

constr - a More Typical Example

Consider the diagram in Figure 15 showing the quantity, $X_{i,j}$, of some product (say lacrosse balls) transported between two plants (i) and three markets (j).[24]

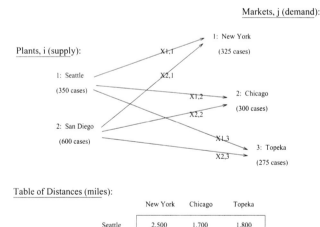

FIGURE 15. Plot of lacrosse ball Transportation

---

[24] This example is based upon an optimization problem found in *GAMS: A User's Guide, Release 2.25*, A. Brooke et al., The Scientific Press, South San Francisco, 1992, pp. 7–32. It in turn is based upon an example in *Linear Programming and Extensions*, G. B. Dantzig, Princeton University Press, Princeton, New Jersey, 1963, pp. 35–42.

The diagram shows the six *variables* in this problem, $X_{i,j}$: the quantity shipped from plant i to market j. One possible, but not optimal, solution is

$$X_{i,j} = \begin{bmatrix} X_{1,1} & X_{1,2} & X_{1,3} \\ X_{2,1} & X_{2,2} & X_{2,3} \end{bmatrix} = \begin{bmatrix} 25 & 200 & 100 \\ 300 & 100 & 175 \end{bmatrix}$$

Also shown in Figure 15 is a table of distances between these cities. Assuming a constant freight rate of $0.09/case/mile, then the shipping cost between plant i and market j (in $/case) is:

$$C_{i,j} = \begin{bmatrix} C_{1,1} & C_{1,2} & C_{1,3} \\ C_{2,1} & C_{2,2} & C_{2,3} \end{bmatrix} = \begin{bmatrix} 225.00 & 153.00 & 162.00 \\ 225.00 & 162.00 & 126.00 \end{bmatrix}$$

The *function* you want to minimize is the total shipping cost, $F(X_{i,j})$:

$$F(X_{i,j}) = \sum_{i=1}^{2} \sum_{j=1}^{3} C_{i,j} X_{i,j}$$

which, for the above non-optimized value of X, totals $158,175.00.

The diagram shows the *supply* at each plant (A_i), and the *demand* at each market (B_j):

$$A_i = \begin{bmatrix} 350 \\ 600 \end{bmatrix}$$
$$B_j = \begin{bmatrix} 325 & 300 & 275 \end{bmatrix}$$

Therefore, this problem has a total of five constraints to be satisfied.

The five constraints are:

(a) you cannot ship more than the supply (A_i) available at each of the two plants:

$$\sum_{j=1}^{3} X_{1,j} \leq A_1 \quad \rightarrow \quad \sum_{j=1}^{3} X_{1,j} - A_1 \leq 0$$

$$\sum_{j=1}^{3} X_{2,j} \leq A_2 \quad \rightarrow \quad \sum_{j=1}^{3} X_{2,j} - A_2 \leq 0$$

(b) you cannot ship less than the demand (B_j) at each of the three markets.

$$\sum_{i=1}^{2} X_{i,1} \geq B_1 \quad \rightarrow \quad -(\sum_{i=1}^{2} X_{i,1} - B_1) \leq 0$$

$$\sum_{i=1}^{2} X_{i,2} \geq B_2 \quad \rightarrow \quad -(\sum_{i=1}^{2} X_{i,2} - B_2) \leq 0$$

$$\sum_{i=1}^{2} X_{i,3} \geq B_3 \quad \rightarrow \quad -(\sum_{i=1}^{2} X_{i,3} - B_3) \leq 0$$

The constr function requires that all constraining inequalities be put in the form of a *less than or equal to zero* inequality; which is why we rewrote the above expressions to fit this requirement.

You can invoke the constr function in MATLAB's *Optimization Toolbox* to solve for a value of $X_{i,j}$ that minimizes the total cost subject to these constraints.[25]

First, you create an M-file to define the function to be minimized and the constraints, e.g., ftrans.m:

```
function [f,g] = ftrans(X)
global A B C;

f    = sum( sum( C .* X ) );

% sum of first row
g(1) =        sum( X(1,:) )  -  A(1);
g(2) =        sum( X(2,:) )  -  A(2);

% sum of first column
g(3) =  -(    sum( X(:,1) )  -  B(1)  );
g(4) =  -(    sum( X(:,2) )  -  B(2)  );
g(5) =  -(    sum( X(:,3) )  -  B(3)  );
```

As mentioned before, constr expects the constraints, g(1) through g(5), to be of the form: g(i) ≤ 0. Look carefully at the exact way they are defined above. The = signs are not part of the inequality, but are the method of assigning the inequalities to elements of the constraint vector, g. Notice that the ≤ 0 part of the inequalities is left off.

Now you can invoke the constr function to solve for $X_{i,j}$, the quantity to ship between each plant (*i*) and market (*j*):

```
global A B C;

A = [350
     600];

B = [325    300    275];

C = 0.09 * [2500    1700    1800
            2500    1800    1400] / 1000;     % Had to scale by
                                              % 1000 (see text).

            X0 = [0   0   0
                  0   0   0];     % Guess.

            options = foptions;
            options(1)  = 0;
            options(2)  = 1e-4;
            options(3)  = 1e-4;
            options(13) = 0;                  % For now.
            options(14) = 6 * 100;

            Xlo = [0   0   0
                   0   0   0];
```

---

[25] constr is actually designed to handle *non-linear* problems, but you can still use it to solve linear problems such as this one. The next section uses a linear solver which is more efficient at finding a solution.

```
                                        Xhi = [];
X = constr('ftrans', X0, options, Xlo, Xhi);

X

[f, g] = ftrans(X)
```

The result of invoking constr resembles:

```
X =
   47.7532   300.0000    -0.0000
  277.2468         0    275.0000

f =
  153.6750

g =
   -2.2468   -47.7532   -0.0000        0        0
```

Observe that all of the elements of g are less than or equal to zero, indicating that all of the given constraints have been satisfied.

An additional constraint on the problem is that the $X_{i,j}$ values must be positive or zero. *Simple bounds* like this *could* be expressed as six further constraints, but constr provides a simpler mechanism of providing upper- and lower-bound matrices for the values in $X_{i,j}$. In the above example, Xlo(i,j) contains a lower bound on the value of X(i,j). Xhi has been specified as a *null matrix* to indicate that there are no upper bounds on the values in X(i,j).

Also note that this example divides $C_{i,j}$ by 1000, so that costs are calculated in thousands of dollars. Therefore the total cost is actually f * 1000 which is $153,675.00 (a savings of $4500). This division was done to avoid the following warning:[26]

```
Warning: Matrix is close to singular or badly scaled.
         Results may be inaccurate. RCOND = 1.075171e-17
```

Such scaling is not necessary when using the lp function, as is shown in the next section.

As mentioned earlier, options(13) specifies the number of *equality* constraints, which in this example is 0 (the default). If you change the above example to set:

```
options(13) = 1;
```

then constr treats the first constraint as an *equality*, i.e.:

$$\sum_{j=1}^{3} X_{1,j} - A_1 = 0$$

meaning that you have to ship *exactly* the number of cases produced from plant 1—no more and no less.

---

[26] Refer to the *Optimization Toolbox User's Guide* under *Practicalities* for further information on warning messages like this.

In this case, the solution is:

```
X =
   50.0000  300.0000         0
  275.0000         0  275.0000

f =
  153.6750

g =
    0.0000  -50.0000         0   -0.0000         0
```

The only market that is affected is New York (j = 1). The total cost (f) did not rise because New York is equidistant from the two plants.

lp - Minimize *Linear* $f(x_i)$, Subject to *Linear* Constraints

lp finds the minimum of a *linear* function of one or more variables, $f(x_i)$, subject to *linear* constraints (both *inequalities* and *equalities*).

The last example was actually linear, so you can solve it using lp, which is more efficient in terms of memory and execution time. When using lp, you must first express your optimization problem in *matrix* form (which is customary in linear programming theory).

The *variables* were stored in a matrix, $X_{i,j}$, for convenience, but now *must* be re-expressed as a (column) vector, x:

$$x = \begin{bmatrix} X_{1,1} \\ X_{2,1} \\ X_{1,2} \\ X_{2,2} \\ X_{1,3} \\ X_{2,3} \end{bmatrix}$$

Our $C_{i,j}$ matrix must also be converted to a (column) vector, c:

$$c = \begin{bmatrix} C_{1,1} \\ C_{2,1} \\ C_{1,2} \\ C_{2,2} \\ C_{1,3} \\ C_{2,3} \end{bmatrix}$$

Then the cost *function* becomes

$$f(x_i) = \sum_{i=1}^{6} c_i x_i,$$

which can then be expressed simply as the vector dot product of the two vectors c and x:

$$f(x) = c^T x.$$

As for the *constraints*, the idea is to express them in the following form:

$$Mx \leq v,$$

where, in this specific example, M is a 5×6 matrix, and v is a five-element vector:

$$\begin{bmatrix} 1 & 0 & 1 & 0 & 1 & 0 \\ 0 & 1 & 0 & 1 & 0 & 1 \\ \hline -1 & -1 & 0 & 0 & 0 & 0 \\ 0 & 0 & -1 & -1 & 0 & 0 \\ 0 & 0 & 0 & 0 & -1 & -1 \end{bmatrix} \begin{bmatrix} X_{1,1} \\ X_{2,1} \\ \hline X_{1,2} \\ X_{2,2} \\ \hline X_{1,3} \\ X_{2,3} \end{bmatrix} \leq \begin{bmatrix} A_1 \\ A_2 \\ \hline -B_1 \\ -B_2 \\ -B_3 \end{bmatrix}$$

The model can now be readily translated into MATLAB and a call to lp finds an x that minimizes $c^T x$, such that $Mx \leq v$ and $x_i \geq 0$:

```
A = [350
     600];

B = [325   300   275];

C = 0.09 * [2500   1700   1800
            2500   1800   1400];

                c = reshape(C, 6, 1);
%                         X11 X21   X12 X22   X13 X23
                M = [  1    0     1    0     1    0
                       0    1     0    1     0    1
                      -1   -1     0    0     0    0
                       0    0    -1   -1     0    0
                       0    0     0    0    -1   -1 ];

                v = [ A(1)
                      A(2)
                     -B(1)
                     -B(2)
                     -B(3) ];

                xlo = zeros(6, 1);

                xhi = [];

                X0 = [0    0    0
                      0    0    0];

                x0 = reshape(X0, 6, 1);

                n = 0;

[x, lambda, how] = lp(c, M, v, xlo, xhi, x0, n);
```

```
X = reshape(x, 2, 3)
how

obj = c' * x              % Same as:    sum( sum( C .* X ) )
con = M*x - v
```

The results resemble:

```
X =
          0.0000   300.0000          0
        325.0000          0   275.0000

how =
      ok

obj =
      153675

con =
       -50.0000
        -0.0000
              0
              0
              0
```

Because you are using the default values for xhi (the upper bound), x0 (the initial guess), and n (the number of equality constraints), the invocation of lp could be shortened to

```
x = lp(c, M, v, xlo);
```

MATLAB's handy reshape function has been used to take the elements of the 2 × 3 matrix C and put them into a 6 × 1 matrix (column vector) c. reshape copies the elements column-wise, so that it behaves as if the following had been issued:[27]

```
c(1) = C(1,1);                              % column 1
c(2) = C(2,1);

c(3) =           C(1,2);                    % column 2
c(4) =           C(2,2);

c(5) =                    C(1,3);           % column 3
c(6) =                    C(2,3);
```

Similarly, reshape is used to convert the resulting vector x to the 2 × 3 matrix X. It behaves as if you had entered:

```
X(1,1)                    = x(1);           % column 1
X(2,1)                    = x(2);

        X(1,2)            = x(3);           % column 2
        X(2,2)            = x(4);

                X(1,3) = x(5);              % column 3
                X(2,3) = x(6);
```

---

[27] reshape itself uses a much more efficient method.

lp returns the solution status in the variable how. In this case, the value returned was ok, which indicates successful solution. If the constraints do not allow *any* solutions, then how is infeasible. On the other hand, if the constraints are not restrictive enough (the minimum value is $-\infty$), then how will be unbounded.

Notes

- You did not have to divide the cost by 1000 this time. Linear programming algorithms are not so susceptible to scaling problems as non-linear programming algorithms are.

- It is instructive to examine the lambda vector (containing the so-called *Lagrange multipliers* at the solution), which was returned by the call to lp.

    ```
    lambda =
             0
        0.0000
      225.0000
      153.0000
      126.0000

             0
             0
             0
        9.0000      <-- corresponds to:    -x(4) <= 0
       36.0000      <-- corresponds to:    -x(5) <= 0
             0
    ```

    The first five elements are associated with the *constraints*, and the next six elements are associated with the *lower bounds*. In cases where the solution actually lies along a line defined by a constraint or bound, the corresponding lambda value will be non-zero.

    For example, in the above solution, both x(4) and x(5) are right on their lower bound (0.0), so both g(9) and g(10) have non-zero values.

- n specifies the number of *equality* constraints, which in this example is 0 (the default). If you change the program to set

    ```
    n = 1;
    ```

    then lp takes the first constraint in the M matrix as an *equality* constraint, i.e.:

    ```
    x(1) + x(3) + x(5) = A(1)
    ```

    meaning that you have to ship *exactly* the supply of cases from plant 1: no more and no less. In this case the solution is:

    ```
    X =
         50.0000   300.0000         0
        275.0000         0   275.0000

    obj =
        153675
    ```

    Again, the only market that is affected is New York (j = 1). As before, the total cost (obj) does not rise because New York is equidistant from the two plants.

Command Listing

V_{res} = constr(fnc, M_{init})
Computes a local minimum of the multi-variable function/constraints fnc in the neighborhood of the values in starting matrix M_{init}. (This command is in the *Optimization Toolbox*.)
Output: A vector of numeric values is assigned to V_{res}.
Argument options: V_{res} = constr(fnc, M_{init}, options) to specify various control parameters of constr. For a listing of the elements of options that are actually used in constr, see the example in the tutorial section. ✦ V_{res} = constr(fnc, M_{init}, options, V_{lower}, V_{upper}) to specify upper and lower bounds on the variables involved.
Additional information: The function fnc can either be found in a standard M-file (`fnc.m`) ✦ The function fnc does not need to be linear—for purely linear functions with linear constraints, use the more efficient lp function. ✦ To compute the *value* of the function at the given minimum, compute eval(str)(V_{res}). ✦ The default values for the optional parameters can be found in the vector *foptions*. ✦ See the tutorial section of this chapter for an example of using this command. ✦ For more information on the workings of constr, see the *Optimization Toolbox User's Guide*.
See also: *fzero, fsolve, fmins, fmin, lp*

num = fmin(fnc, num_a, num_b)
Computes a local minimum of the one-variable function fnc in the interval (num_a, num_b).
Output: A numeric value is assigned to num.
Argument options: num = fmin(fnc, num_a, num_b, options) to specify various control parameters of fmin. For a listing of the elements of options that are actually used in fmin, see the example in the tutorial section. ✦ num = fmin(fnc, num_a, num_b, options, $expr_1$, $expr_2$, ..., $expr_n$) to present the extra parameters $expr_1$ through $expr_n$ to the function fnc.
Additional information: The default values for the optional parameters can be found in the vector *foptions*. ✦ To compute the *value* of the function at the given minimum, compute eval(fnc)(num).
See also: *fzero, fsolve, fmins, constr, lp*

V_{res} = fmins(fnc, V_{init})
Computes a local minimum of the multi-variable function fnc in the neighborhood of the values in starting vector V_{init}.
Output: A vector of numeric values is assigned to V_{res}.
Argument options: V_{res} = fmins(fnc, V_{init}, options) to specify various control parameters of fmins. For a listing of the elements of options that are actually used in fmin, see the example in the tutorial section. ✦ V_{res} = fmins(fnc, V_{init}, options, $expr_1$, $expr_2$, ..., $expr_n$) to present the extra parameters $expr_1$ through $expr_n$ to the function fnc.
Additional information: The default values for the optional parameters can be found in the vector *foptions*. ✦ To compute the *value* of the function at the given minimum, compute eval(fnc)(V_{res}). ✦ For more information on the workings of fmins, see the *Optimization Toolbox User's Guide*.

See also: *fzero, fsolve,* fmin, constr, lp

$V_{res} = lp(V_{cc}, M, V)$

Computes a local minimum of the linear programming problem represented by V_{cc}, the vector of constant coefficient, and M and V, the matrix and vector of coefficients for the linear constraints. (This command is in the *Optimization Toolbox*.)
Output: A vector of numeric values is assigned to V_{res}.
Argument options: $V_{res} = lp(V_{cc}, M, V, V_{lower}, V_{upper}, V_{init})$ to specify upper and lower bounds on the variables involved. V_{init} represents an initial guess at the solution.
✦ $V_{res} = lp(V_{cc}, M, V, V_{lower}, V_{upper}, V_{init}, n)$ to specify that the first n constraints be treated as *equality* constraints. By default, all constraints are treated as inequalities. ✦ $[V_{res},$ name$_1$, name$_2] = lp(V_{cc}, M, V)$ to assign the set of Lagrangian multipliers to name$_1$ and the status of the completed computation to name$_2$. The status can be one of *ok*, *infeasible*, or *unbounded*.
Additional information: For non-linear problems, use the constr function. ✦ To compute the *value* of the function at the given minimum, compute eval(str)(V_{res}).
✦ For an example of how to set such a problem up, see the tutorial at the beginning of this chapter. ✦ For more information on the workings of lp, see the *Optimization Toolbox User's Guide*.
See also: *fzero,* fmins, fmin, constr

Integration and Differentiation

This chapter looks at MATLAB functions for numerically integrating and differentiating functions of one or more variables.

The four MATLAB functions covered are:

- quad integrates a function of one variable.[28]
- quad8 integrates a function of one variable (using a different algorithm).
- dblquad integrates a function of *two* variables.
- polyder differentiates a polynomial or rational polynomial.

Chapter **D01** of the *NAG Toolbox* (available separately) contains several integration (quadrature) functions written in FORTRAN which can be used from within MATLAB.

There are also several methods for integration and differentiation provided in the *Symbolic Toolbox*, which are demonstrated in the *Symbolic Computations* chapter.

Integrating Functions of One Variable, $f(x)$

Consider the following simple function which we wish to integrate for x from 0 to 3:
$$y(x) = e^{(2x)}$$
It can be integrated analytically, to come up with the indefinite integral:
$$Y(x) = \frac{1}{2} e^{(2x)}$$
so the definite integral from 0 to 3 is:

$$\begin{aligned} Y(3) - Y(0) &= \frac{1}{2}(e^6 - 1) \\ &= 201.214\ldots \end{aligned}$$

---

[28] *Quadrature* is synonymous with *definite integration*, hence the names quad and quad8.

Not *all* functions are so easy to integrate analytically, so we must turn to numerical methods to approximate the integral. Compare this exact answer with the numerical approximations afforded by MATLAB's quad and quad8 functions.

The first step is to create an M-file to define the above function. Call it growth.m:

```
function y = growth(x)

    y = exp(2*x);
```

Then, invoke MATLAB's quad function to integrate it:

```
                a = 0;
                 b = 3;
                   tol = 1e-3;
                      trace = [];
Q = quad('growth', a, b, tol, trace)
```

The result is:

```
Q =
   201.2158
```

tol is a *relative tolerance* which specifies how accurate the result should be (default: 10^{-3}). trace set to 1 (or *any* non-zero value) tells quad to plot the value of our growth function at points where it is evaluated during the integration process (default: [] or 0 for no plot).

Using the default values for tol and trace, this example could be abbreviated to:

```
Q = quad('growth', 0, 3)
```

The quad8 command takes the same arguments, making it easy to interchange for quad:

```
Q = quad8('growth', a, b, tol, trace)
```

The result with quad8 is generally more accurate:

```
Q =
   201.2144
```

The difference between quad and quad8 is the numerical method used by each:

- quad uses an adaptive recursive Simpson's rule technique.

- quad8 uses an adaptive recursive Newton-Côtes 8-panel rule.

Generally speaking, quad8 computes the integral with fewer function evaluations than required by quad.

Integrating Functions of Two Variables over a Rectangular Domain

The function dblquad is available to perform double integrations.

Consider the following simple function:

$$z(x, y) = e^{(2x+2y)}.$$

which we wish to integrate for x from 0 to 3 and y from 0 to 3:

$$Q = \int_{x=0}^{3} \left(\int_{y=0}^{3} z(x,y) dy \right) dx$$

The first step is to create an M-file, growdbl.m, to compute the value of the integrand:

```
function zvec = growdbl(yvec, x)

n = length(yvec);

for i=1:n
    zvec(i) = exp(2*x + 2*yvec(i));
```

Notice that the inner variable (in this case, y) has been specified first and contains a vector of values.

Finally, create a main-line program to call dblquad to compute the double integral:

```
[ya, yb,   xa, xb] = deal(0, 3,   0, 3);

tol   = [];            % Use default value
trace = [];            % Use default value

method = 'quad8';      % Over-ride default

%                            inner       outer
%                           vvvvvvv     vvvvvvv
    Q = dblquad('growdbl',   ya, yb,    xa, xb, ...
                ...
                tol, trace, method)
```

The last three arguments to dblquad are optional. The default for method is 'quad', but we over-rode it here by specifying 'quad8'.

The deal function is a utility that simply assigns the variables on the left-hand side to the matching values in the function call.

This results in:

```
Q =
    4.0487e+004
```

polyder - Derivative of a Polynomial or Rational Polynomial

polyder calculates the analytic derivative of polynomials whose coefficients are stored in a vector.

Consider the following polynomial and its derivative:

$$p(x) = x^2 + 2$$
$$p'(x) = 2x$$

As usual, we represent $p(x)$ in MATLAB with the following vector, p:

```
p = [1 0 2];
```

The derivative of p is found using polyder:

```
der_p = polyder(p)
```

which results in:

```
der_p =
     2     0
```

Command Listing

M_{out} = **cumtrapz**(M_y)

Performs cumulative trapezoidal numerical integration on matrix M_y.
Output: A matrix is assigned to M_{out}.
Argument options: M_{out} = cumtrapz(V_x, M_y) to perform the integration with respect to the column vector V_x. ✱ M_{out} = cumtrapz(M_y, int) or M_{out} = cumtrapz(V_x, M_y, int) to integrate along the intth dimension of M_y.
See also: cumsum, trapz

V_{out} = **polyder**(V)

Computes the derivative of the polynomial whose coefficients are represented by vector V.
Output: A vector of coefficients is assigned to V_{out}.
Argument options: V_{out} = polyder(V_a, V_b) to compute the derivative of the product of polynomials V_a and V_b. ✱ [V_n, V_d] = polyder(V_b, V_a) to compute the derivative of the rational polynomial V_b/V_b. The numerator and denominator of the result are assigned to V_n and V_d, respectively.
See also: polyval, diff

num$_{int}$ = **dblquad**(fnc, num$_{in,min}$, num$_{in,max}$, num$_{out,min}$, num$_{out,max}$)

Computes the definite double integral of function fnc, with bounds on the inner integral of num$_{in,min}$ and num$_{in,max}$ and bounds on the outer integral of num$_{out,min}$ and num$_{out,max}$.

Output: A numerical value is assigned to num_{int}.
Additional information: See the entry for quad for details on optional parameters.
* The function fnc must be such that it takes a vector and a scalar and returns a vector.
See also: quad, quad8, quaddemo

num_{int} = **quad(fnc, num_a, num_b)**
Computes the definite integral of function fnc from between lower bound num_a and upper bound num_b.
Output: A numerical value is assigned to num_{int}.
Argument options: num_{int} = quad(fnc, num_a, num_b, num_{tol}) to set the relative error tolerance to num_{tol}. The default value is 1e-3. * num_{int} = quad(fnc, num_a, num_b, num_{tol}, num_{trace}), where num_{trace} is nonzero, to specify that a graph showing the progress of the integration be displayed.
Additional information: The function fnc must be such that it returns a vector when given a vector. * The definite integral is found using an adaptive recursive Simpson's rule. * quad8 is better than quad at handling *soft singularities*.
See also: quad8, quaddemo, dblquad, trapz, diff

num_{int} = **quad8(fnc, num_a, num_b)**
Computes the definite integral of function fnc from between lower bound num_a and upper bound num_b.
Output: A numerical value is assigned to num_{int}.
Argument options: num_{int} = quad8(fnc, num_a, num_b, num_{tol}) to set the relative error tolerance to num_{tol}. The default value is 1e-3. * num_{int} = quad8(fnc, num_a, num_b, num_{tol}, num_{trace}), where num_{trace} is nonzero, to specify that a graph showing the progress of the integration be displayed.
Additional information: The function fnc must be such that it returns a vector when given a vector. * The definite integral is found using a numerical method called *quadrature*. An adaptive recursive Newton-Côtes 8 panel rule is used to implement it. * quad8 does not handle all singularities properly, tending to "guess" after a certain number of recursions. When this happens, a warning message is displayed.
* quad8 is, however, better than quad at handling *soft singularities*.
See also: quad, quaddemo, dblquad, trapz, diff

quaddemo
Runs a demonstration script detailing the workings of the quad and quad8 commands.
Output: Not applicable.
Additional information: Various commands are automatically entered for you.
* Occasionally, you will be prompted to strike any key to continue the demonstration. * For a complete list of demonstrations available on your platform, see the on-line help file for demos.
See also: quad, quad8, dblquad, odedemo, zerodemo, fftdemo, fplotdemo

Ordinary Differential Equations

Introduction

This chapter looks at MATLAB functions for numerically solving ordinary differential equations (ODEs) of first-order and higher.

The four MATLAB functions covered are:

- ode23 solves ODEs using 2nd- and 3rd-order Runge-Kutta formulas.
- ode45 solves ODEs using 4th- and 5th-order Runge-Kutta formulas.[29]
- ode115s solves stiff systems of differential equations.
- odeset to set ODE solution options.

The main focus of this chapter is on *initial value problems* (IVPs), where certain conditions are given at a system's starting time, t_0. We show one example of a *boundary value problem* (BVP), where conditions are specified at a later time, and solve it using "shooting method" techniques (with help from MATLAB's fzero function).

Chapter **D02** of the *NAG Toolbox* (available separately) contains several functions for solving ordinary differential equations written in FORTRAN which can be used from within MATLAB.

There are also several methods for solving ODEs provided in the *Symbolic Toolbox*, which are demonstrated in the *Symbolic Computations* chapter.

MATLAB's *PDEs Toolbox* (available separately) solves *partial* differential equations. As well, the *NAG Toolbox* contains functions for solving PDEs.

Solving a Single 1st-Order ODE (IVP)

Recall the function you were integrating in the last chapter using quad and quad8:

$$y(x) = e^{2x} \qquad Y(x) = \frac{1}{2}e^{2x}.$$

---

[29] There is a function called ode78 on the MATLAB "ftp" site (see the MATLAB *Resources* chapter), which was contributed by MATLAB users.

Now change the variables to ones more commonly used in ODE problems, $x \to t$ and $y(x) \to \frac{d}{dt}x(t)$:

$$\frac{d}{dt}x(t) = e^{2t} \qquad x(t) = \frac{1}{2}e^{2t}.$$

So far, the differential equation depends only on t. It is easily handled using regular integration functions, until you recast it into an equivalent form that depends on $x(t)$ itself:

$$\frac{d}{dt}x(t) = 2x(t).$$

The straightforward analytic solution to this ODE is:

$$x(t) = x(0)e^{2t},$$

where $x(0)$ is the initial condition. Assuming an initial value of $x(0) = \frac{1}{2}$ and evaluating $x(t)$ at $t = 3$, yields 201.714.

In general, ODEs are not so easy to solve analytically and numerical methods *must* be used. The following sections, compare this exact answer with the numerical approximations afforded by MATLAB's ode23 and ode45 functions.

Solving IVPs using ode23

The first step is to create an M-file to define the derivative function. Call it grow.m:

```
function der_x = grow(t, x)

    der_x(1) = 2 * x(1);
```

Since this is an example in *one* variable, both x and der_x are scalars. Therefore, the subscript (1) could be omitted throughout the above code.[30]

Now invoke the MATLAB function ode23 to integrate this differential equation as t goes from 0 to 3:

```
                       tspan = [0, 3];
                          x0 = [1/2];

[t, x] = ode23('grow', tspan, x0);

[m, n] = size(x)
[t(m), x(m)]

plot(t, x, '-', ...
     t, x, 'o')

print -deps dogrow23.eps
```

The result is:

---

[30] We included it here to make the transition to higher order differential equations easier to understand.

```
m =
    23

n =
    1

ans =
    3.0000  200.7214
```

ode23 returns *two* vectors, t and x, containing the solution at unevenly spaced points over the timespan, tspan.

We use size to determine the number of rows, m, in x and t. Then we display the last (m^{th}) element which contains the solution at the end of the timespan. Then the solution is plotted over the entire timespan.

See the section *Setting ODE Solver Options* for details on controlling such options as error tolerances, output options, etc.

Solving IVPs using ode45

In general, ode45 produces more accurate results than does ode23.

Now use ode45 and compare the results.

```
[t, x] = ode45('grow', tspan, x0);
```

The results are:

```
m =
    45

n =
    1

ans =
    3.0000  201.7214
```

Interpolating from the Solution

Figures 16 and 17 plot the solutions found using ode23 and ode45, respectively.

To make the ode45 plot smoother, you can use Matlab's *spline interpolation* functions to estimate intermediate values. Here are the basic commands:

```
step = (tf - t0) / 100;    % Time-step.
ti   = t0 : step : tf;     % 101 ti-values!

pp = spline(t, x);         % Spline coefficients.
xi = ppval(pp, ti);        % 101 xi-values that
                           %   approximate x(ti)!
plot(t,  x,  'o', ...
     ti, xi, '.', ...
     ti, xi, '-')

print -deps grow45sp.eps
```

The two main lines in the above are:

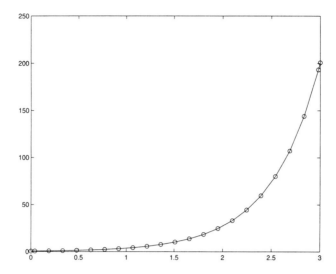

FIGURE 16. Solution to ODE computed by ode23

```
pp = spline(t, x);          % Spline coefficients.
```

which computes the coefficients of the spline curve for the existing points, and

```
xi = ppval(pp, ti);         % 101 xi-values that
                            %   approximate x(ti)!
```

which evaluates that spline curve at the ti values. These two lines can be replaced with just one line:

```
xi = spline(t, x, ti);
```

which is simpler, but not as efficient if you will be doing further interpolations to the same data later on (because you will have to recalculate the coefficients).

Finding t such that $x(t) = a$ (Root-Finding)

Suppose you were interested in finding the time t_c at which $x(t_c) = 50$. Two approaches to this problem are:

1. Indirect root-finding (using spline interpolation)
2. Direct root-finding (using repeated ODE solution)

Indirect Root-Finding (Using Spline Interpolation)

Recall from the previous section that you can interpolate from the solution to find x at intermediate time values:

```
pp = spline(t, x);          % Spline coefficients.

xmid = ppval(pp, 1.5);      % Estimate x(1.5).
```

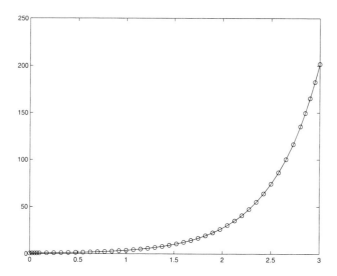

FIGURE 17. Solution to ODE computed by ode45

Also recall that when using fzero to solve for $x(t) = a$ (covered in the *Non-Linear Equations* chapter), you have to create an M-file to evaluate $x(t) - a$. Call this M-file xval.m, and have it use ppval to evaluate the spline at any value of t:

```
function x = xval(t)

global pp;
global x_target;

    x = ppval(pp, t) - x_target;
```

Now pass xval to fzero to find t_c where $x(t_c) = 50$:

```
global pp;
global x_target;

            [t, x] = ode45('grow', [0., 3.], 1/2);
pp = spline(t, x);

x_target = 50;
                           guess = 0.;
t_target = fzero('xval', guess)
```

The result is:

```
t_target =
   2.3026
```

Direct Root-Finding (Using Repeated ODE Solution)

Once again, because you are using fzero, you must create an M-file to evaluate $x(t) - a$. Call this M-file xsol.m, and have it use ode45 to integrate to find x(t) at any value of t:

```
function x = xsol(t)
global x_target;
                            tspan = [0, t];
                               x0 = [1/2];
[ts, xs] = ode45('grow', tspan, x0);

[m, n] = size(xs);
x(1) = xs(m,1) - x_target(1);
```

Now pass xsol to fzero to find t_c, where $x(t_c) - 50 = 0$:

```
global x_target;

x_target(1) = 50;
                           guess = 0.;
t_target = fzero('xsol', guess)
```

The result is:

```
t_target =
    2.3026
```

which is the same answer as before.

As in earlier examples, this is treated as a system with one ODE, easing the transition to larger systems later in this chapter. If you prefer to treat it as a single ODE, then the last line in `xsol.m` can be re-written as:

```
x = xs(m) - x_target;
```

Notes

- A problem with this approach is that it repeatedly solves the ODE from the initial values, (t0, x0), up to fzero's current guess.

Stiff ODEs

ode45 will perform slowly on differential equations that are characterized as being *stiff*.

An example of a stiff ODE is:[31]

$$\frac{dy}{dt} = -100(y - \cos t) - \sin t$$

whose analytic solution is:

$$y(t) = \cos t + ce^{-100t}$$

where $c = y(0) - 1$. For our purposes we will use $y(0) = 2$, so that the analytic solution is:

$$y(t) = \cos t + e^{-100t}$$

---

[31] This example is taken from *Differential Equations: Theory and Applications*, Raymond M. Redheffer, Jones and Bartlett Publishers, Inc., Boston, MA, 1991, p. 628, in Chapter 24 on *Numerical Methods*.

Let us first attempt to solve this differential equation with ode45. First we create an M-file containing our differential equation:

```
function yprime    = stiffex(t, y)

        yprime(1) = -100*(y(1) - cos(t)) - sin(t);
```

Then we solve it using ode45 as follows:

```
                        t0 = 0;
                          tf = 20;
                            y0 = 2;
[t,y] = ode45('stiffex', [t0, tf], y0);
[m,n] = size(y);

[t(m), y(m)]
 m
```

which results in:

```
ans =
   20.0000    0.4080

m =
        2545
```

Although the answer is correct, it took over 2000 steps (an average step-size of less than 0.01) to arrive at that answer.

Contrast this to using ode15s to solve the ODE:

```
                        t0 = 0;
                          tf = 20;
                            y0 = 2;
[t,y] = ode15s('stiffex', [t0, tf], y0);
[m,n] = size(y);

[t(m), y(m)]
 m
```

which results in:

```
ans =
   20.0000    0.4081

m =
         115
```

The exact solution is *cos(20)* which to 4 decimal places is *0.4081*. This solver (ode15s) took just over 100 steps (an average step-size of just under 0.2) to arrive at its answer.

Finally, it is not just computer time that stiff solvers save. By keeping the step-size from becoming too small, cumulative round-off errors are also reduced.

Solving Higher-Order Systems of ODEs

Consider the following *spring-mass-damper* system:

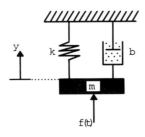

FIGURE 18. Spring-mass-damper system

You can think of the mass m as being a car, which has suspension with stiffness k and shock absorbers with damping coefficient b. The external force $f(t)$ could represent the effects of a bumpy road.

You want to study the effects of varying k and b on the vertical motion of the car over time.

The differential equation describing this model is:[32]

$$m\frac{d^2}{dt^2}y(t) + b\frac{d}{dt}y(t) + ky(t) = f(t)$$

or:

$$\frac{d^2}{dt^2}y(t) + \frac{b}{m}\frac{d}{dt}y(t) + \frac{k}{m}y(t) = \frac{1}{m}f(t)$$

with initial conditions of $x(0) = 1$ and $\frac{d}{dt}x(0) = 0$.

This classic example of a linear second-order ODE arises in the study of many physical systems. Often this equation is re-cast in the following form:

$$\frac{d^2}{dt^2}y(t) + 2\omega_n\zeta\frac{d}{dt}y(t) + \omega_n^2 y(t) = \frac{f(t)}{m}$$

where ω_n (omega_n) is the *natural frequency* without damping and ζ (zeta) is the *damping value* ($\zeta \geq 0$).

Most software packages for solving differential equations require that you re-express the equation as a mathematically equivalent system of first-order equations. MATLAB is no exception, so define a new variable $v(t)$ such that:

$$v(t) = \frac{d}{dt}y(t)$$

Now you have the following system of *first*-order ODEs:

$$\frac{d}{dt}y(t) = v(t)$$
$$\frac{d}{dt}v(t) = \frac{f(t)}{m} - 2\omega_n\zeta v(t) - \omega_n^2 y(t))$$

---

[32]This model is a linear approximation to the non-linear physical processes at work.

with initial conditions of $y(0) = 1$ and $v(0) = 0$.

There are three analytic solutions, depending on the value of ζ. Consider only the *under-damped* case, where $\zeta < 1$.[33] Also assume that $\omega_n = 1$ and $f(t) = 0$, and thereby the analytic solution simplifies to:

$$y(t) = e^{-\zeta t}\left(\frac{\zeta \sin(\sqrt{1-\zeta^2}t)}{\sqrt{1-\zeta^2}} + \cos(\sqrt{1-\zeta^2}t)\right)$$

$$v(t) = -e^{-\zeta t}\frac{\sin(\sqrt{1-\zeta^2}t)}{\sqrt{1-\zeta^2}}$$

For example, with $\zeta = 0$ (no damping), after 10π seconds, $y(10\pi) = y(0) = 1$ and $v(10\pi) = v(0) = 0$.

Figure 19 plots $y(t)$ for various values of ζ between 0 and 1.

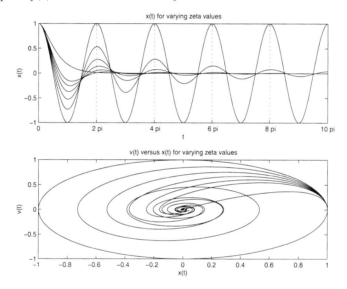

FIGURE 19. Analytic solution to 2nd-order spring-mass-damper problem

Solve this problem numerically using MATLAB's ode45 function and compare with the analytic solution above. The first step is to create an M-file to define the two ODEs, called smd.m (for spring-mass-damper):

```
function der_yv = smd(t, yv)

global m omega_n zeta;        % "t" a vector?

    y = yv(1);                % Extract y and v
    v = yv(2);                %----------------

%---------------------------------------------------
```

---

[33]In the *critically-damped* case, $\zeta = 1$, and in the *over-damped* case, $\zeta > 1$.

```
% Compute derivative of y(t) and v(t):
%------------------------------------------------
    der_y = v;
    der_v = 0/m - (2 * omega_n * zeta) * v ...
              - (omega_n)^2 * y;

%------------------------------------------------
    der_yv(1) = der_y;      % Store in der_yv
    der_yv(2) = der_v;      %-----------------

end
```

Then invoke ode45 to integrate this system of differential equations:

```
global m omega_n zeta;

m       = 1;
omega_n = 1;
zeta    = 0.1;       % "0" means no damping

                tspan = [0, 10*pi];

                    y0 = 1;
                    v0 = 0;
                    yv0 = [y0;
                           v0];

[t, yv] = ode45('smd', tspan, yv0);

y = yv(:,1);           % Extract y and v from
v = yv(:,2);           % columns 1 and 2 of yv

[m, n] = size(yv)
[t(m), y(m) v(m)]      % Solution at "tf".

subplot(2,1,1);
    plot(t, y, '-', ...
         t, y, 'o')           % Trajectory
    title('Position vs Time')

subplot(2,1,2);
    plot(y, v, '-', ...
         y, v, 'o')           % Phase-plane
    title('Velocity vs Position')

print -deps dosmd45.eps
```

The result is:

```
m =
    34

n =
     2

ans =
    31.4159    0.0437    0.0077
```

The values returned in ans represent 10*pi, y(10*pi), and v(10*pi), respectively.

The initial conditions for y and v must be entered in the two-element *column vector*, $yv0$. During its computation, ode45 stores the values of y and v (computed at various time steps) in the *rows* of the matrix yv:

```
t0 =      0.0000                 yv0 = [ 1.0000 ]
                                       [ 0.0000 ]   ---+
                                                       |
t = [    0.0000  ]       yv = [ 1.0000   0.0000 ]   <--+
    [      .     ]            [   .         .   ]
    [      .     ]            [   .         .   ]
    [   31.4159  ]            [ 0.0437   0.0077 ]
```

The program then extracts the values of y and v from yv:

```
y = [ 1.0000 ]        v = [ 0.0000 ]
    [   .    ]            [   .    ]
    [   .    ]            [   .    ]
    [ 0.0437 ]            [ 0.0077 ]
```

The two plots created (which are put one above the other with the subplot command) are the plots of position over time in one graph and the *phase-plane diagram* (v versus y) in the other. See Figure 20.

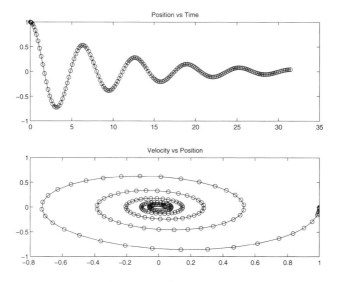

FIGURE 20. Numeric solution to 2nd-order spring-mass-damper problem

Finding $v(t_0)$ such that $v(t_1) = v_1$ (BVP)

In the previous example, both conditions are given at t_0, namely: $y(t_0) = 1$ and $v(t_0) = 0$.

Suppose that instead of knowing the initial position, $y(t_0)$, you know a later position, $y(2\pi) = 1$. This means that somehow you must determine the initial position, $y(t_0)$. (You also know the initial velocity, $v(t_0) = 0$.)

One way of finding a missing initial condition (from a condition at a later time) is the so-called "shooting method", which uses a non-linear equation solver, such as MATLAB's fzero function, to try different values of $y(t_0)$ until one is found where the ODE solution passes through $y(t_1)$.

The first step is to create an M-file, which given a value for $y_0 = y(t_0)$, computes the solution to the ODEs at $t = t_1$. The M-file then returns the distance between the computed position, $y(t_1)$, and the desired position, y_1. Call the M-file y1dist:

```
function dist = y1dist(y0)

global t0     v0;
global t1 y1;

    [t, yv] = ode45('smd', t0, t1, [y0
                                    v0]);

    [m, n] = size(yv);
                            %---------------------
    y = yv(:, 1);           % Or, more efficiently:
                            %
    dist = y1 - y(m);       % dist = y1 - yv(m,1);
                            %---------------------
```

Since the basic system being examined hasn't changed, you can still use the previously defined M-file for smd.

Notice in the following code that zeta = 0.3, so that a lot of damping takes place. Hence, you can expect that $y(t_0)$ will be much larger than $y(t_1) = 1$.

With problems like this, it helps to have a good initial guess for $y(t_0)$ to pass to the non-linear equation solver (fzero). In order to come up with that guess, plot the value of y1dist for various values of y0:

The plot is in Figure 21.

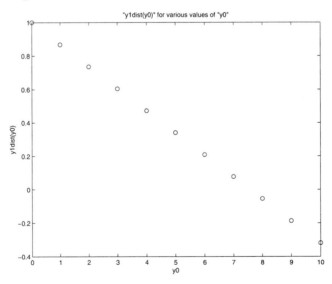

FIGURE 21. Sample starting points for BVP form of spring-mass-damper problem

From the graph, choose $y_0 = 7$ for the initial guess.

Finally, pass y1dist to fzero to find a more accurate value for y0.

```
        global m;                m       = 1;
        global omega_n;          omega_n = 1;
        global zeta;             zeta    = 0.3;

        %-------------------------------------------------
        % Solve BVP for unknown "y0 = y(t0)"
        %-------------------------------------------------

        global t0    v0;         t0 = 0;       v0 = 0;
        global t1 y1;            t1 = 2*pi;    y1 = 1;

                                 y0_guess = 7.;
        y0 = fzero('y1dist', y0_guess)

        %-------------------------------------------------
        % Now can solve IVP for "yf = y(tf)"
        %-------------------------------------------------

                                 tspan = [t0, 10*pi];
                                 yv0 = [y0;
                                        v0];
        [t, yv] = ode45('smd', tspan, yv0);

        y = yv(:,1);             % Extract y and v from
        v = yv(:,2);             % columns 1 and 2 of yv

        [m, n] = size(yv)
        [t(m), y(m) v(m)]        % Solution at "tf".

        subplot(211);
            plot(t, y, '-', ...
                 t, y, 'o')      % Trajectory
            title('Position vs Time')

        subplot(212);
            plot(y, v, '-', ...
                 y, v, 'o')      % Phase-plane
            title('Velocity vs Position')

        print -deps doy1dist.eps
```

Notice that once y0 is found, ode45 is called to compute the solution through to t_f. The solution to the IVP is plotted in Figure 22.

The results of the above code are:

```
        y0 =
            7.5818

        m =
            106

        n =
            2

        ans =
            31.4159    -0.0001    0.0006
```

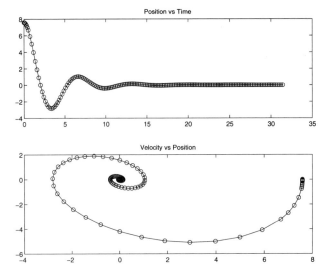

FIGURE 22. Numeric solution to BVP form of spring-mass-damper problem

Notes

- The value of y0 found by fzero can be checked using the analytic solution to our original IVP (where $y(0)$ was 1). Comparing results at $t = t_1$:

```
y1 - (y0 * y_exact(2*pi))
```

results in a difference very close to zero:

```
ans =
   8.0528e-07
```

Setting ODE Solver Options

All of the ODE solving functions allow you to set options at the end of their parameter sequences. For example, to specify a function to display intermediate output), invoke ode45 as follows:

```
[t, yv] = ode45('smd', tspan, yv0, 'OutputFcn', 'odeplot')
```

To change this value for the remainder of the session, use odeset before any subsequent invocations of ode23, ode45, ode15s, etc.

```
odeset('OutputFcn', 'odeplot')
```

If you enter the command odeset without any parameters, MATLAB prints out a list of the current option values for solving ODEs.

The default values for the options are:

```
        AbsTol: [ positive scalar or vector {1e-6} ]
           BDF: [ on | {off} ]
        Events: [ on | {off} ]
   InitialStep: [ positive scalar ]
      Jacobian: [ on | {off} ]
     JConstant: [ on | {off} ]
      JPattern: [ on | {off} ]
          Mass: [ on | {off} ]
  MassConstant: [ on | off ]
      MaxOrder: [ 1 | 2 | 3 | 4 | {5} ]
       MaxStep: [ positive scalar ]
     OutputFcn: [ string ]
     OutputSel: [ vector of integers ]
        Refine: [ positive integer ]
        RelTol: [ positive scalar {1e-3} ]
         Stats: [ on | {off} ]
    Vectorized: [ on | {off} ]
```

See the command listing for odeset for details on the other options.

Command Listing

[V_t, V_x] = **ode113**(fnc, [num$_{start}$, num$_{end}$], V_{init})

Solves the system of ordinary differential equations represented by function fnc for time ranging between num$_{start}$ and num$_{end}$. Initial conditions for the system are provided in vector V_{init}.

Output: Two vectors, V_t and V_x, return the sampled values of time and the solution to the system at those times, respectively.

Argument options: [V_t, V_x] = ode113(fnc, [num$_{start}$, num$_{end}$], V_{init}, V_{opt}) to use the optional values found in structured option vector V_{opt}. See odeset and odeget for details on options.

Additional information: The function fnc must take two input parameters, a scalar time and a state vector, and it must return a column vector of state derivatives. For more information and an example, see the tutorial section at the start of this chapter and the on-line help file for odefile. ♣ The solution is found using a variable order method. ♣ If integration cannot be performed over the entire time range given, an error message is returned. ♣ See the on-line help file for more information about this command.

See also: ode23, ode45, odeset, odeget

[V_t, V_x] = **ode15s**(fnc, [num$_{start}$, num$_{end}$], V_{init})

Solves the system of stiff differential equations represented by function fnc for time ranging between num$_{start}$ and num$_{end}$. Initial conditions for the system are provided in vector V_{init}.

Output: Two vectors, V_t and V_x, return the sampled values of time and the solution to the system at those times, respectively.

Argument options: [V_t, V_x] = ode23s(fnc, [num$_{start}$, num$_{end}$], V_{init}, V_{opt}) to use the optional values found in structured option vector V_{opt}. See odeset and odeget for details on options.

Additional information: See the entry for ode113 and the on-line help file for more information about this command.
See also: ode113, ode23s, odeset, odeget

[V_t, V_x] = ode23(fnc, [num$_{start}$, num$_{end}$], V_{init})
Solves the system of ordinary differential equations represented by function fnc for time ranging between num$_{start}$ and num$_{end}$. Initial conditions for the system are provided in vector V_{init}.
Output: Two vectors, V_t and V_x, return the sampled values of time and the solution to the system at those times, respectively.
Argument options: [V_t, V_x] = ode23(fnc, [num$_{start}$, num$_{end}$], V_{init}, V_{opt}) to use the optional values found in structured option vector V_{opt}. See odeset and odeget for details on options.
Additional information: The function fnc must take two input parameters, a scalar time and a state vector, and it must return a column vector of state derivatives. For more information and an example, see the tutorial section at the start of this chapter and the on-line help file for odefile. ✦ The solution is found using a second/third order Runge-Kutta formula. ✦ In most cases, ode45 is more accurate and more efficient than ode23. ✦ If integration cannot be performed over the entire time range given, an error message is returned. ✦ See the on-line help file for more information about this command.
See also: ode23s, ode45, ode113, odeset, odeget

[V_t, V_x] = ode23s(fnc, [num$_{start}$, num$_{end}$], V_{init})
Solves the system of stiff differential equations represented by function fnc for time ranging between num$_{start}$ and num$_{end}$. Initial conditions for the system are provided in vector V_{init}.
Output: Two vectors, V_t and V_x, return the sampled values of time and the solution to the system at those times, respectively.
Argument options: [V_t, V_x] = ode23s(fnc, [num$_{start}$, num$_{end}$], V_{init}, V_{opt}) to use the optional values found in structured option vector V_{opt}. See odeset and odeget for details on options.
Additional information: See the entry for ode23 and the on-line help file for more information about this command.
See also: ode23, ode15s, odeset, odeget

[V_t, V_x] = ode45(fnc, [num$_{start}$, num$_{end}$], V_{init})
Solves the system of ordinary differential equations represented by function fnc for time ranging between num$_{start}$ and num$_{end}$. Initial conditions for the system are provided in vector V_{init}.
Output: Two vectors, V_t and V_x, return the sampled values of time and the solution to the system at those times, respectively.
Argument options: [V_t, V_x] = ode45(fnc, [num$_{start}$, num$_{end}$], V_{init}, V_{opt}) to use the optional values found in structured option vector V_{opt}. See odeset and odeget for details on options.
Additional information: The function fnc must take two input parameters, a scalar time and a state vector, and it must return a column vector of state derivatives. For

more information and an example, see the tutorial section at the start of this chapter and the on-line help file for odefile. ♣ The solution is found using a fourth/fifth order Runge-Kutta formula. ♣ If integration cannot be performed over the entire time range given, an error message is returned. ♣ See the on-line help file for more information about this command.
See also: ode23, ode113, odeset, odeget

odedemo
Runs a demonstration script detailing how to solve ordinary differential equations in MATLAB.
Output: Not applicable.
Additional information: Various commands are automatically entered for you. ♣ Occasionally, you will be prompted to strike any key to continue the demonstration. ♣ For a complete list of demonstrations available on your platform, see the on-line help file for demos.
See also: ode23, ode45, quaddemo, zerodemo, fftdemo, fplotdemo

M = odeget(V_{opt}, str)
Returns the value in ODE options vector V_{opt} corresponding to option name str.
Output: A value is assigned to M.
Additional information: These option values are used in the ODE solvers ode23, ode45, etc. ♣ See the on-line help file for odeset for details about individual options.
See also: ode23, ode45, ode113, ode23s, ode15s, odeset

V_{opt} = odeset(str_1, M_1, ..., str_n, M_n)
Creates an ODE options vector where option str_1 through str_n are set to values M_1 through M_n, respectively.
Output: A structured vector is assigned to V_{opt}.
Argument options: V_{opt} = odeset(V_{old}, str_1, M_1, ..., str_n, M_n) to start with the values in option structure V_{old}. ♣ V_{opt} = odeset(V_{old}, V_{new}) to start with the values in option structure V_{old} and replace any overlaps with the values in V_{new}.
Additional information: These option vectors are used in the ODE solvers ode23, ode45, etc. ♣ See the on-line help file for details about individual options.
See also: ode23, ode45, ode113, ode23s, ode15s, odeget

Mathematical Functions and Operators

Introduction

Operators

MATLAB contains many different operators for manipulating both number and matrices. Descriptions of these operators are spread throughout the tutorial sections of the other chapters in this book, including the chapters MATLAB *Quick Start*, *Linear Algebra*, and *Symbolic Computations*. Read these sections and the individual command entries for more details.

Standard Functions

MATLAB contains dozen of predefined commands for computing values of well-known (and lesser-known) mathematical functions at particular points. Among these are included such functions as sine, arctangent, exponential, and the error function, named in such a way as to be readily recognized by most users.

There are two standard ways of invoking these commands: with individual values or with matrices of values.

When providing individual values, there are often different results produced depending on whether the input is real-valued or complex.

```
    realang = asin(.5432)
    compang = asin(-.1234 + .3456i)
realang =
    0.5742
compang =
  -0.1168 + 0.3413i
```

Sometimes, real-valued input results in a complex result.

```
    asin(2.341)
ans =
    1.5708 - 1.4946i
```

Most of the commands in this chapter, when passed a matrix of numeric values, return a similarly sized matrix containing the result of passing each individual member of the original matrix through the given command.

```
M = [1,2,3;4,5,6;7,8,9];

Ms = sqrt(M)
```

Ms =

    1.0000   1.4142   1.7321
    2.0000   2.2361   2.4495
    2.6458   2.8284   3.0000

Some functions, however, perform operations upon matrices *as matrices*, not as individual elements. Most of these command names end in the letter m, such as sqrtm and expm.

```
Ms2 = sqrtm(M)

Me = expm(M)
```

Ms2 =

    0.4498 + 0.7623i   0.5526 + 0.2068i   0.6555 - 0.3487i
    1.0185 + 0.0842i   1.2515 + 0.0228i   1.4844 - 0.0385i
    1.5873 - 0.5940i   1.9503 - 0.1611i   2.3134 + 0.2717i

Me =

  1.0e+006 *

    1.1189   1.3748   1.6307
    2.5339   3.1134   3.6929
    3.9489   4.8520   5.7552

Command Listing

$M_{new} = M_1 + M_2$

Adds together matrices M_1 and M_2.
Output: A matrix is assigned to M_{new}.
Argument options: $M_{new} = M + num$ or $M_{new} = num + M$ to add the scalar value num to each element of M. ♣ $num_{new} = num_1 + num_2$ to add together two scalar values.
Additional information: M_1 and M_2 must be of identical dimensions.
See also: -, *, /, \, ^, '

$M_{new} = M_1 - M_2$

Subtracts matrix M_2 from matrix M_2.
Output: A matrix is assigned to M_{new}.
Argument options: $M_{new} = M - num$ to subtract the scalar value num from each element of M. ♣ $M_{new} = num - M$ to subtract the values in M from the matrix of scalar values num. ♣ $num_{new} = num_1 - num_2$ to subtract num_2 from num_1.
Additional information: M_1 and M_2 must be of identical dimensions.
See also: +, *, /, \, ^, '

$M_{new} = M_1 * M_2$
Multiplies, in a linear algebra context, matrices M_1 and M_2.
Output: A matrix is assigned to M_{new}.
Argument options: $M_{new} = M * num$ or $M_{new} = num * M$ to multiply the scalar value num by each element of M. ✤ $num_{new} = num_1 * num_2$ to multiply two scalar values.
Additional information: The number of columns in M_1 must be the same as the number of rows in M_2. ✤ To multiply corresponding elements of two matrices together, use the .* operator.
See also: +, -, /, \, ^, ', .*

$M_{new} = M_1 .* M_2$
Multiplies together the corresponding elements of matrices M_1 and M_2.
Output: A matrix is assigned to M_{new}.
Argument options: $M_{new} = M .* num$ or $M_{new} = num .* M$ to multiply the scalar value num by each element of M. ✤ $num_{new} = num_1 .* num_2$ to multiply two scalar values.
Additional information: M_1 and M_2 must be of identical dimensions. ✤ This multiplication differs from the standard linear algebra definition of multiplication of matrices. See the entry for * for more information.
See also: +, -, *, ./, .\, .^

$M_{new} = M_{sq} \backslash M$
Performs *matrix left division* on square matrix M_{sq} and matrix M.
Output: A matrix is assigned to M_{new}.
Argument options: $M_{new} = M_1 \backslash M_2$, when M_1 is not square, to find a least-squares solution to a system of equations. See the examples in the Linear Equations chapter or the on-line help file for more details. ✤ $num_{new} = num_1 \backslash num_2$ to return the value num_2/num_1.
Additional information: When a square matrix is given, the result is equivalent to $inv(M_{sq}) * M$. ✤ The dimensions of the matrices used must be consistent with the related matrix multiplication performed. ✤ If there are any problems with the dimension or conditioning of input matrices, an error message is displayed.
See also: +, -, /, ^, ', .\

$M_{new} = M_1 .\backslash M_2$
Divides the elements of matrix M_2 by the corresponding elements of matrix M_1.
Output: A matrix is assigned to M_{new}.
Argument options: $M_{new} = M .\backslash num$ or $M_{new} = num .\backslash M$ to divide using the scalar value num. ✤ $num_{new} = num_1 .\backslash num_2$ to return the value num_2/num_1.
Additional information: M_1 and M_2 must be of identical dimensions.
See also: +, -, \, .*, ./, .^

$M_{new} = M / M_{sq}$
Performs *matrix right division* on square matrix M_{sq} and matrix M.
Output: A matrix is assigned to M_{new}.
Argument options: $num_{new} = num_1/num_2$ to return the value num_1/num_2.
Additional information: The result is equivalent to $M * inv(M_{sq})$. ✤ The dimensions of the matrices used must be consistent with the related matrix multiplication per-

formed. ✦ If there are any problems with the dimension or conditioning of input matrices, an error message is displayed.
See also: +, -, \, ^, ', ./

$M_{new} = M_1 ./ M_2$
Divides the elements of matrix M_1 by the corresponding elements of matrix M_2.
Output: A matrix is assigned to M_{new}.
Argument options: $M_{new} = M ./ num$ or $M_{new} = num ./ M$ to divide using the scalar value num. ✦ $num_{new} = num_1 ./ num_2$ to return the value $num_1 ./ num_2$.
Additional information: M_1 and M_2 must be of identical dimensions.
See also: +, -, /, .*, .\, .^

$M_{new} = M_{sq}$ ^ num
Calculates the square matrix M_{sq} to the power num.
Output: A matrix is assigned to M_{new}.
Argument options: $M_{new} = num$ ^ M, to raise the scalar num to the matrix power M using eigenvalues and eigenvectors. ✦ $num_{new} = num_1$ ^ num_2 to take the power of two scalar values.
Additional information: If num is a positive integer, then simple matrix multiplication is done. ✦ If num is a negative integer, then the inverse of M_{sq} is taken first. ✦ For non-integer values of num, eigenvalues and eigenvectors are used.
See also: +, -, *, /, \, ', .^

$M_{new} = M_1 .^ M_2$
Takes the elements of matrix M_1 to the powers of the corresponding elements in matrix M_2.
Output: A matrix is assigned to M_{new}.
Argument options: $M_{new} = M .^ num$ to take the num^{th} power of each element of M. ✦ $num_{new} = num_1 .^ num_2$ to take the power of two scalar values.
Additional information: M_1 and M_2 must be of identical dimensions.
See also: +, -, ^, .*, ./, .\, .^

$M_{new} = M'$
Calculates the transpose of matrix M.
Output: A matrix is assigned to M_{new}.
Additional information: If the elements of M are complex, the elements in the transposed matrix are the complex conjugates of the original values. To retain the original values, use the .' operator.
See also: +, -, *, /, \, ^, .'

$M_{new} = M.'$
Calculates the array transpose of matrix M.
Output: A matrix is assigned to M_{new}.
See also: '

$M_{bool} = M_1 < M_2$
Determines if the elements of matrix M_1 are less than the elements of matrix M_2.

Output: A matrix of zeroes and ones is assigned to M_{bool}.
Argument options: $M_{bool} = M < num$ or $M_{bool} = num < M$ to compare a scalar value to each element of a matrix. ✤ $num_{bool} = num_1 < num_2$ to compare two scalar values.
Additional information: M_1 and M_2 must be of identical dimensions. ✤ If an element satisfies the relation, a 1 is placed in the corresponding element of M_{bool}. Otherwise, the element is 0. ✤ Only the *real* parts of the elements are compared.
See also: <=, >, >=, ==, ~=, &, -,~, xor

$M_{bool} = M_1 <= M_2$

Determines if the elements of matrix M_1 are less than or equal to the elements of matrix M_2.
Output: A matrix of zeroes and ones is assigned to M_{bool}.
Argument options: $M_{bool} = M <= num$ or $M_{bool} = num <= M$ to compare a scalar value to each element of a matrix. ✤ $num_{bool} = num_1 <= num_2$ to compare two scalar values.
Additional information: M_1 and M_2 must be of identical dimensions. ✤ If an element satisfies the relation, a 1 is placed in the corresponding element of M_{bool}. Otherwise, the element is 0. ✤ Only the *real* parts of the elements are compared.
See also: <, >, >=, ==, ~=, &, -,~, xor

$M_{bool} = M_1 > M_2$

Determines if the elements of matrix M_1 are greater than the elements of matrix M_2.
Output: A matrix of zeroes and ones is assigned to M_{bool}.
Argument options: $M_{bool} = M > num$ or $M_{bool} = num > M$ to compare a scalar value to each element of a matrix. ✤ $num_{bool} = num_1 > num_2$ to compare two scalar values.
Additional information: M_1 and M_2 must be of identical dimensions. ✤ If an element satisfies the relation, a 1 is placed in the corresponding element of M_{bool}. Otherwise, the element is 0. ✤ Only the *real* parts of the elements are compared.
See also: <, <=, >=, ==, ~=, &, -,~, xor

$M_{bool} = M_1 >= M_2$

Determines if the elements of matrix M_1 are greater than or equal to the elements of matrix M_2.
Output: A matrix of zeroes and ones is assigned to M_{bool}.
Argument options: $M_{bool} = M >= num$ or $M_{bool} = num >= M$ to compare a scalar value to each element of a matrix. ✤ $num_{bool} = num_1 >= num_2$ to compare two scalar values.
Additional information: M_1 and M_2 must be of identical dimensions. ✤ If an element satisfies the relation, a 1 is placed in the corresponding element of M_{bool}. Otherwise, the element is 0. ✤ Only the *real* parts of the elements are compared.
See also: <, <=, >, ==, ~=, &, -,~, xor

$M_{bool} = M_1 == M_2$

Determines if the elements of matrix M_1 are equal to the elements of matrix M_2.
Output: A matrix of zeroes and ones is assigned to M_{bool}.
Argument options: $M_{bool} = M == num$ or $M_{bool} = num == M$ to compare a scalar value to each element of a matrix. ✤ $num_{bool} = num_1 == num_2$ to compare two scalar values.

Additional information: M_1 and M_2 must be of identical dimensions. ♣ If an element satisfies the relation, a 1 is placed in the corresponding element of M_{bool}. Otherwise, the element is 0. ♣ Both *real* and *complex* parts of the elements are compared.
See also: $<, <=, >, >=, \tilde{=}, \&, -, \tilde{\ }$, xor

$M_{bool} = M_1 \tilde{=} M_2$

Determines if the elements of matrix M_1 are not equal to the elements of matrix M_2.
Output: A matrix of zeroes and ones is assigned to M_{bool}.
Argument options: $M_{bool} = M \tilde{=}$ num or M_{bool} = num $\tilde{=} M$ to compare a scalar value to each element of a matrix. ♣ $num_{bool} = num_1 \tilde{=} num_2$ to compare two scalar values.
Additional information: M_1 and M_2 must be of identical dimensions. ♣ If an element satisfies the relation, a 1 is placed in the corresponding element of M_{bool}. Otherwise, the element is 0. ♣ Both *real* and *complex* parts of the elements are compared.
See also: $<, <=, >, >=, ==, \&, -, \tilde{\ }$, xor

$M_{log} = M_1 \,\&\, M_2$

Performs a logical *and* on the elements of matrices M_1 and M_2.
Output: A matrix of zeroes and ones is assigned to M_{log}.
Argument options: $num_{log} = num_1 \,\&\, num_2$ to compare two scalar values.
Additional information: M_1 and M_2 must be of identical dimensions. ♣ A value of 0 is interpreted as FALSE; every other value is interpreted as TRUE. ♣ These logical operators have the lowest precedence in expressions.
See also: $-, \tilde{\ }$, xor, $<, <=, >, >=, ==$

$M_{log} = M_1 \mid M_2$

Performs a logical *or* on the elements of matrices M_1 and M_2.
Output: A matrix of zeroes and ones is assigned to M_{log}.
Argument options: $num_{log} = num_1 \mid num_2$ to compare two scalar values.
Additional information: M_1 and M_2 must be of identical dimensions. ♣ A value of 0 is interpreted as FALSE; every other value is interpreted as TRUE. ♣ These logical operators have the lowest precedence in expressions.
See also: $\&, \tilde{\ }$, xor, $<, <=, >, >=, ==$

$M_{log} = \tilde{\ } M$

Performs a logical *not* on the elements of matrix M.
Output: A matrix of zeroes and ones is assigned to M_{log}.
Argument options: $num_{log} = \tilde{\ }num$ to perform a logical *not* on numeric value num.
♣ A value of 0 is interpreted as FALSE; every other value is interpreted as TRUE.
♣ These logical operators have the lowest precedence in expressions.
See also: $\&, \mid$, xor, $<, <=, >, >=, ==$

$num_{new} = \mathbf{abs}(num)$

Computes the absolute value of num.
Output: If num is a real value, its absolute value is assigned to num_{new}. If num is a complex value, the *norm* (magnitude) of the value is assigned to num_{new}.

Argument options: M_{new} = abs(M) to compute the absolute values of each of the elements of matrix M. A matrix of real values is returned. ✸ V = abs(str) to return a vector containing the ACSII values of the characters in str.
Additional information: Strings are represented internally in vector form.
See also: angle

ang = **acos**(num)

Computes the inverse cosine of num.
Output: If num is a real value in the range $[-1, 1]$, a real value (in radians) is assigned to ang. If num is a complex value or a real value outside of the range $[-1, 1]$, a complex value is assigned to ang.
Argument options: M_{new} = acos(M) to compute the inverse cosine of each of the elements of matrix M.
See also: asin, atan, cos, acosh

ang = **acosh**(num)

Computes the inverse hyperbolic cosine of num.
Output: The appropriate angle is assigned to ang.
Argument options: M_{new} = acosh(M) to compute the inverse hyperbolic cosine of each of the elements of matrix M.
See also: asinh, atanh, cosh, acos

ang = **acot**(num)

Computes the inverse cotangent of num.
Output: If num is a real value, a real value (in radians) is assigned to ang. If num is a complex value, a complex value is assigned to ang.
Argument options: M_{new} = acot(M) to compute the inverse cotangent of each of the elements of matrix M.
See also: asec, acsc, cot, acoth

ang = **acoth**(num)

Computes the inverse hyperbolic cotangent of num.
Output: The appropriate angle is assigned to ang.
Argument options: M_{new} = acoth(M) to compute the inverse hyperbolic cotangent of each of the elements of matrix M.
See also: asech, acsch, coth, acot

ang = **acsc**(num)

Computes the inverse cosecant of num.
Output: If num is a real value in the range $(-1, 1)$ or a complex value, a complex value is assigned to ang. If num is a real value outside of the range $(-1, 1)$, a real value (in radians) is assigned to ang.
Argument options: M_{new} = acsc(M) to compute the inverse cosecant of each of the elements of matrix M.
See also: asec, acot, csc, acsch

ang = **acsch**(num)
Computes the inverse hyperbolic cosecant of num.
Output: The appropriate angle is assigned to ang.
Argument options: M_{new} = acsch(M) to compute the inverse hyperbolic cosecant of each of the elements of matrix M.
See also: asech, acoth, csch, acsc

num_{new} = **airy**(num)
Computes the Airy function value of num.
Output: An Airy function value is assigned to num_{new}.
Argument options: num_{new} = airy(int, num), where int is either 1, 2, or 3, to return the special derivative of the Airy function. ✤ [num_{new}, int_{err}] = airy(int, num) to return an error flag in int_{num}. ✤ M_{new} = airy(M) to compute the Airy function values of each of the elements of matrix M. A matrix of values is returned.
See also: besselJ, bessely, besseli, besselk

ang = **angle**(cmplx)
Computes the phase angle of complex value cmplx.
Output: A value (in radians) is assigned to ang.
Argument options: M_{new} = angle(M) to compute the phase angles of each of the elements of matrix M.
Additional information: The magnitude of a complex value can be found with abs.
✤ If a real value is passed to ang, a value of 0 or 3.1416 is returned, depending on whether the value passed is positive or negative.
See also: abs

ang = **asec**(num)
Computes the inverse secant of num.
Output: If num is a real value in the range $(-1, 1)$ or a complex value, a complex value is assigned to ang. If num is a real value outside of the range $(-1, 1)$, a real value (in radians) is assigned to ang.
Argument options: M_{new} = asec(M) to compute the inverse secant of each of the elements of matrix M.
See also: acsc, acot, sec, asech

ang = **asech**(num)
Computes the inverse hyperbolic secant of num.
Output: The appropriate angle is assigned to ang.
Argument options: M_{new} = asech(M) to compute the inverse hyperbolic secant of each of the elements of matrix M.
See also: acsch, acoth, sech, asec

ang = **asin**(num)
Computes the inverse sine of num.
Output: If num is a real value in the range $[-1, 1]$, a real value (in radians) is assigned to ang. If num is a complex value or a real value outside of the range $[-1, 1]$, a complex value is assigned to ang.

Argument options: M_{new} = asin(M) to compute the inverse sine of each of the elements of matrix M.
See also: acos, atan, sin, asinh

ang = **asinh**(num)
Computes the inverse hyperbolic sine of num.
Output: The appropriate angle is assigned to ang.
Argument options: M_{new} = asinh(M) to compute the inverse hyperbolic sine of each of the elements of matrix M.
See also: acosh, atanh, sinh, asin

ang = **atan**(num)
Computes the inverse tangent of num.
Output: If num is a real value, a real value (in radians) is assigned to ang. If num is a complex value, a complex value is assigned to ang.
Argument options: M_{new} = atan(M) to compute the inverse tangent of each of the elements of matrix M.
See also: atan2, acos, asin, tan, atanh

ang = **atan2**(num_n, num_d)
Computes the four-quadrant inverse tangent of the value represented by the numerator and denominator values num_1 and num_2, respectively.
Output: A real value is returned.
Argument options: M_{new} = atan2(M_1, M_2) to compute the inverse tangent of each of the values represented by elements of M_1 and M_2, matrices of equal dimensions.
Additional information: This command is similar to the atan command, except that entering the numerator and denominator separately allows you to know which quadrant the result is in.
See also: acos, asin, tan, atanh

ang = **atanh**(num)
Computes the inverse hyperbolic tangent of num.
Output: The appropriate angle is assigned to ang.
Argument options: M_{new} = atanh(M) to compute the inverse hyperbolic tangent of each of the elements of matrix M.
See also: acosh, asinh, tanh, atan

V_{new} = **besselh**(n, V)
Computes the complex Hankel order n of the elements in vector V.
Output: A vector of complex values is assigned to V_{new}.
Additional information: n must be an integer.
See also: airy, besseli, besselj, besselk, bessely,

V_{new} = **besseli**(num, V)
Computes the modified Bessel function of the first kind of order num of the elements in vector V.
Output: A vector is assigned to V_{new}.

Additional information: num must be non-negative.
See also: airy, besselh, besselj, besselk, bessely

V_{new} = **besselj**(num, V)
Computes the Bessel function of the first kind of order num of the elements in vector V.
Output: A vector is assigned to V_{new}.
Additional information: num must be non-negative.
See also: airy, besselh, besseli, besselk, bessely

V_{new} = **besselk**(num, V)
Computes the modified Bessel function of the second kind of order num of the elements in vector V.
Output: A vector is assigned to V_{new}.
Additional information: num must be non-negative.
See also: airy, besselh, besseli, besselj, bessely

V_{new} = **bessely**(num, V)
Computes the Bessel function of the second kind of order num of the elements in vector V.
Output: A vector is assigned to V_{new}.
Additional information: num must be non-negative.
See also: airy, besselh, besseli, besselj, besselk

num = **beta**(num_1, num_2)
Computes the Beta function value at num_1 and num_2.
Output: A real value is assigned to num.
Argument options: M_{new} = beta(M_1, M_2) to compute the Beta function values of the elements of M_1 and M_2, two matrices of equal dimensions.
Additional information: Both num_1 and num_2 must be real and non-negative.
♣ beta(num_1, num_2) is defined as (gamma(num_1) * gamma(num_2)) / gamma(num_1 + num_2).
See also: betainc, betaln, gamma

num = **betainc**(num_1, num_2, num_3)
Computes the incomplete Beta function value at num_1, num_2 and num_3.
Output: A real value is assigned to num.
Additional information: num_1 must a real value in the range [0, 1], and both num_2 and num_3 must be real and non-negative. ♣ For the definition of the incomplete Beta function, see the on-line help file.
See also: beta, betaln, gammainc

num_{new} = **betaln**(num_1, num_2)
Computes the logarithm of the Beta function value at num_1 and num_2.
Output: If num_1 and num_2 are real values, a real value is assigned to num_{new}. If either num_1 or num_2 is a complex value, a complex value is assigned to num_{new}.
Argument options: M_{new} = betaln(M_1, M_2) to compute the logarithm of the Beta function values of the elements of M_1 and M_2, two matrices of equal dimensions.

Additional information: betaln(num$_1$, num$_2$) is defined as (gammaln(num$_1$) + gammaln(num$_2$)) - gammaln(num$_1$ + num$_2$).
See also: beta, betainc, gammaln

int = ceil(num)
Rounds the numeric value num up to the next highest integer.
Output: An integer value is assigned to int.
Argument options: M$_{new}$ = ceil(M) to round each of the elements of matrix M to their next highest integer values and assign the result to M$_{new}$.
See also: floor, round, fix

V = colperm(M)
Computes a permutation vector of the columns of M such that they are ordered from left to right with respect to the number of nonzero entries in each column.
Output: A permutation vector is assign to V.
Additional information: Reordering a matrix in this fashion can be helpful in several matrix calculations including LU factorization and Cholesky factorization.
See also: colmmd, chol, lu, symrcm

cmplx$_{new}$ = conj(cmplx)
Computes the complex conjugate of complex value cmplx.
Output: A complex value is assigned to cmplx$_{new}$.
Argument options: M$_{new}$ = conj(M) to compute the complex conjugate of each element of matrix M and assign the result to M$_{new}$.
Additional information: The complex conjugate of cmplx equals real(cmplx) - i*imag(cmplx).
See also: imag, real

num = cos(ang)
Computes the cosine of ang.
Output: If ang is a real value (in radians), a real value is assigned to num. If ang is a complex value, a complex value is assigned to num.
Argument options: M$_{new}$ = cos(M) to compute the cosine of each of the elements of matrix M.
See also: sin, tan, cosh, acos

num = cosh(ang)
Computes the hyperbolic cosine of ang.
Output: The appropriate value is assigned to num.
Argument options: M$_{new}$ = cosh(M) to compute the hyperbolic cosine of each of the elements of matrix M.
See also: sinh, tanh, cos, acosh

num = cot(ang)
Computes the cotangent of ang.
Output: If ang is a real value (in radians), a real value is assigned to num. If ang is a complex value, a complex value is assigned to num.

Argument options: M_{new} = cot(M) to compute the cotangent of each of the elements of matrix M.
See also: sec, csc, coth, acot

num = **coth**(ang)
Computes the hyperbolic cotangent of ang.
Output: The appropriate value is assigned to num.
Argument options: M_{new} = coth(M) to compute the hyperbolic cotangent of each of the elements of matrix M.
See also: sech, csch, cot, acoth

V_{cp} = **cross**(V_1, V_2)
Computes the vector cross product of the three-element vectors V_1 and V_2.
Output: A three-element vector is assigned to V_{cp}.
Argument options: M_{cp} = cross(M_1, M_2), where M_1 and M_2 are $3 \times n$ matrices of identical dimensions, to calculate the cross products of the columns of the matrices and assign the results to the columns of M_{cp}.
Additional information: To find the dot product, use dot.
See also: dot, cumprod

num = **csc**(ang)
Computes the cosecant of ang.
Output: If ang is a real value (in radians), a real value is assigned to num. If ang is a complex value, a complex value is assigned to num.
Argument options: M_{new} = csc(M) to compute the cosecant of each of the elements of matrix M.
See also: sec, cot, csch, acsc

num = **csch**(ang)
Computes the hyperbolic cosecant of ang.
Output: The appropriate value is assigned to num.
Argument options: M_{new} = csch(M) to compute the hyperbolic cosecant of each of the elements of matrix M.
See also: sech, coth, csc, acsch

V_{cp} = **cumprod**(V)
Computes the cumulative products of the elements of vector V.
Output: A vector of the same length as V is assigned to V_{cp}.
Argument options: M_{cp} = cumprod(M) to calculate the cumulative products of the columns of matrix M and assign the result to M_{cp}.
Additional information: To simply find the product of an entire vector or matrix column, use the prod command.
See also: cumsum, prod

V_{cs} = **cumsum**(V)
Computes the cumulative sums of the elements of vector V.
Output: A vector of the same length as V is assigned to V_{cs}.

Argument options: M_{cs} = cumsum(M) to calculate the cumulative sums of the columns of matrix M and assign the result to M_{cs}.
Additional information: To simply find the sum of an entire vector or matrix column, use the sum command.
See also: cumprod, sum

M = diag(V)

Creates a matrix with the elements of vector V on the main diagonal and zeroes elsewhere.
Output: A matrix is assigned to M.
Argument options: M = diag(V, n) to create a matrix with the elements of vector V on the nth superdiagonal and zeroes elsewhere. ♣ M = diag(V, -n) to create a matrix with the elements of vector V on the nth subdiagonal and zeroes elsewhere. ♣ V = diag(M) to extract the elements of the main diagonal of matrix M and assign them the vector V. ♣ V = diag(M, n) to extract the elements of the nth superdiagonal of M. ♣ V = diag(M, -n) to extract the elements of the nth subdiagonal of M.
Additional information: All vectors returned by diag are *column vectors*.
See also: tril, triu

V_{new} = diff(V)

Computes the differences between adjacent elements of vector V.
Output: A vector of length length(V) - 1 is assigned to V_{new}.
Argument options: V_{new} = diff(V, n) to find the nth difference vector. This is found with recursive calls to diff. ♣ M_{new} = diff(M) or M_{new} = diff(M, n) to take the differences down the columns of matrix M. The result is a matrix with fewer rows than M and is assigned to M_{new}.
Additional information: The difference of each adjacent pair is taken. ♣ Using diff(V_y) ./ diff(V_x) will find the approximate derivative.
See also: sum, prod, gradient, del2

num = dot(V_1, V_2)

Computes the dot (scalar) product of vectors V_1 and V_2, which are of equal length.
Output: A scalar value is assigned to num.
Argument options: V = dot(M_1, M_2), where M_1 and M_2 are $m \times n$ matrices, to calculate the dot products of the columns of the matrices and assign the results to an n-element row vector, V.
Additional information: To find the cross product, use cross.
See also: cross, prod

[num_{sn}, num_{cn}, num_{dn}] = ellipj(num_u, num_m)

Computes the values for the elliptic functions *sn*, *cn*, and *dn* given the numeric values num_u and num_m.
Output: Three numeric values are assigned to num_{sn}, num_{cn}, and num_{dn}.
Argument options: [M_{sn}, M_{cn}, M_{dn}] = **ellipj**(M_u, M_m) to find the elliptic function values for the corresponding elements of matrices M_u and M_m. These two input matrices must be of the same dimensions.

Additional information: num_m must be between 0 and 1, inclusive. ✦ For more information on the formulae defining the elliptic functions, see the on-line help page.
See also: ellipke

[num_K, num_E] = **ellipke**(num_m)
Computes the values for the elliptic integrals of the first and second kind, given the numeric value num_m.
Output: The elliptic integral of the first kind is assigned to num_K. The elliptic integral of the second kind is assigned to num_E.
Argument options: [M_K, M_E] = **ellipke**(M_m) to find the elliptic integral values for the elements of matrix M_u.
Additional information: num_m must be between 0 and 1, inclusive. ✦ For more information on the formulae defining the elliptic integrals, see the on-line help page.
See also: ellipj

num_e = **erf**(num)
Computes the error function value at num.
Output: A real value is assigned to num_e.
Argument options: M_e = erf(M) or V_e = erf(V) to compute the error function values at each element of a matrix or vector.
Additional information: For a definition of the error function, see the on-line help file.
See also: erfc, erfcx, erfinv

num_e = **erfc**(num)
Computes the complementary error function value at num.
Output: A real value is assigned to num_e.
Argument options: M_e = erfc(M) or V_e = erfc(V) to compute the complimentary error function values at each element of a matrix or vector.
Additional information: By definition, erfc(num) = 1 - erf(num). ✦ For a definition of the error function, see the on-line help file.
See also: erf, erfcx, erfinv

num_e = **erfcx**(num)
Computes the scaled complementary error function value at num.
Output: A real value is assigned to num_e.
Argument options: M_e = erfcx(M) or V_e = erfcx(V) to compute the scaled complimentary error function values at each element of a matrix or vector.
Additional information: By definition, erfcx(num) = e^{x^2} erfc(num). ✦ For a definition of the error function, see the on-line help file.
See also: erf, erfc, erfinv

num = **erfinv**(num_e)
Computes the inverse error function value at num_e.
Output: A real value is assigned to num.
Argument options: M_e = erfinv(M) or V_e = erfinv(V) to compute the inverse error function values at each element of a matrix or vector.

Additional information: For a definition of the inverse error function, see the on-line help file.
See also: erf, erfc, erfcx

M = eye(n)
Creates the n × n identity matrix.
Output: A square matrix is assigned to M.
Argument options: M = eye(m, n) to create an m × n matrix with ones on the main diagonal (starting in the upper-left corner) and zeroes elsewhere.
Additional information: When working with other matrices, using eye in combination with the size command can be very helpful.
See also: speye, ones, zeros, rand

num_e = exp(num)
Computes the exponential function value at num.
Output: If num is real, a real value is assigned to num_e. If num is complex, a complex value is assigned to num_e.
Argument options: M_e = exp(M) to compute the exponential values of each of the elements of matrix M.
Additional information: Matrix exponentials (which are different from exp(M)) are computed with expm. ♣ If M contains any values that are too large, then this command may overflow.
See also: log, log10, expm

num_{ei} = expint(num)
Computes the exponential integral at num.
Output: A numerical value is assigned to num_{ei}.
Argument options: M_{ei} = expint(M) to compute the exponential integral of each of the elements of matrix M.
Additional information: The exponential integral at x is defined as $\int_x^\infty \frac{e^{-t}}{t} dt$. ♣ Both the input and output may be either real or complex valued.
See also: exp, expm

M_e = expm(M)
Computes the matrix exponential of matrix M.
Output: A matrix of equal dimensions to M is assigned to M_e.
Additional information: To find the element-by-element exponential values, use exp. ♣ For most matrices, logm is the inverse operation to expm.
See also: logm, funm, sqrtm, exp

V_p = factor(int)
Computes the prime factorization of positive integer int.
Output: A vector of positive integer values is assigned to V_p.
See also: primes, isprime

int = fix(num)
Rounds the numeric value num to an integer by truncation.

Output: An integer value is assigned to int.
Argument options: M_{new} = fix(M) to round each of the elements of matrix M and assign the result to M_{new}.
See also: floor, ceil, round

M_{new} = **fliplr(M)**

Reverses the order of the columns of matrix M.
Output: A matrix of equal size to M is assigned to M_{new}.
See also: flipud, reshape, rot90

M_{new} = **flipud(M)**

Reverses the order of the rows of matrix M.
Output: A matrix of equal size to M is assigned to M_{new}.
See also: fliplr, reshape, rot90

int = **floor(num)**

Rounds the numeric value num down to the next lowest integer.
Output: An integer value is assigned to int.
Argument options: M_{new} = floor(M) to round each of the elements of matrix M to their next lowest integer values and assign the result to M_{new}.
See also: ceil, round, fix

M_e = **funm(M, 'fnc')**

Computes the function fnc of matrix M.
Output: A matrix of equal dimensions to M is assigned to M_e.
Additional information: funm does not work element by element, but on the matrix as a whole. ✢ There are several MATLAB functions with specialized matrix equivalents. Included are expm, logm, and sqrtm. ✢ When the results may be inaccurate, a warning message is displayed.
See also: logm, expm, sqrtm, sin, cos

num_{new} = **gamma(num)**

Computes the Gamma function value at num.
Output: If num is a real value, a real value is assigned to num_{new}. If num is a complex value, a complex value is assigned to num_{new}.
Argument options: M_{new} = beta(M) to compute the Gamma function values of the elements of matrix M.
Additional information: gamma(num + 1) is defined as n! or prod(1:n).
See also: gammainc, gammaln, gamma, prod

num = **gammainc(num_1, num_2)**

Computes the incomplete Gamma function value at num_1 and num_2.
Output: If num_1 and num_2 are real values, a real value is assigned to num. If either num_1 or num_2 is a complex value, a complex value is assigned to num.
Argument options: M_{new} = betainc(M_1, M_2) to compute the Gamma function values of the elements of M_1 and M_2, two matrices of equal dimensions.

Additional information: For the definition of the incomplete Gamma function, see the on-line help file.
See also: gamma, gammaln, betainc

num_{new} = **gammaln**(num)

Computes the logarithm of the Gamma function value at num.
Output: If num is a real value, a real value is assigned to num_{new}. If num is a complex value, a complex value is assigned to num_{new}.
Argument options: M_{new} = betaln(M) to compute the logarithm of the Gamma function values of the elements of matrix M.
See also: gamma, gammainc, betaln

$posint_{gcd}$ = **gcd**(int_1, int_2)

Computes the *greatest common divisor* of integer values int_1 and int_2.
Output: A positive integer value is assigned to $posint_{gcd}$.
Additional information: This command *cannot* be used with matrix input, or an error message is displayed.
See also: lcm

[M_x, M_y] = **gradient**(M)

Computes a numerical approximation to the gradient field of M.
Output: The partial derivatives in the two variables (directions) are assigned to matrices M_x and M_y.
Argument options: [M_x, M_y] = gradient(M, V_x, V_y) to compute the divided differences with respect to x and y vectors V_x and V_y. ✦ [M_x, M_y] = gradient(M, num_{dx}, num_{dy}) to divide the horizontal and vertical differences by num_{dx} and num_{dy}, respectively. ✦ M_c = gradient(M) to return the horizontal and vertical results as real and imaginary components of complex values, respectively, in M_c. ✦ V_{out} = gradient(V) to compute a numerical approximation to the first derivative of vector V. ✦ V_{out} = gradient(V, V_y) to compute divided differences. ✦ V_{out} = gradient(V, num_{dy}) to divide the difference by num_{dy}.
See also: diff, del2, quiver

i

Represents the square root of -1.
Output: Not applicable.
Additional information: i is an alias used primarily for representing complex values. ✦ You can override the value of i if you wish to use this variable for other purposes (e.g., as a loop index). ✦ The same value is also denoted by j.
See also: j, real, imag, conj

num = **imag**(cmplx)

Determines the imaginary part of complex numeric value cmplx.
Output: A real value is assigned to num.
Argument options: M_{new} = imag(M) to determine the complex parts of each element in matrix M and assign the results to M_{new}.
See also: i, j, real, conj

Inf
Represents positive infinity.
Output: Not applicable.
Argument options: -Inf to represent negative infinity.
Additional information: Any computation that leads to a numeric value too large to represent in floating-point notation results in Inf. ♣ Inf/Inf and Inf - Inf are both equal to NaN.
See also: NaN, isinf, dbstop

int = isequal(expr$_1$, expr$_2$)
Determines if expr$_1$ and expr$_2$ are equal.
Output: If expr$_1$ and expr$_2$ are equal, a value of 1 is assigned to int. Otherwise, a value of 0 is assigned to int.
Argument options: int = isequal(M$_1$, M$_2$) to determine if matrices M$_1$ and M$_2$ are of equal dimension and have identical elements.
Additional information: To return a 0 or 1 for each element of the matrices, use the == syntax.
See also: isfinite, isinf, isnan

int = isfinite(expr)
Determines if expr is finite.
Output: If expr is not Inf, -Inf or NaN, a value of 1 is assigned to int. Otherwise, a value of 0 is assigned to int.
Argument options: M$_{new}$ = isfinite(M) to determine if each element of matrix M is finite. An equally sized matrix of zeroes and ones is assigned to M$_{new}$.
Additional information: This command can be used in conjunction with any or all.
See also: isinf, isnan, any, all

int = isinf(expr)
Determines if expr equals Inf or -Inf.
Output: If expr is Inf or -Inf, a value of 1 is assigned to int. Otherwise, a value of 0 is assigned to int.
Argument options: M$_{new}$ = isinf(M) to determine if each element of matrix M is equal to Inf or -Inf. An equally sized matrix of zeroes and ones is assigned to M$_{new}$.
Additional information: This command can be used in conjunction with any or all.
See also: isfinite, isnan, any, all

int = isprime(int$_p$)
Determines if int$_p$ is prime.
Output: If int$_p$ is prime, a value of 1 is assigned to int. Otherwise, a value of 0 is assigned to int.
Argument options: M$_{out}$ = isprime(M$_p$) to determine if each individual elements of M$_p$ is prime.
See also: primes, factor

int = isnan(expr)
Determines if expr equals NaN (Not a Number).

Output: If expr is NaN, a value of 1 is assigned to int. Otherwise, a value of 0 is assigned to int.
Argument options: M_{new} = isnan(M) to determine if each element of matrix M is equal to NaN. An equally sized matrix of zeroes and ones is assigned to M_{new}.
Additional information: This command can be used in conjunction with any or all.
See also: isfinite, isinf, any, all

int = **isnumeric**(expr)

Determines if expr is a numeric expression.
Output: If expr is numeric, a value of 1 is assigned to int. Otherwise, a value of 0 is assigned to int.
Argument options: int = isnumeric(M) to determine if matrix M is made up of nothing but numeric elements.
Additional information: To return a 0 or 1 for each element of the matrices, use the == syntax.
See also: *numeric*, isfinite, isinf, isnan

j

Represents the square root of -1.
Output: Not applicable.
Additional information: j is an alias used primarily for representing complex values.
♦ You can override the value of j if you wish to use this variable for other purposes (e.g., as a loop index). ♦ The same value is also denoted by i.
See also: i, real, imag, conj

posint$_{lcm}$ = **lcm**(posint$_1$, posint$_2$)

Computes the *lowest common multiple* of positive integer values posint$_1$ and posint$_2$.
Output: A positive integer value is assigned to posint$_{lcm}$.
Additional information: This command *cannot* be used with matrix or negative integer input, or an error message is displayed.
See also: gcd

M = **legendre**(n, V)

Computes the Legendre function value of degrees 0 through n of the *m* real values in vector V.
Output: An n+1 × m matrix is returned. Every value V(i) that satisfies abs(V(i)) < 1 returns a column of real values. Otherwise, a column of complex values is returned.
Additional information: n must be an integer value of less than 257. Any negative values are rounded to 0.

V = **linspace**(num$_1$, num$_2$, int)

Creates a vector consisting of int elements, the values of which are linearly spaced between num$_1$ and num$_2$, inclusive.
Output: A vector is assigned to V.
Argument options: V = linspace(num$_1$, num$_2$) to create a vector with 100 elements.
Additional information: The same result can also be had by using the colon operator (:).

See also: logspace

num_l = **log(num)**
Computes the natural logarithm of num.
Output: If num is real and non-negative, a real value is assigned to num_l. Otherwise, a complex value is assigned to num_l.
Argument options: M_l = log(M) to compute the natural logarithm of each of the elements of matrix M.
Additional information: Matrix logarithms (which are different from log(M)) are computed with logm.
See also: logm, log10, exp

num_l = **log2(num)**
Computes the logarithm base 2 of num.
Output: If num is real and non-negative, a real value is assigned to num_l. Otherwise, a complex value is assigned to num_l.
Argument options: $[num_f, num_e]$ = log2(num) to return two numeric values such that num = num_f * 2^num_e. ✦ M_l = log2(M) to compute the logarithm base 2 of each of the elements of matrix M.
See also: log, log10, pow2, num2hex

num_l = **log10(num)**
Computes the common logarithm (to base 10) of num.
Output: If num is real and non-negative, a real value is assigned to num_l. Otherwise, a complex value is assigned to num_l.
Argument options: M_l = log10(M) to compute the common logarithm of each of the elements of matrix M.
See also: log

M_l = **logm(M)**
Computes the matrix logarithm of matrix M.
Output: A matrix of equal dimensions to M is assigned to M_l.
Additional information: To find the element-by-element logarithmic values, use exp.
✦ If M contains any zero eigenvalues, then M_l will contain some infinity values.
✦ For most matrices, logm is the inverse operation to expm.
See also: expm, funm, sqrtm, log, eig

V = **logspace(num_1, num_2, int)**
Creates a vector consisting of int elements, the values of which are logarithmically spaced between 10^num_1 and 10^num_2, inclusive.
Output: A vector is assigned to V.
Argument options: V = logspace(num, pi, int) to create the values between 10^num and pi. ✦ V = logspace(num_1, num_2) to create a vector with 50 elements.
See also: linspace

expr = **max(V)**
Determines the maximum value of the elements of vector V.

Output: If V contains a value of NaN, then NaN is assigned to expr. Otherwise, a numeric value is assigned to expr.
Argument options: V = max(M) to assign a row vector containing the maximum element in each column of matrix M to V. ✤ M_{max} = max(M_1, M_2) to determine a matrix that takes the larger values from the corresponding elements of similarly sized matrices M_1 and M_2 and assign it to M_{max}. ✤ [V_{max}, V_i] = max(M) to assign the maximum values to row vector V_{max} and the corresponding row indices from M to V_i. The first occurrence of the maximal value is used.
Additional information: Using max recursively determines the largest element in an entire matrix. ✤ If an element is complex, its magnitude (as found with abs) is used.
See also: min, sort, isnan

expr = min(V)

Determines the minimum value of the elements of vector V.
Output: If V contains a value of NaN, then NaN is assigned to expr. Otherwise, a numeric value is assigned to expr.
Argument options: V = min(M) to assign a row vector containing the minimum element in each column of matrix M to V. ✤ M_{min} = min(M_1, M_2) to determine a matrix that takes the smaller values from the corresponding elements of similarly sized matrices M_1 and M_2 and assign it to M_{min}. ✤ [V_{min}, V_i] = min(M) to assign the minimum values to row vector V_{min} and the corresponding row indices from M to V_i. The first occurrence of the minimal value is used.
Additional information: Using min recursively determines the smallest element in an entire matrix. ✤ If an element is complex, its magnitude (as found with abs) is used.
See also: max, sort, isnan

num = mod(num_x, num_y)

Computes the value num_x mod num_y.
Output: A numeric value is assigned to num.
Argument options: M = mod(M_x, M_y), where M_x and M_y are matrices of real values, to compute the mod values element by element.
See also: rem

NaN

Represents an undefined numerical result (or Not a Number).
Output: Not applicable.
Additional information: Any computation that leads to an undefined value results in NaN. ✤ some situations that produce NaN include division by zero and multiplication of infinity by zero. ✤ NaN can also be used as input to commands. ✤ For more information and cases, see the on-line help file.
See also: Inf, isnan, dbstop

M = nchoosek(V, int)

Creates a matrix, each row of which contains int elements chosen from V.
Output: A int-column matrix is returned.
Additional information: This function is very slow for values greater than 15.
See also: perms

n = **nextpow2**(num)
Computes the value of the lowest integer n such that $2^n \geq abs(num)$.
Output: A non-negative integer value is assigned to n.
Argument options: n = nextpow2(V) to compute the lowest integer n for the *length* of the vector V.
Additional information: nextpow2 is particularly useful in FFT calculations.
See also: fft, pow2, log2, length

M = **ones**(n)
Creates the n × n matrix whose every element equals 1.
Output: A square matrix is assigned to M.
Argument options: M = ones(m, n) to create an m × n matrix of ones. ♣ M = ones(size(M_{old})) to create a matrix of ones with identical dimensions to M_{old}.
Additional information: When working with other matrices, using eye in combination with the size command can be very helpful.
See also: eye, zeros, rand

M = **perms**(V)
Creates a matrix, each row of which contains a permutation of the elements in V.
Output: A matrix is returned.
Additional information: This function is very slow for values greater than 15.
See also: nchoosek

pi
Provides a floating-point approximation to π.
Output: Not applicable.
Additional information: Because MATLAB is a numerical system, pi is not exactly equal to π.
See also: i, j, Inf

V_p = **primes**(int)
Generates all the prime numbers less than or equal to int.
Output: A vector of positive integer values is assigned to V_p.
See also: factor, isprime

num_p = **pow2**(num)
Computes the value for 2 to the power num.
Output: A numeric value of the same type as num is assigned to num_p.
Argument options: M_p = pow2(M_r, M_i), where M_r is a matrix of real values and M_i is a matrix of integers, to compute the matrix of values corresponding to M_r .* (2 .^ M_i).
Additional information: This function corresponds to the IEEE function *scalbn*.
See also: log2, exp, realmax, realmin

num = **prod**(V)
Computes the product of all the elements of vector V.
Output: A numeric value is assigned to V_{cp}.

Argument options: V_r = prod(M) to calculate the products of the elements of the columns of matrix M and assign the result to row vector V_r.
Additional information: To find the product of all the elements in a matrix, use prod recursively.
See also: cumprod, sum

V = randperm(n)
Creates a random permutation of the integers from 1 to n.
Output: A vector containing the elements 1 through n in random order is assigned to V.
See also: rand, randn

[int_n, int_d] = rat(num)
Approximates numerical value num by a simple rational fraction.
Output: Integer values representing the numerator and denominator are assigned to int_n and int_d, respectively.
Argument options: [M_n, M_d] = rat(M_{num}) to approximate M_{num}, a matrix of numeric values by two matrices, M_n and M_d, containing the respective numerators and denominators. ✦ [int_n, int_d] = rat(num, num_{tol}) or [M_n, M_d] = rat(M_{num}, num_{tol}) to set the approximation tolerance to num_{tol}. ✦ rat(num) or rat(M_{num}) to display the complete continued fraction representation of num.
Additional information: The approximations are produced by truncating continued fraction expansions. ✦ To return a string containing the entire fraction, use rats.
See also: rats

str = rats(num)
Approximates numerical value num by a simple rational fraction.
Output: A string representing the approximation is assigned to str.
Argument options: M_{str} = rats(M_{num}) to approximate M_{num}, a matrix of numeric values and assign the resulting strings to matrix M_{str}. ✦ str = rats(num, num_{ch}) or M_{str} = rats(M_{num}, num_{ch}) to set the maximum number of characters in each element to num_{ch}. The default value is 13.
Additional information: The approximations are produced by truncating continued fraction expansions. ✦ This string representation of values is the same as is produced when format is set to *rat*. ✦ If a value cannot be represented by a string of the appropriate size, and its value is important to the result, an asterisk is printed in its place. ✦ To return integers representing the numerator and denominator of the fraction, use rat.
See also: rat, format, eval

num = real(cmplx)
Determines the real part of complex numeric value cmplx.
Output: A real value is assigned to num.
Argument options: M_{new} = imag(M) to determine the real parts of each element in matrix M and assign the results to M_{new}.
See also: i, j, imag, conj

num = **realmax**

Represents the largest floating-point value representable on your computer.
Output: A real value is assigned to num.
Additional information: On computers following the IEEE standards, realmax is approximately $1.7977e+308$.
See also: realmin, pow2, log2

num = **realmin**

Represents the smallest floating-point value representable on your computer.
Output: A real value is assigned to num.
Additional information: On computers following the IEEE standards, realmin is approximately $2.2251e-308$.
See also: realmax, pow2, log2

num_r = **rem**(num_x, num_y)

Computes the remainder when num_x is divided by num_y.
Output: A real value is assigned to num_r.
Argument options: To determine the integer part of num_x/num_y, use the fix function.
See also: mod, fix, sign

M_{new} = **reshape**(M, m, n)

Reshapes the matrix M into an m × n matrix, if possible.
Output: If m * n equals the number of elements of M, a new matrix is assigned to M_{new}. Otherwise, an error message is returned.
Argument options: M_{new} = reshape(M, num_1, ..., num_n) to reshape an n-dimensional matrix.
Additional information: Elements are taken from M and entered into M_{new} proceeding from top to bottom and left to right (i.e., columnwise).
See also: squeeze, shiftdim, fliplr, flipud, rot90

M_{new} = **rot90**(M)

Rotates the m × n matrix M counterclockwise by 90deg.
Output: An n × m matrix is assigned to M_{new}.
Argument options: M_{new} = rot90(M, int) to rotate M counterclockwise by int*90 degrees.
See also: fliplr, flipud, reshape

int = **round**(num)

Rounds the numeric value num to the nearest integer.
Output: An integer value is assigned to int.
Argument options: M_{new} = round(M) to round each of the elements of matrix M to their nearest integer values and assign the result to M_{new}.
Additional information: Values ending in .5 are rounded *away* from zero.
See also: ceil, round, fix

num = **sec**(ang)

Computes the secant of ang.

Output: If ang is a real value (in radians), a real value is assigned to num. If ang is a complex value, a complex value is assigned to num.
Argument options: M_{new} = sec(M) to compute the secant of each of the elements of matrix M.
See also: csc, cot, sech, asec

num = sech(ang)
Computes the hyperbolic secant of ang.
Output: The appropriate value is assigned to num.
Argument options: M_{new} = sech(M) to compute the hyperbolic secant of each of the elements of matrix M.
See also: csch, coth, sec, asech

int = sign(num)
Determines the sign of numeric value num.
Output: If num > 0, then 1 is assigned to int. If num = 0, then 0 is assigned to int. If num < 0, then −1 is assigned to int.
Argument options: M_{new} = sign(M) to determine the sign of the elements of matrix M and assign the result to M_{new}.
Additional information: If num is a complex value, then the result equals num/abs(num).
See also: abs, real

num = sin(ang)
Computes the sine of ang.
Output: If ang is a real value (in radians), a real value is assigned to num. If ang is a complex value, a complex value is assigned to num.
Argument options: M_{new} = sin(M) to compute the sine of each of the elements of matrix M.
See also: cos, tan, sinh, asin

num = sinh(ang)
Computes the hyperbolic sine of ang.
Output: The appropriate value is assigned to num.
Argument options: M_{new} = sinh(M) to compute the hyperbolic sine of each of the elements of matrix M.
See also: cosh, tanh, sin, asinh

V_{new} = sort(V)
Sorts the elements of vector V in ascending order.
Output: A vector with the same number of elements as V is assigned to V_{new}.
Argument options: [V_{new}, V_i] = sort(V) to also assign a vector with the correspondingly sorted indices of V to V_i. In effect, then V_{new} = V(V_i). ♣ M_{new} = sort(M) or [M_{new}, M_i] = sort(M) to sort the columns of matrix M in a similar manner.
Additional information: If any elements are complex, then the function abs is run on them before they are sorted. ♣ If two elements being sorted are of identical value, the relative ordering of their indices is preserved.

See also: min, max, mean, median

M_{new} = **sortrows(M)**
Sorts the rows of matrix M in ascending order.
Output: A matrix is assigned to M_{out}.
Argument options: $[M_{new}, V_i]$ = sortrows(V) to also assign a vector with the correspondingly sorted row indices of M to V_i. ✤ M_{new} = sortrows(M, V_c) to sort the rows of matrix M by the columns in V_c.
Additional information: If any elements are complex, then the function abs is run on them before they are sorted.
See also: min, max, sort

num_{sq} = **sqrt(num)**
Computes the square root of num.
Output: If num is a real non-negative value, a real value is assigned to num_{sq}. If num is a complex value or a real negative value, a complex value is assigned to num_{sq}.
Argument options: M_{sq} = sqrt(M) to compute the square root of each of the elements of matrix M. ✤ For matrix square roots, use sqrtm.
See also: sqrtm, exp, log

M_{sq} = **sqrtm(M)**
Computes a matrix square root of matrix M.
Output: A matrix of equal dimensions to M is assigned to M_{sq}.
Additional information: To find the element-by-element square root values, use sqrt. ✤ The square root is such that $M_{sq} * M_{sq} = M$. ✤ Most matrices have more than one possible square root. If M is positive definite and symmetric, then the square root chosen is the unique positive definite one. ✤ For more information on how to determine all the square roots of M, see the on-line help file.
See also: logm, funm, expm, sqrt

num = **sum(V)**
Computes the sum of all the elements of vector V.
Output: A numeric value is assigned to V_{cp}.
Argument options: V_r = sum(M) to calculate the sums of the elements of the columns of matrix M and assign the result to row vector V_r.
Additional information: To find the sum of all the elements in a matrix, use sum recursively.
See also: cumsum, prod, trace

num = **tan(ang)**
Computes the tangent of ang.
Output: If ang is a real value (in radians), a real value is assigned to num. If ang is a complex value, a complex value is assigned to num.
Argument options: M_{new} = tan(M) to compute the tangent of each of the elements of matrix M.
See also: cos, sin, tanh, atan

num = **tanh**(ang)
Computes the hyperbolic tangent of ang.
Output: The appropriate value is assigned to num.
Argument options: M_{new} = tanh(M) to compute the hyperbolic tangent of each of the elements of matrix M.
See also: cosh, sinh, tan, atanh

M = **zeros**(n)
Creates the n × n matrix whose every element equals 0.
Output: A square matrix is assigned to M.
Argument options: M = zeros(m, n) to create an m × n matrix of zeros. ♣ M = zeros(size(M_{old})) to create a matrix of zeros with identical dimensions to M_{old}.
Additional information: While MATLAB automatically generates the storage needed for growing matrices, it is generally a good idea to initialize a matrix with zeroes *before* you begin allocating its elements. This tends to improve overall efficiency.
See also: eye, ones, rand

Symbolic Computations

While we have decided not to go into great detail about most of the Toolboxes available for MATLAB, we did feel it important to give an overview of a relatively new addition to the MATLAB application family, the *Symbolic Math Toolbox*.

Traditionally, MATLAB has restricted its attention to numeric computations (as the rest of this handbook will attest). For many years, *all* computational software restricted itself to the numeric realm. In the last fifteen years, however, more and more attention has been paid to designing, creating, and utilizing systems that perform *symbolic* calculations.

Symbolic computations are basically defined as ones that involve *unassigned* variables (e.g., the variable x, which is not given a numeric value). Also, in most symbolic systems, standard arithmetic operations can be performed with *exact* results (e.g., 3 divided by 7 is exactly $\frac{3}{7}$, not .428571...), and floating-point approximations can be found to any number of digits (e.g., 3000 decimal places of π).

We think that the marriage of the accuracy and flexibility of symbolic computation with the speed and ease of numeric computation is very exciting and will open many previously uninvestigated doors in the future. The importance of the combination of MATLAB's numeric routines with those of a symbolic program goes beyond simply using the symbolic system *through* MATLAB's interface, residing more in the use of *combined symbolic and numeric* routines to solve particular problems that before have only used one type of system or the other.

How the Connection is Accomplished

To write a complete symbolic computational system takes years upon years of research and coding, and has successfully been done by several software companies. The designers of MATLAB, instead of reinventing the wheel, have instead created an easy-to-use connection between their product and the symbolic computation package Maple.[34]

You need not be overly concerned with *how* this connection was implemented, merely with how to best take advantage of it. There are two different levels of the *Symbolic Math Toolbox* available to you:

---

[34] Copyright Waterloo Maple Inc.

- The basic *Symbolic Math Toolbox* provides about fifty built-in MATLAB commands that interface to specific Maple commands as well as a syntax for calling any other symbolic command within Maple's large standard library. This is the version you are most likely to be using, and the one we spend the most time on here.
- The *Extended Symbolic Math Toolbox* has all the capabilities of the basic package, as well as access to Maple programming structures, input/output capabilities, and commands in specialized packages.

MATLAB Functions and Operators that Support Symbolic Operands

If you use any of the following MATLAB functions with symbolic arguments, MATLAB will perform symbolic calculations. For details on the workings of these individual functions and operators, please refer to their command listings elsewhere in this handbook (primarily in the *Functions* and *Linear Equations* chapters).

Operators

```
  +    -    *    /    \
  ^   .*   .\   ./   .^
  '   .'   :   ==   ~=
```

Mathematical Functions

Trigonometric Functions

```
acos    acosh   acot    acoth   acsc
acsch   asec    asech   asin    asinh
atan    atanh   cos     cosh    cot
coth    csc     csch    sec     sech
sin     sinh    tan     tanh
```

Other Mathematical Functions

```
abs     besseli   besselj   besselk   bessely
conj    erf       exp       gamma     imag
log     real      sqrt
```

Linear Equations Functions

```
det     diag   eig    expm   inv
jordan  null   poly   rank   rref
svd     tril   triu
```

Other Functions

char	disp	display	double	eval
ezplot	isreal	prod	size	sum

Creating Symbolic Expressions Within MATLAB

Now that you know most of the differences between the syntaxes of Maple and MATLAB, try creating some simple Maple expressions from within your MATLAB session.

```
sin(2*x)
```

MATLAB tries to interpret this as a purely MATLAB entity. Since the variable x is as yet undefined, and MATLAB cannot deal with variables without values, the following error message is returned.

```
??? Undefined function or variable 'x'.
```

In order to bypass this automatic evaluation step, enclose the symbolic expression in string quotes.

```
'sin(2*x)'
```

The result is now

```
ans =

sin(2*x)
```

Of course, symbolic expression can contain more than just one unknown.

```
'4*x^3*y^2 - cos(z/r)'
```

What about creating larger MATLAB structures, such as **symbolic matrices**, containing several symbolic expressions? The MATLAB command sym provides a syntax nearly identical to that for entering numerical matrices. The command

```
M = sym('[a, b; c, d]')
```

produces the result

```
M =

[a,b]
[c,d]
```

Calling Symbolic Computation Commands

M, as created in the previous section, can then be passed to other commands available in the *Symbolic Math Toolbox*.

```
inv(M)
```

The result is:

```
ans =

[ d/(a*d-b*c),  -b/(a*d-b*c)]
[-c/(a*d-b*c),   a/(a*d-b*c)]
```

These commands can also perform computation on matrices containing only numeric elements.

```
Mn = sym('[1, 2; 3, 4]')
```

```
inv(Mn)
```

returns

```
ans =

[  -2,    1]
[ 3/2, -1/2]
```

Notice, however, that the resulting elements are in *exact* form, as discussed earlier, not in the floating-point approximation you would expect from as output from standard MATLAB procedures.[35]

Of course, symbolic expressions can be entered as input to symbolic commands directly, without assignment to variable names first, as in the differentiation

```
diff('2*x^3')
```

which returns

```
ans =

6*x^2
```

One thing you need to be aware of is that not all parameters to symbolic commands need to be entered in string syntax. In many commands, string and non-string elements are mixed. For example, to differentiate $\sin(2*x)$ with respect to x four times, the following syntax is used:

```
diff('sin(2*x)', 'x', 4)
```

and the result is

```
ans =

16*sin(2*x)
```

The dual nature of standard MATLAB commands as both commands and functions also holds true for symbolic commands. For example, the call to diff seen earlier can also be entered as:

---

[35] For more details on what type of input is valid for various symbolic commands, and what type of output can be expected in each case, see the individual command entries later in this chapter.

```
diff 2*x^3
```

with the same result.

Notice that there is no need for string quotes here. Unfortunately, if you want to separate sections of your symbolic expressions with spaces (as is perfectly valid Maple syntax), you must reintroduce the quotes or MATLAB thinks you are entering multiple parameters.

```
int sin(x) * cos(x)
```

results in three parameters being sent to the int command, which produces the following error message.

```
??? Error using ==> maple
syntax error before: ..cos(x));

Error in ==> /.../toolbox/symbolic/int.m
On line 35  ==>        r = maple('int',f,[x '=' symrat(a) '..' symrat(b)]);
```

On the other hand,

```
int sin(x)*cos(x)
```

is interpreted correctly

```
ans =

1/2*sin(x)^2
```

Converting Data Between the Two Products

The commands in the *Symbolic Math Toolbox* can compute exact answers from rational matrices

```
M3 = sym('[1,3,2;4,3,1;5,6,2]');
Minv = inv(M3)

Minv =

[    0,   2/3,  -1/3]
[ -1/3,  -8/9,   7/9]
[    1,     1,    -1]
```

but can this exact (rational) answer be used in further standard MATLAB numeric calculations?

The entire matrix can be converted to standard numeric format, assuming that none of its elements contain unknown variables, using the double command.

```
format long
double(Minv)

ans =

                  0   0.66666666666667  -0.33333333333333
 -0.33333333333333  -0.88888888888889   0.77777777777778
  1.00000000000000   1.00000000000000  -1.00000000000000
```

We used format long to cause MATLAB to display answers to 15 digits, which is as good as it gets in numeric systems. Contrast this to evaluating Minv to arbitrary precision within Maple using the vpa function. The desired number of digits is entered as the second parameter.

```
M20 = vpa(Minv, 20)

M20 =

[                         0,    .66666666666666666667,   -.33333333333333333333]
[-.33333333333333333333,    -.88888888888888888889,    .77777777777777777778]
[                        1.,                       1.,                      -1.]
```

Don't make the mistake of thinking that these elements are now in standard MATLAB format. (The presence of the brackets is your clue.)

For the special case of polynomials, the commands polytosym and symtopoly are provided.

```
poly2sym([1 -3 4])
poly2sym([4 2 5 3 -7], 'y')
sym2poly('x^4 - 3*x^2 + x + 6')
```

Your answers are

```
ans =
x^2-3*x+4

ans =
4*y^4+2*y^3+5*y^2+3*y-7

ans =
     1     0    -3     1     6
```

Parameter Types Specific to This Chapter

M_s	a matrix of symbolic expressions
M_{vpa}	a matrix of variable precision numeric values
num_{vpa}	a variable precision numeric value
$poly_s$	a symbolic polynomial
symcond	a symbolic conditional equation (represented as a string)
symexpr	a symbolic expression (represented as a string)
symfun	a string containing a Maple command name
V_s	a vector of symbolic expressions
V_{vpa}	a vector of variable precision numeric values

Command Listing

$poly_{s,out}$ = collect($poly_{s,in}$, str)
Collects terms for variable str in symbolic polynomial $poly_{s,in}$.

Output: A collected symbolic polynomial is assigned to poly$_{s,out}$.
Argument options: poly$_{s,out}$ = collect(poly$_{s,in}$) to collect poly$_{s,in}$ with respect to the default variable, x. ♣ M$_{s,out}$ = collect(M$_{s,in}$, str) to collect each individual symbolic polynomial in symbolic matrix M$_s$ with respect to variable str.
Additional information: str need not be the *only* variable in the given symbolic polynomials. To determine what variables are present, use the symvar command.
See also: expand, factor, horner

M$_s$ = colspace(M)

Computes a column space for the numeric matrix M.
Output: A symbolic matrix is assigned to M$_s$.
Argument options: M$_{s,out}$ = colspace(M$_{s,in}$) to compute a column space for the symbolic matrix M$_{s,in}$.
Additional information: To convert a symbolic matrix (with no unknowns) to standard numeric form, use the numeric command.
See also: null, double, orth

symexpr = compose(symexpr$_1$, symexpr$_2$)

Performs functional composition such that symexpr$_2$ replaces the symbolic variable in symexpr$_1$.
Output: A symbolic expression is assigned to symexpr.
Argument options: symexpr = compose(symexpr$_1$, ..., symexpr$_n$) to compose n expressions. ♣ symexpr = compose(symexpr$_1$, ..., symexpr$_n$, str) to assumed all expressions have domain variable str. ♣ symexpr = compose(symexpr$_1$, ..., symexpr$_n$, str$_1$, ..., str$_n$) to use str$_1$ as domain variable for symexpr$_1$, str$_2$ as domain variable for symexpr$_2$, and so on.
Additional information: If symexpr$_1$ is f and symexpr$_2$ is g, then the result represents $f(g(x))$. ♣ For examples of use, see the on-line help.
See also: finverse, subs

M$_{out}$ = cosint(M$_{in}$)

Computes the cosine integral function at each element of numeric matrix M$_{in}$.
Output: A numeric matrix is assigned to M$_{out}$.
Argument options: M = cosint(M$_s$) to compute the function at each element of symbolic matrix M$_s$. A numeric matrix is returned.
Additional information: The cosine integral of x is sometimes notated as $Ci(x)$. ♣ If a symbolic matrix is used, it must be free of unknown variables.
See also: sinint, mfun

symexpr$_{out}$ = diff(symexpr$_{in}$, str)

Computes the derivative of symbolic expression symexpr$_{in}$, with respect to the variable str.
Output: A symbolic expression is assigned to symexpr$_{out}$.
Argument options: symexpr$_{out}$ = diff(symexpr$_{in}$) to differentiate with respect to the only variable, as found by symvar, in one-variable expression symexpr$_{in}$. ♣ symexpr$_{out}$ = diff(symexpr$_{in}$, str, n) to compute the nth derivative with respect to str. ♣ M$_{s,out}$ = diff(M$_{s,in}$, str) to compute the derivative of each element of symbolic matrix M$_s$.

* V_{out} = diff(V_{in}) to a vector containing the difference between adjacent elements of numeric vector V_{in}. * M_{out} = diff(M_{out}) to perform the difference operation on the columns of numeric matrix M_{out}.
Additional information: The nth difference vector or matrix can also be found by adding n as a parameter. * See the *Linear Equations* chapter for details on the numerical diff, function which computes differences between columns of a *numeric* matrix.
See also: int, gradient, jacobian, del2

digits(int)
Sets the current number of digits used in variable precision arithmetic to n.
Output: Not applicable.
Argument options: int = digits to assign the current number of digits to int. * digits to just display the current number of digits.
Additional information: The number of digits used in variable precision arithmetic is set within Maple by the global variable Digits.
See also: vpa

symexpr$_{out}$ = dsolve(symexpr$_{in}$)
Solves the specially formulated ordinary differential equation, symexpr$_{in}$, for the appropriate dependent variable.
Output: A symbolic expression is assigned to symexpr$_{out}$. If more than one solution is found, they are returned in a symbolic vector of appropriate size.
Argument options: symexpr$_{out}$ = dsolve(symexpr$_{in}$, str) to use str as the independent variable. x is the default. * symexpr$_{out}$ = dsolve(symexpr$_{in}$, symcond, str) to solve the differential equation symexpr$_{in}$ under the initial condition symcond. * symexpr$_{out}$ = dsolve(symexpr$_{in,1}$, ..., symexpr$_{in,n}$, symcond$_1$, ..., symcond$_m$, str) to solve n differential equations under m initial conditions. * [symexpr$_1$, ..., symexpr$_n$] = dsolve(symexpr$_{in}$, symcond, str) to assign the n results to the individual names symexpr$_1$ through symexpr$_n$.
Additional information: To denote a differentiation, use the character D. For example, Dy represents $\frac{dy}{dx}$. Higher order derivatives are created similarly. For example, D4s represents $\frac{d^4s}{dx^4}$. Examples of complete differential equations can be found in the on-line help page. * Both symexpr$_{in}$ and symcond can also contain *multiple* symbolic expressions. Simple create one large string, where the individual symbolic expressions are separated by commas. * Read the entire on-line help page for more information and examples.
See also: solve, linsolve, ode23, ode45, ode113, ode15s, ode23s

symexpr$_{out}$ = expand(symexpr$_{in}$)
Expands the symbolic expression symexpr$_{in}$.
Output: A collected symbolic expression is assigned to symexpr$_{out}$.
Argument options: $M_{s,out}$ = expand($M_{s,in}$) to expand each individual symbolic expression in symbolic matrix M_s.
Additional information: str need not be the *only* variable in the given symbolic polynomials. To determine what variable are present, use the symvar command.
See also: collect, factor

ezplot(symexpr)

Plots the one-variable function symexpr.
Output: A two-dimensional plot is created.
Argument options: ezplot(symexpr, [num$_{min}$, num$_{max}$]) to specify the "x-range" of the plot. ✽ ezplot(symexpr, [num$_{min}$, num$_{max}$], hndl) to use the figure hndl instead of the current figure.

symexpr$_{out}$ = factor(symexpr$_{in}$)

Factors the symbolic expression symexpr$_{in}$.
Output: A factored symbolic expression is assigned to symexpr$_{out}$.
Argument options: $M_{s,out}$ = factor($M_{s,in}$) to factor each individual symbolic expression in symbolic matrix M_s. ✽ symexpr = factor(int) or $M_{s,out}$ = factor(M) to perform prime factorization on a single integer or a matrix of integer values.
See also: collect, expand, simplify

symexpr$_{out}$ = finverse(symexpr$_{in}$)

Determines the functional inverse of symexpr$_{in}$, a symbolic expression of exactly one variable.
Output: A symbolic expression is assigned to symexpr$_{out}$.
Argument options: symexpr$_{out}$ = finverse(symexpr$_{in}$, str) find the function inverse of symexpr. a multi-variable expression, with respect to the variable str.
Additional information: To compose two symbolic functions, use the compose command. ✽ The symvar command can be used to help determine whether a symbolic expression has only one unknown variable.
See also: compose, solve

funtool(symexpr)

Initiates a functional calculator tool.
Output: Not applicable.
Additional information: This tool manipulates symbolic expressions of one variable.

poly$_{s,out}$ = horner(poly$_{s,in}$, str)

Convert poly$_{s,in}$, a polynomial in typical symbolic, form, into Horner form.
Output: A symbolic polynomial is assigned to poly$_{s,out}$.
Argument options: $M_{s,out}$ = horner($M_{s,in}$) to convert each individual symbolic polynomial in symbolic matrix M_s into Horner form.
See also: expand, factor

symexpr$_{out}$ = int(symexpr$_{in}$, str)

Computes the indefinite integral of symbolic expression symexpr$_{in}$, with respect to the variable str.
Output: A symbolic expression is assigned to symexpr$_{out}$.
Argument options: symexpr$_{out}$ = int(symexpr$_{in}$) to integrate with respect to the only variable, as found by symvar, in one-variable expression symexpr$_{in}$. ✽ int to integrate the previous result with respect to its only variable. ✽ symexpr$_{out}$ = int(symexpr$_{in}$, str, symexpr$_a$, symexpr$_b$) or symexpr$_{out}$ = int(symexpr$_{in}$, str, num$_a$, num$_b$) to compute the definite

integral of symexpr$_{in}$ from over the bounds given as the last two parameters. ✣ M$_{s,out}$ = int(M$_{s,in}$, str) to compute the integral of each element of symbolic matrix M$_s$.
Additional information: The nth difference vector or matrix can also be found by adding n as a parameter.
See also: quad, quad8, dblquad, diff, symsum

M$_{s,out}$ = jacobian(M$_{s,in}$, V$_s$)

Computes the Jacobian matrix for one-column symbolic matrix M$_{s,in}$ over the variables contained in symbolic vector V$_s$.
Output: A symbolic matrix is assigned to M$_{s,out}$.
Additional information: The $(i,j)^{th}$ elements of M$_{s,out}$ is computed by $\frac{\partial M_{s,in}(i)}{\partial V_s(j)}$.
See also: diff

M$_s$ = jordan(M)

Computes the Jordan canonical form of numeric matrix M.
Output: A symbolic matrix is assigned to M$_s$.
Argument options: [M$_{s,e}$, M$_{s,j}$] = jordan(M) to also return the eigenvectors of M in symbolic matrix M$_{s,e}$. ✣ M$_{s,out}$ = jordan(M$_{s,in}$) or [M$_{s,e}$, M$_{s,j}$] = jordan(M$_{s,in}$) to compute the Jordan form and/or eigenvectors of a square symbolic matrix, M$_{s,in}$.
Additional information: To convert a symbolic matrix (with no unknowns) to standard numeric form, use the numeric command. ✣ If a symbolic matrix is entered, the elements must all be integers or simple integer ratios.
See also: eig, poly

M$_{out}$ = lambertw(M$_{in}$)

Computes Lambert's W function at each element of numeric matrix M$_{in}$.
Output: A numeric matrix is assigned to M$_{out}$.
Argument options: M = lambertw(M$_s$) to compute the function at each element of symbolic matrix M$_s$.
See also: mfun

symexpr$_{out}$ = maple(symexpr$_{in}$)

Sends symexpr$_{in}$, a symbolic expression in Maple syntax, directly to the Maple kernel for evaluation.
Output: A symbolic expression is assigned to symexpr$_{out}$.
Argument options: symexpr$_{out}$ = maple(symfun, expr$_1$, ..., expr$_n$) to call the Maple function symfun with the n parameters expr$_1$ through expr$_n$. These parameters are automatically converted to symbolic expression, if necessary. ✣ [symexpr$_{out}$, int] = maple(symexpr$_{in}$) to return the error status of the given Maple computation in int. If an error has occurred, int is a positive integer and the error message is in symexpr$_{out}$. Otherwise, int is 0. ✣ maple('traceon') to turn on automatic tracing. Besides computing the result, subsequent results from Maple are preceded by tracing information. ✣ maple('traceoff') to turn off automatic tracing.
Additional information: Depending on which version of the *Symbolic Math Toolbox* you have purchased, different numbers and levels of Maple commands will be available. ✣ To view Maple's on-line help files, use the mhelp command. ✣ When

automatic tracing is on, *every* call to the Maple kernel is traced, not just those caused by direct calls to the maple command.
See also: mhelp, procread, mfun

M = mfun(symfun, expr$_1$, expr$_2$, ..., expr$_n$)

Sends numeric expressions expr$_1$ through expr$_n$ to the Maple-known function symfun.
Output: A numeric matrix is assigned to M.
Additional information: Only up to four numeric parameters, including single values and matrices, are allowed. ♣ Among the functions available through mfun are Euler numbers and polynomials, the Reimann Zeta function, the Logarithmic Integral, and several orthogonal polynomials. For a complete list of all the functions as well as information on the proper types of parameters, see help mfunlist. ♣ For more latitude on which Maple functions can be called, use the maple command.
See also: maple, mhelp

mhelp(symfun)

Displays the Maple on-line help information for the Maple function symfun.
Output: A help page is displayed.
Additional information: Besides Maple function names, other keywords (e.g., list, library, etc.) have on-line information.
See also: help, maple

[symexpr$_n$, symexpr$_d$] = numden(symexpr)

Computes the numerator and denominator of the symbolic expression, symexpr.
Output: Relatively prime polynomials with integer coefficients are assigned to symexpr$_n$ and symexpr$_d$.
Argument options: [int$_n$, int$_d$] = numden(num) to convert numeric value num to a ratio of integer values. ♣ [M$_{s,n}$, M$_{s,n}$] = numden(M$_s$) or [M$_n$, M$_d$] = numden(M) to convert each element of the input matrix into its numerator and denominator.

poly$_s$ = poly2sym(poly, str)

Converts poly, a polynomial in standard MATLAB form, to a polynomial in variable str in Maple symbolic form.
Output: A symbolic polynomial is assigned to poly$_s$.
Argument options: poly$_s$ = poly2sym(poly) to use the default variable name x.
Additional information: If the coefficients of poly are not integer values, they are converted to rationals using the symrat command.
See also: sym2poly, polyval

pretty(symexpr)

Displays the symbolic expression symexpr in a two-dimensional ASCII format which resembles typeset math.
Output: Not applicable.
Argument options: pretty(M$_s$) to print the symbolic matrix M$_s$ in a similar form.
♣ pretty(symexpr, int to use a character width of int when producing the pretty output. The default is 79. ♣ pretty to display the previous result in pretty form.

procread(filename)
Reads a Maple procedure, written in Maple's own programming language, from the file filename.
Output: Not applicable.
Additional information: In order to use this function, you must have the Extended *Symbolic Math Toolbox* installed. ♣ Once the new Maple procedure has been loaded, it can be accessed through the maple command. ♣ The Maple procedure can be of any size. ♣ For more information on how to program in Maple, see a Maple-specific textbook or *The Maple Handbook*.
See also: maple

rsums(symexpr)
Creates a figure which interactively approximates the integral of symbolic expression symexpr.
Output: A figure window is created.
Additional information: The integral is approximated from 0 to 1. The value of the Riemann sum is displayed above the graph. ♣ Initially, 10 ranges are used in the sum. A slider allows you to increase or decrease this number.

$symexpr_{out}$ = simple($symexpr_{in}$)
Simplifies $symexpr_{in}$ by applying various algebraic manipulations, so long as they shorten the overall length of the expression.
Output: A symbolic expression is assigned to $symexpr_{out}$.
Argument options: simple(symexpr) to display information about each manipulation that is performed during the simplification. ♣ [$symexpr_{out}$, str] = simple($symexpr_{in}$) to assign a description of the simplification procedure to str. ♣ simple(M_s) to simplify each individual element of M_s. ♣ simple to simplify the previous result.
Additional information: The simplify command is different than the simple command. simplify uses Maple's built-in simplification rules to simplify an expression. There is no guarantee that the result of simplify will be shorter than the original expression.
See also: simplify, factor, expand, collect

$symexpr_{out}$ = simplify($symexpr_{in}$)
Simplifies $symexpr_{in}$ by applying various algebraic manipulations.
Output: A symbolic expression is assigned to $symexpr_{out}$.
Argument options: simple(M_s) to simplify each individual element of M_s. ♣ simple to simplify the previous result.
Additional information: The simplify command is different than the simple command. simplify uses Maple's built-in simplification rules to simplify an expression, while simple tries to shorten the length of the expression.
See also: simple, factor, expand, collect

M_{out} = sinint(M_{in})
Computes the sine integral function at each element of numeric matrix M_{in}.
Output: A numeric matrix is assigned to M_{out}.
Argument options: M = sinint(M_s) to compute the function at each element of symbolic matrix M_s. A numeric matrix is returned.

Additional information: The sine integral of x is sometimes notated as $Si(x)$. ✸ If a symbolic matrix is used, it must be free of unknown variables.
See also: cosint, mfun

symexpr$_{out}$ = **solve**(symexpr$_{in}$)
Solves the symbolic equation symexpr$_{in}$ for its only variable.
Output: A symbolic expression is assigned to symexpr$_{out}$. If more than one solution is found, they are returned in a symbolic vector of appropriate size.
Argument options: symexpr$_{out}$ = solve(symexpr$_{in}$, str) to use solve symexpr$_{in}$ for the variable str. ✸ symexpr$_{out}$ = solve(symexpr$_{in,1}$, ..., symexpr$_{in,n}$, str$_1$, ..., str$_n$) to solve n equations for n variables. ✸ [V$_{s,1}$, ..., V$_{s,n}$] = solve(symexpr$_{in,1}$, ..., symexpr$_n$, str$_1$, ..., str$_n$) to return solution for the n variables as individual column vectors.
Additional information: If the = syntax is not used in symexpr$_{in}$, then the expression is automatically set equal to 0. ✸ Each symexpr$_{in}$ can also contain *multiple* symbolic equations. Simple create one large string, where the individual equations are separated by commas. ✸ If the result of solve contains calls to RootOf, the allvalues command can be used to get specific values. ✸ Read the entire on-line help page for more information and examples.
See also: dsolve, linsolve, finverse, fsolve, fzero

symexpr$_{out}$ = **subs**(symexpr$_{in}$, symexpr$_{new}$, symexpr$_{old}$)
Replaces each occurrence of symexpr$_{old}$ found in symexpr$_{in}$ with symexpr$_{new}$.
Output: A symbolic expression is assigned to symexpr$_{out}$.
Argument options: symexpr$_{out}$ = subs(symexpr$_{in}$, symexpr$_{new}$) to replace each occurrence of the only variable in symexpr$_{in}$ with symexpr$_{new}$. ✸ M$_{out}$ = subs(M$_{in}$, symexpr$_{new}$, symexpr$_{old}$) to make the substitution throughout the elements of a symbolic matrix.
Additional information: The subs command only performs syntactical substitution. That is, the expression being searched for is not found unless it is a proper subexpression of the original expression. See mhelp(subs) for more information.
See also: simplify, mhelp

M$_s$ = **sym**(symexpr)
Creates a symbolic matrix from a single symbolic expression symexpr, containing symbolic elements separated by commas, and rows separated by semicolons.
Output: A symbolic matrix is assigned to M$_s$.
Argument options: M$_s$ = sym(M) to convert the standard numeric matrix M to a symbolic matrix. Each element of M is converted to its exact rational format.
Additional information: To convert variable-free symbolic matrices back to standard numeric form, use the numeric command. ✸ To perform matrix operation upon symbolic matrices, use the symop command.
See also: double, vpa

poly = **sym2poly**(poly$_s$)
Convert the symbolic polynomial poly$_s$ to its standard numeric coefficient vector form.
Output: A standard polynomial vector is assigned to poly.

Additional information: poly$_s$ must contain only one variable. ✦ The coefficients of poly$_s$ are converted to numeric values using the numeric command.
See also: poly2sym, double, polyval

symexpr$_{out}$ = **symsum**(symexpr$_{in}$, str)
Computes the indefinite summation of symbolic expression symexpr$_{in}$, with respect to the variable str.
Output: A symbolic expression is assigned to symexpr$_{out}$.
Argument options: symexpr$_{out}$ = symsum(symexpr$_{in}$) to sum with respect to the only variable, as found by symvar, in one-variable expression symexpr$_{in}$. ✦ symexpr$_{out}$ = diff(symexpr$_{in}$, str, symexpr$_a$, symexpr$_b$) to compute the definite sum from symexpr$_a$ to symexpr$_b$. ✦ symsum to perform indefinite summation on the previous result.
See also: int

symexpr$_{out}$ = **taylor**(symexpr$_{in}$, str)
Computes the Taylor series expansion of symbolic expression symexpr$_{in}$, about str=0.
Output: A symbolic expression is assigned to symexpr$_{out}$.
Argument options: symexpr$_{out}$ = taylor(symexpr$_{in}$) to find the expansion with respect to the only variable, as found by symvar, in one-variable expression symexpr$_{in}$.
✦ symexpr$_{out}$ = taylor(symexpr$_{in}$, int) or symexpr$_{out}$ = taylor(symexpr$_{in}$, str, int) to compute the expansion to order int.
Additional information: Any calls to the function O in the result represent order terms for the expansion.

M$_{vpa}$ = **vpa**(M$_s$)
Converts the elements of symbolic matrix M$_{vpa}$ to variable precision values.
Output: A variable precision matrix is assigned to M$_{vpa}$.
Argument options: expr$_{vpa}$ = vpa(symexpr) to convert a single symbolic expression to variable precision. ✦ M$_{vpa}$ = vpa(M$_s$, n) to use n digits of precision. The default is 16. vpa to convert the previous result.
Additional information: Maple can perform floating-point computation to nay number of digits. ✦ To alter the number of digits used for subsequent calls to vpa, alter the value of digits. ✦ To convert from variable precision values to standard MATLAB values, use the numeric command.
See also: double, digits

M$_{out}$ = **zeta**(M$_{in}$)
Computes the Riemann Zeta function at each element of numeric matrix M$_{in}$.
Output: A numeric matrix is assigned to M$_{out}$.
Argument options: M = zeta(M$_s$) to compute the function at each element of symbolic matrix M$_s$. A numeric matrix is returned.
Additional information: If a symbolic matrix is used, it must be free of unknown variables.
See also: mfun

Graphing Points and Curves

This chapter looks at plotting 2-D and 3-D curves from data points or function values.

The MATLAB functions covered are:

- plot and plot3 plot 2-D/3-D curves from *data points*.
- fplot plots 2-D curves from *function values*.
- loglog, semilogx, and semilogy plot 2-D curves against *logarithmic axes*.
- polar plots 2-D curves against *polar axes*.
- hist, bar, barh, stairs, and rose plot 2-D histograms and bar graphs.
- patch, fill, and fill3 shade beneath curves (and other areas).
- compass, feather, and quiver plot arrows from complex elements.

See the *Graphics Properties* chapter for details on customizing graphs (e.g., multiple y-axes, etc.).

Several MATLAB toolboxes contain specialized graphing functions, for example:

- Bode and Nyquist plots in the *Control Toolbox*.
- Probability plots in the *Statistics Toolbox*.

`plot, plot3` - Plot 2D/3D Curves from *Data Points*

The plot function creates *x-y* graphs from points contained in vectors. For example, consider the points:

```
x = [ -pi : pi/2 : pi ]
y = sin(x)
```

which produces:

```
x =
    -3.1416   -1.5708        0    1.5708    3.1416

y =
    -0.0000   -1.0000        0    1.0000    0.0000
```

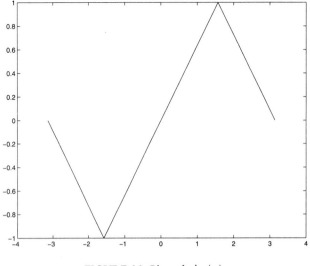

FIGURE 23. Plot of $\sin(x)$

You can plot these (x, y) points simply by using:

```
plot(x,y)
```

which produces the plot shown in Figure 23.

Notice that MATLAB labels the tick marks and automatically scales the axes to fit the data.[36]

You can also plot multiple sets of points on the same axes, e.g.:

```
y2 = cos(x)
plot(x,y,  x,y2)
```

which produces the plot shown in Figure 24.

Plotting Columns of Matrices

In cases such as the above where two sets of points share the same x-values, the two sets of y-values can be rows (or columns) in a matrix. For example:

```
ys = [y
      y2]
```

results in ys containing:

```
ys =
    -0.0000   -1.0000        0    1.0000    0.0000
    -1.0000    0.0000   1.0000    0.0000   -1.0000
```

and

---

[36] You can, however, override the scaling, tick mark placement, and labeling (see the *Graphics Properties* chapter).

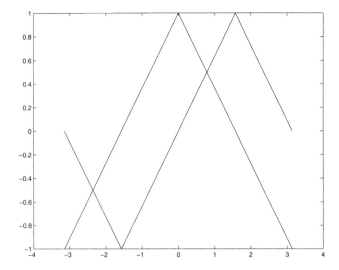

FIGURE 24. Plot of $\sin(x)$ and $\cos(x)$

```
plot(x, ys)
```

produces a plot identical to the previous plot.

Adding Axis Labels, Titles, Legends, and Grid Lines

You can add x- and y-axis labels and a title to the plot.

```
xlabel('This is the xlabel')
ylabel('This is the ylabel')

title('This is the title')
```

These commands must be entered *after* a plot command has produced a figure. You can also add grid lines to the plot using the grid command.

```
grid on          /or/        grid('on')
```

MATLAB puts the grid lines wherever there are tickmarks on the axes.[37]

You can also add a legend to the plots using the legend function:

```
legend('y = sin(x)', ...
       'y2 = cos(x)', ...
       -1)
```

The -1 option tells legend to plot the legend *outside* the figure's axes. By default, legend plots the legend *inside* the axes, trying not to obscure any data points.[38]

---

[37] On logarithmic axes, MATLAB will put in intermediate lines if there is room.

[38] You can always click on the legend with the left mouse button, and drag the legend to a new location.

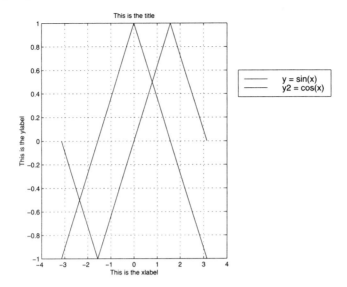

FIGURE 25. Plot of $\sin(x)$ and $\cos(x)$, with axis labels, title, legend and grid lines

It is important to note that in the above call to legend, the two descriptive strings had to be in the same order that the two lines were plotted.

The result of these customizations is shown in Figure 25.

Varying Line Types, Markers, and Colors

By default MATLAB plots points with a solid line style and no markers. You can plot points using other line styles and/or markers. For example, you can plot the two curves as follows:

```
plot(x, y, '--', ...
     x, y2, 'o')
```

Dashed lines are used for the first set of points, and *circles* for the second set. The results are shown in Figure 26. Notice that the legend still correctly reflects the line type or marker.

If you are creating this plot on a color screen, MATLAB plots the two curves in different colors. You will also notice that the colors are correctly reflected in the legend.

A sample of all available line types and markers is shown in Figure 27.

You can mix markers and line styles for the same set of data points by combining their identifiers.

```
plot(x, y, '>--', ...
     x, y2, 'o-')
```

as shown in Figure 28.

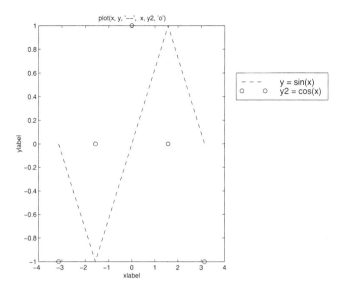

FIGURE 26. Plot of $\sin(x)$ with dashed line between points, and $\cos(x)$ with circle marker at points

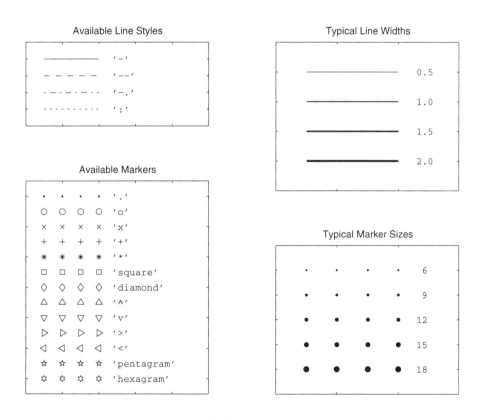

FIGURE 27. Available line styles and markers

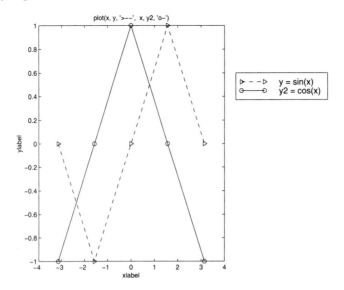

FIGURE 28. Plot of cos(x) and sin(x) with both markers *and* line styles.

Varying Line Width and Marker Size

By default, the width of lines is 0.5 points, where 1 point ≈ $\frac{1}{72}$ inch. You can specify the width of the line using:

```
plot(x, y, '--', 'LineWidth', 4.0)     % Default is 0.5
```

By default, the height of markers is 6 points. You can specify the height of the marker using:

```
plot(x, y2, 'o', 'MarkerSize', 12)     % Default is 6.
```

The result is shown in Figure 29.

Notice that the legend *does* reflect both the line width, and the marker size.

The complete program to produce this plot is as follows:

```
x = [ -pi : pi/2 : pi ];
y = sin(x);
y2 = cos(x);

plot(x, y,   '--', 'LineWidth', 8 * 0.5)

hold on
plot(x, y2, 'o',  'MarkerSize', 2 * 6)
hold off

title('LineWidth: 8 * 0.5    MarkerSize: 2 * 6')
xlabel('xlabel');   ylabel('ylabel');

legend('y = sin(x)', ...
       'y2 = cos(x)', ...
       -1)

print -deps doplot6.eps
```

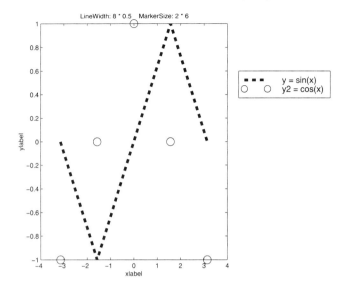

FIGURE 29. Plot of $\sin(x)$ and $\cos(x)$, varying line width and marker size

Because of the extra options 'LineWidth' and 'MarkerSize', it is necessary to make two separate calls to plot, which won't take multiple lines when these types of options are used. Ordinarily, each new invocation of plot *replaces* the previous plot on the screen.

The command hold on tells MATLAB to hold on to what is already on the screen, so that the next plot is *added* to the graph. Once the last plot has been added to this graph, issue the command hold off to tell MATLAB to replace the entire graph when you next invoke plot.

Varying Color

MATLAB chooses a different color for each curve on the screen.[39]

You can specify which color is to be used for each curve. For example:

```
plot(x, y, 'r--', ...
     x, y2, 'bo')
```

Red ('r') dashed lines ('-') are used for the first set of points, and blue ('b') circles ('o') are used for the second set.

When multiple plot lines are added to the same set of axes, MATLAB automatically picks different colors for the lines, cycling through a preset *color order* as listed below.[40]

The default color order is returned by entering

---

[39] However, the curves are *all* printed in black unless you specify otherwise. See the *Graphics Properties* chapter for details.

[40] We recommend specifying the color you want for each individual line.

```
get(gca, 'ColorOrder')
```

which returns

```
ans =

         0         0    1.0000    <--- blue    = 'b'
         0    0.5000         0    <--- green   = 'g'
    1.0000         0         0    <--- red     = 'r'
         0    0.7500    0.7500    <--- cyan    = 'c'
    0.7500         0    0.7500    <--- magenta = 'm'
    0.7500    0.7500         0    <--- yellow  = 'y'
    0.2500    0.2500    0.2500    <--- grey
```

When MATLAB gets to the eighth curve it cycles through the colors again, beginning with the first color (blue).

Apart from these seven colors, you can also specify:

```
         0         0         0    <--- black   = 'k'
    1.0000    1.0000    1.0000    <--- white   = 'w'
```

or *any* color if you know its "red-green-blue" (RGB) code.[41] The previous plot can be reproduced using RGB color notation as follows.

```
plot(x, y,  'LineStyle', '--', 'Color', [1 0 0])
hold on
plot(x, y2, 'LineStyle', 'o',  'Color', [0 0 1])
hold off
```

One shade of red is given by:

```
Dark Red:         [0.40  0.  0.]
```

For shades of grey, the red, green, and blue values are all equal. For example, medium gray is given by:

```
Medium Gray:      [0.5  0.5  0.5]
```

Let's plot our two sets of points using two different shades of gray:

```
plot(x, y,  'LineStyle', '-', 'Color', [0.75 0.75 0.75])

hold on
plot(x, y2, 'LineStyle', '-', 'Color', [0.5  0.5  0.5])
hold off
```

The results are shown in Figure 30.

---

[41] If you know a color by its "hue-saturation-value" code, you can use the hsv2rgb function to convert to RGB.

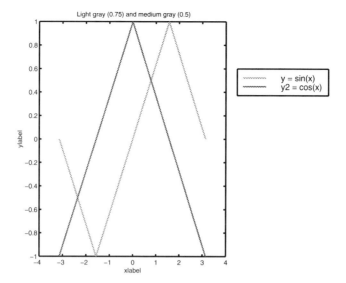

FIGURE 30. Sample plot using 2 shades of gray

`plot3` - Points and Curves in 3-D.

plot3 plots points and curves in 3-D. In its most common form, it is very similar to using plot, except you must also specify the *z* coordinates of the points.

For example, the following plots two curves in 3-D:

```
x = [ -pi : pi/2 : pi ];
y = [ -pi : pi/2 : pi ];

z = sin(x)
z2 = cos(x)

%-------------------------------------------------

plot3(x, y, z,   '--', ...
      x, y, z2, 'o')

view([10, 35])

%-------------------------------------------------

title('plot3(x, y, z,  ''--'',   x, y, z2, ''o'')')

xlabel('xlabel');
ylabel('ylabel');
zlabel('zlabel');

legend('z = sin(x),   y = x', ...
       'z2 = cos(x),  y = x', ...
       -1)

print -deps doplot9.eps
```

Besides plot3, the only other "new" function is view, which controls the viewpoint:

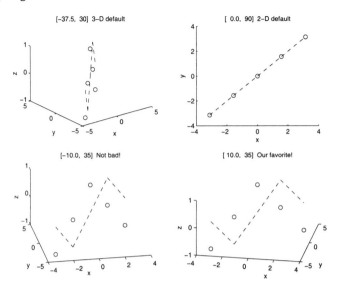

FIGURE 31. Four typical viewpoints

```
view( [azimuth, zenith] )
```

where:

- azimuth is the angle in degrees, counter-clockwise from the *negative* y-axis. The default is -37.5 deg.[42]

- zenith is the angle in degrees above the *xy*-plane. Default is 30 deg.

An interesting point is that MATLAB actually treats 2-D plots as a special case of 3-D plots in which the viewpoint is set to:

```
azimuth = 0.0
zenith  = 90.
```

If you want to, you can change the viewpoint of a 2-D plot to anything you like. Try it and see.

Figure 31 shows four typical viewpoints.

`fplot` - Plot 2-D Curves from *Function Values*

fplot provides a quick method for plotting a function. The first step is to create an M-file to define the function. For example, `myfun.m` could contain:

---

[42] Most software follows the mathematical convention of measuring the azimuth from the *positive* x-axis. However, once you get used to MATLAB's convention, it is actually easier to work with.

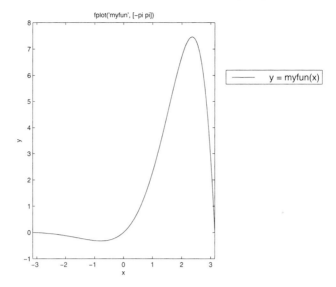

FIGURE 32. Plotting a function using fplot

```
function y = myfun(x)
y = x.^2 - 2;
```

Then to plot the function between $-\pi$ and $+\pi$ you issue:

```
fplot('myfun', [-pi pi])
```

The result is shown in Figure 32.

You *can* specify a line style in calls to fplot, so you could issue: fplot('myfun', [-pi pi], '-.')

fplot does not offer all the flexibility of plot. However, you can use fplot to compute good points for plotting, and then use plot to plot them in whatever style you like.

For example:

```
[xf, yf] = fplot('myfun', [-pi pi]);
plot(xf, yf)
```

You can pass an expression involving the variable *x* to fplot. In this example, we could simply issue: fplot('sin(x) .* exp(x)', [-pi pi])

`loglog`, `semilogx`, and `semilogy` - Plot 2-D Curves against *Logarithmic Axes*

MATLAB allows you to make any of the three available axes logarithmic. There are two ways of doing this, and this section illustrates both ways.

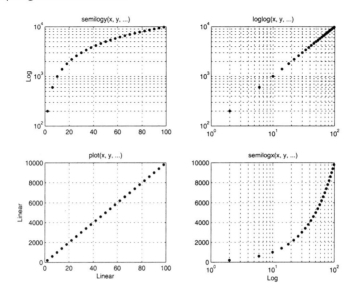

FIGURE 33. Four combinations of linear and logarithmic axes

Figure 33 plots $y = 100x$ using the four combinations of linear/logarithmic x-/y-axis styles.

Each of loglog, semilogx, and semilogy takes the same arguments as plot. The commands to produce the figure are:

```
x = [ 2 : 4 : 98 ];
y = x .* 100;

subplot(2,2,1)
semilogy(x, y, '.', 'MarkerSize', 12)
title('semilogy(x, y, ...)')
% xlabel('Linear')
ylabel('Log')
grid on

            subplot(2,2,2)
            loglog(x, y, '.', 'MarkerSize', 12)
            grid on
            title('loglog(x, y, ...)')
            % xlabel('Log')
            % ylabel('Log')

subplot(2,2,3)
plot(x, y, '.', 'MarkerSize', 12)
grid on
title('plot(x, y, ...)')
xlabel('Linear')
ylabel('Linear')

            subplot(2,2,4)
            semilogx(x, y, '.', 'MarkerSize', 12)
            grid on
            title('semilogx(x, y, ...)')
            xlabel('Log')
            % ylabel('Linear')
```

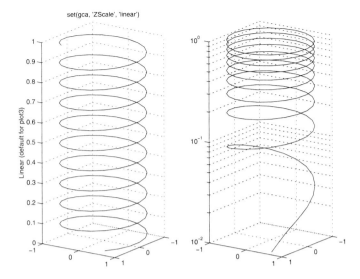

FIGURE 34. 3-D plot with logarithmic z-axis

```
print -deps dolog1.eps
```

The subplot command allows you to combine more than one set of axes in the same MATLAB figure. The first two parameters specify how many axes the figure is to have down and across, respectively. (Here, you want a 2 × 2 "matrix" of axes.) The last parameter specifies which of these axes is currently being worked on, reading from left to right, top to bottom, starting at 1.

Plotting 3-D Curves against Logarithmic Axes

MATLAB does not have 3-D equivalents of loglog, semilog, etc.; however, logarithmic axes can be readily defined using plot3. Figure 34 illustrates a logarithmic z-axis.

The commands to produce this figure are:

```
t = [ .01 : .005 : .99 ];

x = cos(20 * pi * t);
y = sin(20 * pi * t);
z = t;

subplot(1,2,1)
    plot3(x, y, z, '-')           %/////////
    set(gca, 'ZScale', 'linear')  % Default
                                  %\\\\\\\\\
    view([125, 7])
    grid on

    title('set(gca, ''ZScale'', ''linear'')')
    zlabel('Linear (default for plot3)')

subplot(1,2,2)
    plot3(x, y, z, '-')           %////////////
```

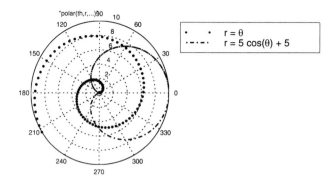

FIGURE 35. Polar plot of $r = \theta$ and $r = 5\cos(\theta) + 5$

```
set(gca, 'ZScale', 'log')    % Log Z-axis
                             %\\\\\\\\\\\
view([125, 7])
grid on

title('set(gca, ''ZScale'', ''log'')')
zlabel('Log')

print -deps dolog2.eps
```

Notice that we first invoke plot3 to produce linear axes:

```
plot3(x, y, z, '-')
```

and then change the *z*-axis to logarithmic scaling using:

```
set(gca, 'ZScale', 'log')
```

Similarly, we could change the *x*- and/or *y*-axis scaling produced by plot3 (or plot) to logarithmic scaling using:

```
set(gca, 'XScale', 'log')    % Log X-axis

set(gca, 'YScale', 'log')    % Log Y-axis
```

Refer to the *Graphics Properties* chapter for more information on using set to modify axis attributes, etc.

polar - Plot 2-D Curves against *Polar Axes*

MATLAB allows you to produce *polar plots* using:

```
polar(th, r)
```

where th is a vector of *angles* (in radians), and r is a vector of corresponding *radius* values.

Figure 35 plots $r = th$ and $r = 5\cos(th) + 5$.

The commands to produce the figure are:

```
th = [0.0 : 0.1 : 10.0 ];              % Angles in RADIANS.

r1 = th;                                % Radius values.
r2 = 5*cos(th) + 5;                     % Another set.

%------------------------------------------------

h(1) = polar(th, r1, 'k.');

set(h(1), 'Markersize', 2 * 6)

%------------------------------------------------

hold on

h(2) = polar(th, r2, 'k-.');
set(h(2), 'LineWidth', 4 * 0.5)

hold off

%------------------------------------------------

title('"polar(th,r,...)"                               ')

legend(h,   'r = \theta',                   ...
            'r = 5 cos(\theta) + 5',        ...
       -1)

print -deps polar1.eps
```

Notes

- We had to make two separate calls to polar to produce this plot—unlike plot, the polar function does not accept multiple sets of data in one invocation.

- We also had to change the 'MarkerSize' and 'LineStyle' using set—unlike plot, the polar function does not accept such properties within its parameter sequence.[43]

- legend does not (currently) automatically pick up the line style and marker types from polar-plots, so we had to specify them in the invocation of legend.

- Notice that we used the
theta syntax to add the θ character to the legend. MATLAB5.1 supports the use of Greek and other TEXcharacter codes.

Another way to produce polar plots is to use regular linear axes, as shown in Figure 36.

The commands to produce this plot are as follows:

```
th = [0.0 : 0.1 : 10.0 ];              % Angles in RADIANS.

r1 = th;                                % Radius values.
r2 = 5*cos(th) + 5;                     % Another set.
```

---

[43] Refer to the *Graphics Control* chapter for more information on using set to modify line attributes, etc.

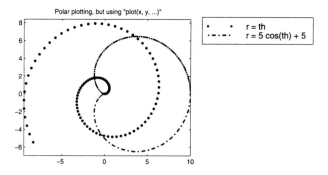

FIGURE 36. Polar plot using regular linear axes

```
%--------------------------------------------------
                                    %///////////////////
[x1, y1] = pol2cart(th, r1);        % Polar-to-Cartesian
[x2, y2] = pol2cart(th, r2);        %    co-ordinates.
                                    %\\\\\\\\\\\\\\\\\\\
%--------------------------------------------------
plot(x1, y1, 'r.', 'Markersize', 2 * 6)

hold on
plot(x2, y2, 'b-.', 'LineWidth',  4 * 0.5)
hold off
                                    %////////////////////////////////
axis('equal')                       % (1 unit in x) = (1 unit in y)
                                    %\\\\\\\\\\\\\\\\\\\\\\\\\\\\\\\\

title('Polar plotting, but using "plot(x, y, ...)"')

%--------------------------------------------------

legend('r = th',             ...
       'r = 5 cos(th) + 5',  ...
       -1)

print -deps polar2.eps
```

This example makes use of MATLAB's pol2cart function to convert polar (th, r) coordinates to cartesian (x, y) coordinates.

Notice that it also specifies:

```
axis('equal')
```

so that one unit in the x direction equals one unit in the y direction. Otherwise circles (e.g., $r = \cos(\theta)$) would appear as ovals.

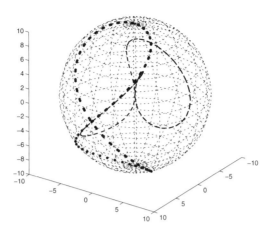

FIGURE 37. Spherical plot produced using plot3 on cartesian axes

Plotting 3-D Curves against *Spherical Axes*

MATLAB does not have functions *specifically* for plotting in spherical coordinates. However, you can readily convert spherical (θ, ϕ, r) coordinates to cartesian (x, y, z) coordinates using sph2cart, and then plot the 3-D data points using plot3.

Figure 37 illustrates plotting data given in spherical coordinates.

The commands to produce this plot are as follows:

```
az = [0.0 : 0.1 : 10.0 ];          % Azimuth angles
                                   %    in RADIANS

el = az;                           % Zenith angles

r1 = 10*ones(size(az));            % Radius values
r2 = 10*cos(az);                   % Another set

%-------------------------------------------------
                                   %///////////////
[x1, y1, z1] = sph2cart(az, el, r1);  %   Spherical-to-
[x2, y2, z2] = sph2cart(az, el, r2);  %   cartesian coords
                                   %\\\\\\\\\\\\\\\

%-------------------------------------------------

plot3(x1, y1, z1, 'r.',   'Markersize', 2 * 6)

hold on

plot3(x2, y2, z2, 'b--', 'LineWidth', 3 * 0.5)

%- - - - - - - - - - - - - - - - - - - - - - - -
                                   %///////////////////////
```

```
            [Xs, Ys, Zs] = sphere(20);          % "sphere(20)" produces
                                                %    21x21 X,Y,Z matrices
                                                %    representing 'unit'
                                                %    sphere (radius = 1).
                                                %
            h = surf(10*Xs, 10*Ys, 10*Zs);      % "surf(...)" plots it.
                                                %\\\\\\\\\\\\\\\\\\\\\\\\
            set(h, 'FaceColor', 'none')                 % Transparent
            set(h, 'EdgeColor', [0.0  0.5  0.0])        % Dark green
            set(h, 'LineStyle', ':')                    % Dotted lines

            hold off

            %-------------------------------------------------------------

            view([125, 25]);
                                                %////////////////////////
            axis('equal')                       % (1 unit in x) = (1 unit in y)
                                                %               = (1 unit in z)
                                                %\\\\\\\\\\\\\\\\\\\\\\\\
            title('Spherical plotting using "plot3(x, y, z, ...)"')

            %-------------------------------------------------------------

            legend('r.',   'r = th',           ...
                   'b--',  'r = 10 cos(th)',   ...
                   -1)

            print -deps polar3.eps
```

sphere(20) produces the (x, y, z) coordinates for the surface of a sphere in a form usable by surf, which plots the surface.

As you may have noticed, the above example combines the results of two calls to plot3 with a call to surf. Generally, plots created with different plotting commands (both 2-D and 3-D) can be combined on one set of axes.

`hist`, `bar`, `barh`, and `rose` - Plot 2-D Histograms and Bar Graphs

MATLAB can produce histograms in both cartesian and polar coordinates. Consider the following simple set of data stored in y:

```
            y = [-27  -23   -15   15   23  27]
```

A plot of these y-values against themselves is shown in Figure 38 (top left).

Now group the data into six different bins, defined by the range of y-values. Then count the number falling in each range. The result is as follows:

```
            min(y) < y <= -20              n = 2
              -20  < y <= -10              n = 1
              -10  < y <=   0              n = 0
                0  < y <=  10              n = 0
               10  < y <=  20              n = 1
               20  < y <= max(y)           n = 2
```

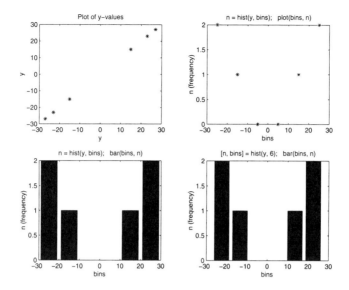

FIGURE 38. Sample data and histogram

You can describe the bins using their end-points:

```
ranges = [ min(y)    -20    -10    0    +10    +20    max(y) ]
n      = [            2      1     0     0      1      2            ]
```

MATLAB's hist function can do the counting and compute n for us. But first, you must describe the bins by their *mid-points*, rather than by their *end-points*.

```
ranges = [ min(y)    -20    -10    0    +10    +20    max(y) ]
bins   = [        -25      -15   -5   +5     +15    +25            ]
```

Now invoke hist to count the number in each bin. The complete set of commands issued in this example is:

```
y      = [-27   -23   -15                15    23    27 ]
bins   = [      -25   -15    -5    5     15    25 ]
n      = hist(y, bins)
```

which results in:

```
n =
            2     1     0     0     1     2
```

Use the plot function to plot the frequencies (n) against the bin values:

```
plot(bins, n)
```

as shown in Figure 38 (top right).

However, the bar function in MATLAB produces an easier to interpret graph, by plotting bars around the bin values:

```
bar(bins, n)
```

The result is shown in Figure 38 (bottom left).

You can also produce *horizontal* bars with the barh function. For example,

```
barh(bins, n)
```

would produce the same plot but with the bars running left to right, not up and down.

If you invoke hist with no *output* arguments, then hist computes the frequencies, and calls bar. For example:

```
hist(y, bins)
```

produces the same result as the bar plot in the bottom-left corner of Figure 38.

If you invoke hist without specifying bins, then it computes and returns ten bin values for you. For example:

```
[n, bins] = hist(y)
```

results in:

```
n =
     2    0    1    0    0    0    0    1    0    2
bins =
   -24.3 -18.9 -13.5  -8.1  -2.7   2.7   8.1  13.5  18.9  24.3
```

You can also specify the number of bins that you want (say 6). For example:

```
[n, bins] = hist(y, 6)
bar(bins, n)
```

results in:

```
n =
     2    1    0    0    1    2
bins =
   -22.5 -13.5  -4.5   4.5  13.5  22.5
```

The plot of this histogram is shown in Figure 38 (bottom right).

Notice that each bin that hist(y, 6) returns has the same *uniform* width. In this example they are all nine units wide. Also notice that the bins *exactly* enclose all the y-values, no more and no less. Hence, the last midpoint (at 22.5) is exactly 4.5 less than max(y) (at 27).

You can also pass your own bins of *non-uniform* width to hist. For example, change the last bin value from 25 to 30:

```
bins2 = [   -25   -15   -5    5    15    30 ]
n = hist(y, bins2);
bar(bins2, n)
```

The resulting histogram is shown in Figure 39.

This histogram corresponds to the following y-ranges:

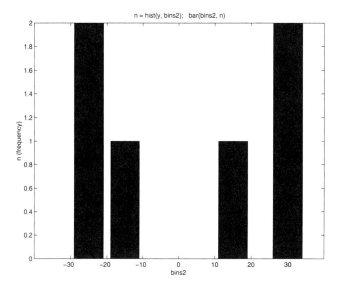

FIGURE 39. Histogram with bins of **non-uniform** width

```
ranges = [ -30    -20    -10    0    10    22.5    37.5 ]
```

meaning that all the bins are 10 units wide, except that the last two which are both 12.5 units wide. Notice, as well, that the second last bin value (15) is *not* at the *mid-point* of a range.

So, as you can see, the hist function: (1) computes end-points that are midway between the supplied bin values, and then (2) counts the numbers in the resulting ranges.

rose is the polar equivalent of hist and bar together. The vectors passed to rose are given in radians between 0 and 2π. As with hist you can specify: (1) the number of bins, or (2) the actual bins themselves. In this example, we specify the number of bins as 12 (the default is 20):

```
thetas = [ 260 270 280   170 190   90 ] * (pi/180);
[tt, rr] = rose(thetas, 12);
```

The output can then be passed directly to polar for plotting: (If no output arguments are given, rose invokes polar.)

```
polar(tt, rr, '-')
```

The result is shown in Figure 40.

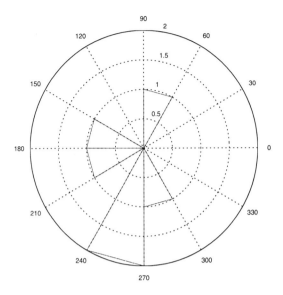

FIGURE 40. "Rose plot" - polar histogram

`patch`, `fill`, and `fill3` - Shade Beneath Curves (and Other Areas)

MATLAB allows you to fill defined areas with solid colors. A common example is filling in the area under bars in a histogram. We use bar to compute the bar lines without plotting them, and use plot to plot them:

```
n = hist(y, bins)          % compute frequencies
     [bb, nn] = bar(bins, n)     % compute bar coords
plot(bb, nn, '--')         % plot dashed bars
```

Now, replace the call to plot with a call to fill. For example, to plot red bars you would issue:

```
fill(bb, nn, 'r')          % 'r' for red
```

or, alternatively, you would specify the RGB code for the color:

```
fill(bb, nn, [1 0 0])      % [red, green, blue]
```

The result is shown in Figure 41 (left side).

If you are following this example on your computer, you will notice that part(s) of the *x*-axis has been obscured. One simple way to fix this is to have plot draw the top and sides of the bars on the same graph:

```
hold on
plot(bb, nn, 'r-')
hold off
```

The result is shown in Figure 41 (right side).

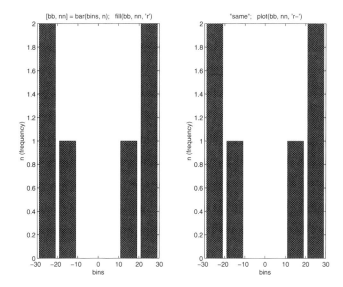

FIGURE 41. Using fill to shade a histogram

`compass`, `feather`, and `quiver` - Plot Arrows from Complex Elements

Complex elements can be plotted with arrows, using one of these three MATLAB functions:

- compass plots arrows all emanating from one point ($x=0$, $y=0$).
- feather plots arrows emanating from the x-axis (equally-spaced).
- quiver plots arrows emanating from various (x, y) locations.

Consider the following matrix of complex numbers, DZ, formed from DX and DY:

```
A = [ 1/4   5/4
      2/4   6/4
      3/4   7/4 ] * pi;

R = A / pi;

DX = R .* cos(A);
DY = R .* sin(A);

DZ = DX + i*DY       % where "i" is sqrt(-1)
```

which results in:

```
DZ =
   0.1768 + 0.1768i   -0.8839 - 0.8839i
   0.0000 + 0.5000i   -0.0000 - 1.5000i
  -0.5303 + 0.5303i    1.2374 - 1.2374i
```

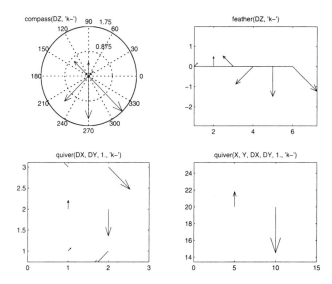

FIGURE 42. Plotting complex matrix elements using compass, feather, and quiver

Now let's plot DZ using each of compass, feather, and quiver. The results are shown in Figure 42.

The commands to produce the plots are as follows:

```
    subplot(221)
        compass(DZ, 'k-')
% or:   compass(DX, DY, 'k-')

        axis('equal')           % Not needed
                                % with "compass"

        title('compass(DZ, ''k-'')')

%-------------------------------------------------

    subplot(222)
        feather(DZ, 'k-')
% or:   feather(DX, DY, 'k-')

        axis('equal')
        title('feather(DZ, ''k-'')')

%-------------------------------------------------

    subplot(223)
                    rel = 1.
        quiver(DX, DY, rel, 'k-')

        axis([ 0   2+1    0   3+1 ])
        axis('equal')
        title('quiver(DX, DY, 1., ''k-'')')

%-------------------------------------------------

    subplot(224)
        [X, Y] = meshgrid( [5 10], [10,
```

```
                            20,
                            30] );
                   rel = 1.
    quiver(X, Y, DX, DY, rel, 'k-');

    axis([ 0   10+5    0   30+10 ])
    axis('equal')
    title('quiver(X, Y, DX, DY, 1., ''k-'')')

print -depsc complex1.eps
```

Notes

- In all of the above plots we have specified:

    ```
    axis('equal')
    ```

    to tell MATLAB that axes should be scaled such that one unit in x is equal to one unit in y. This causes complex numbers such as $1 + 1i$ to be represented by a vector drawn at 45 deg, rather than skewed. Furthermore, complex numbers of equal magnitude are plotted with vectors of equal length.

    In the case of compass, equal axes is actually the default, so specifying axis('equal') is redundant.

- Both compass and feather plot the vectors using the magnitude of the complex number as the length. However, quiver scales all the vectors so that they extend no further than their nearest neighbor's origin.

    You can change the *relative size* of the vectors, by changing rel. For example, setting rel = 2.0 doubles the length of the vectors in the above plots.

    With compass and feather you have no control over where the origin of the vectors is. compass plots all vectors from (x=0, y=0). feather plots all vectors along the x-axis, starting at x=1. In the case of matrices, feather plots in "column order", plotting the first column, then the second, and so on.

    However, quiver *does* give you control over the (x,y) location of the origin of the vectors. In the plot on the bottom left of Figure 42, we use the default of (x=j, y=i), where (i,j) are the row and column indices of the elements. In the plot on the bottom right, we define our own x and y coordinates for each element.

- The quiver plot on the left might suit your needs better if the numbering on the y-axis is reversed (for example, if you were trying to visualize the *pattern* of complex numbers in a matrix). This can be achieved by adding the following line *after* the invocation of quiver:

    ```
    axis('ij')         %%     Equivalent to:
                       %% set(gca, 'YDir', 'reverse')
    ```

    However, this would also have the effect of flipping the vectors upside down! To counteract that you could pass quiver the negative of your y-values:

    ```
    quiver(DX, -DY, rel, '-')
    ```

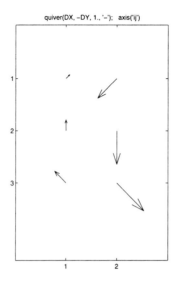

FIGURE 43. Plotting a complex 3 × 2 matrix using quiver

A further refinement for the matrix plot would be to make the lengths of the x- and y-axis proportional to the number of columns and rows in the matrix. For example, if the matrix is square, you can make the axes the same length as each other simply by using:

```
axis('square')
```

In this case, we have a 3 × 2 matrix, so we want the x-axis to be $\frac{2}{3}$ the length of the y-axis. This is just a little more complicated to achieve:[44]

```
x_over_y = get(gca, 'AspectRatio')
             [m,n] = size(DZ);
x_over_y(1) = n/m;
set(gca, 'AspectRatio', x_over_y)
```

Furthermore, it is better to restrict the axes tick marks to actual row and column numbers:

```
set(gca, 'XTick', [1:n], ...
         'YTick', [1:m])
```

Putting all of these commands together produces the result shown in Figure 43.[45]

---

[44] The vector x_over_y returned by set below actually has *two* elements. The second element controls the ratio of x and y data units. For example, it was set to 1.0 when we issued: axis('equal') in the above examples, so that one unit in x would equal one unit in y.

[45] Refer to the *Graphics Properties* chapter for more information on using set to modify axis attributes, etc.

Command Listing

area(V_y)
Displays a filled area plot defined by the y-values in vector V_y.
Output: A two-dimensional plot is displayed.
Argument options: area(V_x, V_y), where V_x and V_y are vectors of identical length, to plot the values of V_y at the corresponding horizontal locations in V_x. The values in V_x must be evenly spaced and ascending. ✤ area(M) or area(M_1, M_2) to plot a series of filled areas corresponding to the columns of the given matrices, stacked one on top of the other on the same set of axes. ✤ area(V_y, num) to filled areas at a y-level of num. Default is num $= 0$. ✤ V_{hndl} = area(V_x, V_y) to return the vector of patch object handles necessary to create the filled area graph with the plot command.
Additional information: If only one vector is passed to area, the other vector is assumed to contain, in order, the elements 1 through n, where n is the length of the given vector. ✤ If only one matrix is passed to area, each column of the other matrix is assumed to contain, in order, the elements 1 through n, where n is the number or rows of the given matrix.
See also: bar, plot, hist

bar(V)
Displays a bar graph where the height of the bars corresponds to the value of the elements of vector V.
Output: A two-dimensional plot is displayed.
Argument options: bar(V_x, V_y), where V_x and V_y are vectors of identical length, to plot the values of V_y at the corresponding horizontal locations in V_x. The values in V_x must be evenly spaced and ascending. ✤ bar(M) or bar(M_1, M_2) to plot a series of bar graphs corresponding to the columns of the given matrices, staggered one on top of the other on the same set of axes. ✤ bar(M, str) to supply an optional value to control the plot. Available values are *'grouped'*, and *'stacked'*. ✤ V_{hndl} = bar(V_x, V_y) to return the vector of patch object handles necessary to create the bar graph with the plot command.
Additional information: If only one vector is passed to bar, the other vector is assumed to contain, in order, the elements 1 through n, where n is the length of the given vector. ✤ If only one matrix is passed to bar, each column of the other matrix is assumed to contain, in order, the elements 1 through n, where n is the number or rows of the given matrix.
See also: barh, bar3, bar3h, plot, area, stairs, stem, hist, errorbar

barh(V)
Displays a horizontal bar graph where the length of the bars corresponds to the value of the elements of vector V.
Output: A two-dimensional plot is displayed.
Argument options: barh(V_x, V_y), where V_x and V_y are vectors of identical length, to plot the values of V_y at the corresponding horizontal locations in V_x. The values in V_x must be evenly spaced and ascending. ✤ barh(M) or bar(M_1, M_2) to plot a series of bar graphs corresponding to the columns of the given matrices, staggered one on

top of the other on the same set of axes. ♦ barh(M, str) to supply an optional value to control the plot. Available values are *'grouped'*, and *'stacked'*. ♦ V$_{hndl}$ = barh(V$_x$, V$_y$) to return the vector of patch object handles necessary to create the bar graph with the plot command.
Additional information: If only one vector is passed to barh, the other vector is assumed to contain, in order, the elements 1 through *n*, where *n* is the length of the given vector. ♦ If only one matrix is passed to barh, each column of the other matrix is assumed to contain, in order, the elements 1 through *n*, where *n* is the number or rows of the given matrix.
See also: bar, bar3, bar3h, plot, stairs, stem, hist, errorbar

[num$_\theta$, num$_r$] = cart2pol(num$_x$, num$_y$)

Converts the cartesian coordinate specified by *x* and *y* values num$_x$ and num$_y$ into polar coordinates.
Output: A numeric value representing angle (in radians) is assigned to num$_\theta$ and a numeric value representing radius is assigned to num$_r$.
Argument options: [M$_\theta$, M$_r$] = cart2pol(M$_x$, M$_y$), where M$_x$ and M$_y$ are matrices of identical dimensions, to convert each corresponding pair of *x* and *y* values. ♦ [M$_\theta$, M$_r$, M$_z$] = cart2pol(M$_x$, M$_y$, M$_z$) to transform values in three-dimensional cartesian coordinates into cylindrical coordinates.
Additional information: Use the polar command to plot the resulting values.
See also: polar, pol2cart, cart2sph, sph2cart

comet(V$_x$, V$_y$)

Displays an "animated" comet plot of the values in V$_x$ versus those in V$_y$.
Output: A two-dimensional plot is displayed.
Argument options: comet(V$_y$) to plot the comet plot of the real values in V$_y$ against their indices. That is, V$_x$ = [1 2 3 ...]. ♦ comet to show an example of a comet plot. ♦ comet(V$_x$, V$_y$, num), where num is a value between 0 and 1, to specify that the tail of the comet is num * length(V$_y$). The default value for num is 0.1.
Additional information: A comet plot draws values of V$_y$ as V$_x$ increases. A "tail," which eventually fades to another color, is left by the comet as it progresses. ♦ comet takes no more than three input parameters.
See also: comet3, plot, movie

comet3(V$_x$, V$_y$, V$_z$)

Displays an "animated" three-dimensional comet plot of the values represented by V$_x$, V$_y$, and V$_z$.
Output: A three-dimensional plot is displayed.
Argument options: comet3(V$_z$) to plot the comet plot of the real values in V$_z$ against standard values for V$_x$ and V$_y$. ♦ comet3 to show an example of a three-dimensional comet plot. ♦ comet(V$_x$, V$_y$, V$_z$, num), where num is a value between 0 and 1, to specify that the tail of the comet is num * length(V$_y$). The default value for num is 0.1.
Additional information: A comet plot draws values of V$_z$ as V$_x$ and V$_y$ increase. A "tail," which eventually fades to another color, is left by the comet as it progresses. ♦ comet3 takes no more than four input parameters.
See also: comet, plot3, movie

compass(V_{cmplx})

Displays the compass plot of V_{cmplx}, a vector of complex elements.
Output: A two-dimensional plot is displayed.
Argument options: compass(V_x, V_y), where V_x and V_y are vectors of identical length, to plot the equivalent compass plot with values $V_x + i*V_y$. ✤ compass(M_{cmplx}) or compass(M_x, M_y) to plot the elements from matrices instead of vectors.
See also: rose, feather, quiver, polar

errorbar(V_x, V_y, V_e)

Displays an errorbar graph of the function defined by x and y vectors V_x and V_y and error levels defined by vector V_e.
Output: A two-dimensional plot is displayed.
Argument options: errorbar(M_x, M_y, M_e) to plot a series of errorbar graphs corresponding to the columns of the given matrices.
Additional information: The vectors V_x, V_y, and V_e (or the corresponding matrices) must all be of the same size. ✤ The individual error bars are drawn both above and below the corresponding ($V_x(i)$, $V_y(i)$) points a distance of $V_e(i)$.
See also: plot, bar, std

feather(V_{cmplx})

Displays the feather plot of V_{cmplx}, a vector of complex elements.
Output: A two-dimensional plot is displayed.
Argument options: feather(V_x, V_y), where V_x and V_y are vectors of identical length, to plot the equivalent feather plot with values $V_x + i*V_y$. ✤ feather(M_{cmplx}) or feather(M_x, M_y) to plot the elements from matrices instead of vectors.
See also: rose, compass, quiver

fill(V_x, V_y, [num_r, num_g, num_b])

Displays a polygon, whose vertices are defined by the x and y values in corresponding elements of vectors V_x and V_y, with the color defined by RGB values num_r, num_g, and num_b.
Output: A two-dimensional plot is displayed.
Argument options: fill(V_x, V_y, str), where str is one of 'r', 'g', 'b', 'c', 'm', 'y', 'w', or 'k', to use one of the eight predefined colors. ✤ fill(V_x, V_y, V_c), where V_c is a vector of identical size to V_x and V_y, to color each vertex with the corresponding color from V_c and interpolate the colors within the polygon. ✤ fill(M_x, M_y, V_c) or fill(M_x, M_y, M_c) to create one polygon plot for each column of M_x and M_y.
Additional information: All vectors and matrices passed to fill must be of appropriately related sizes. ✤ If the last vertex specified does not equal the first vertex specified, then the connection is made automatically.
See also: fill3, patch

fill3(V_x, V_y, V_z, [num_r, num_g, num_b])

Displays a polygon, whose vertices are defined by the x, y, and z values in corresponding elements of vectors V_x, V_y and V_z, with the color defined by RGB values num_r, num_g, and num_b.
Output: A three-dimensional plot is displayed.

Argument options: fill3(V_x, V_y, V_z, str), where str is one of 'r', 'g', 'b', 'c', 'm', 'y', 'w', or 'k', to use one of the eight predefined colors. ✤ fill3(V_x, V_y, V_z, V_c), where V_c is a vector of identical size to V_x V_y, and V_z, to color each vertex with the corresponding color from V_c and interpolate the colors within the polygon. ✤ fill3(M_x, M_y, M_z, V_c) or fill3(M_x, M_y, M_z, M_c) to create one polygon plot for each column of M_x, M_y, and M_z.
Additional information: All vectors and matrices passed to fill3 must be of identical size. ✤ If the last vertex specified does not equal the first vertex specified, then the connection is made automatically.
See also: fill, patch

fplot(str, [num_a, num_b])

Displays a two-dimensional plot of the one-variable function represented by string str over x-axis values from num_a to num_b.
Output: A two-dimensional plot is displayed.
Argument options: fplot('[fnc_1, ..., fnc_n]', [num_a, num_b]) to plot n different functions on the same set of axes. fplot(str, [num_{xa}, num_{xb}, num_{ya}, num_{yb}], str_c) to limit the y-axis to displaying values from num_{ya} to num_{yb}. ✤ fplot(str, [num_a, num_b], str_c) to plot the function according to the color and type specification string str_c. For more information, see the entries for LineStyle and Color. ✤ fplot(str, [num_a, num_b], posint) to use a minimum of posint sample points over the range. ✤ fplot(str, [num_a, num_b], $posint_1$, ang, $posint_2$) to specify that at most $posint_2$ subdivisions be taken in sections of the plot with rapidly changing angle. Default is 20. ✤ [V_x, V_y] = fplot(str, [num_a, num_b]) to assign the relevant vectors to V_x and V_y. No plot is displayed until plot is called. ✤
Additional information: str can either be the name of a function defined in a .m file or an actual one-variable Matlab function enclosed in quotes (e.g., 'cos(x)').
See also: plot, LineStyle, feval

fplotdemo

Runs a demonstration script detailing the workings of the fplot command.
Output: Not applicable.
Additional information: Various commands are automatically entered for you.
✤ Occasionally, you will be prompted to strike any key to continue the demonstration. ✤ For a complete list of demonstrations available on your platform, see the on-line help file for demos.
See also: fplot, quaddemo, odedemo, zerodemo, fftdemo

gplot(M_{adj}, M_{pos})

Displays the two-dimensional graph theory graph defined by adjacency matrix M_{adj} and node position matrix M_{pos}.
Output: A two-dimensional plot is displayed.
Argument options: gplot(M_{adj}, M_{pos}, str) to draw the lines in the plot according to the color and type specification string str. For more information, see the entries for LineStyle and Color.
Additional information: The individual elements of adjacency matrix M_{adj}(i, j) equal 1 if there is an edge joining nodes i and j, and equal 0 if there is no such edge. ✤ The node position matrix M_{pos} is an $n \times 2$ matrix containing x and y positions in the figure for each node of the graph.

See also: plot, spy

hist(V_y)
Displays a 10-bin histogram of the elements of vector V_y.
Output: A two-dimensional plot is displayed.
Argument options: hist(V_y, n) to draw the n-bin histogram. ✦ hist(V_y, V_x) to plot the histogram of the elements in V_y with respect to the bins specified in V_x. ✦ [V_f, V_x] = hist(V_y) to assign the frequency counts and bin specifications to V_f and V_x, respectively. No plot is displayed, but the result can be shown using bar.
See also: bar, rose, stairs

loglog(V_x, V_y)
Displays the two-dimensional plot, with logarithmic scaling, of the values in vectors V_x versus V_y.
Output: A two-dimensional plot is displayed.
Argument options: loglog(V_x, V_y, str) to draw the log-log plot with line style specified by str. ✦ loglog(V_{x1}, V_{y1}, str_1, ..., V_{xn}, V_{yn}, str_n) to draw the n log-log plots on the same set of axes.
Additional information: If there are any positive values in V_y, then all negative values in V_y are ignored. Otherwise, negative values are plotted. ✦ To create a log plot where only one of the axes is scaled, use semilogx or semilogy. ✦ For more information on line styles and multiple plots, see the listing for plot.
See also: semilogx, semilogy, plot

pie(V)
Displays a pie chart where the arc length of individual slices correspond to the values of the elements of vector V.
Output: A two-dimensional plot is displayed.
Argument options: pie(V, V_{ex}), where V_{ex} and V are vectors of identical length, to "pull" out those slices where V_{ex} is non-zero. ✦ V_{hndl} = pie(V) to return the vector of patch and text object handles necessary to create the pie chart.
Additional information: If the elements of V sum to greater than 1, then each element is divided by that sum to determine its share of the pie. If the sum is less than 1, then each element is treated as a percentage, and an incomplete pie is created.
See also: pie3, plot

pie3(V)
Displays a "raised" pie chart where the arc length of individual slices correspond to the values of the elements of vector V.
Output: A three-dimensional plot is displayed.
Argument options: pie3(V, V_{ex}), where V_{ex} and V are vectors of identical length, to "pull" out those slices where V_{ex} is non-zero. ✦ V_{hndl} = pie3(V) to return the vector of patch, surface, and text object handles necessary to create the pie chart.
Additional information: If the elements of V sum to greater than 1, then each element is divided by that sum to determine its share of the pie. If the sum is less than 1, then each element is treated as a percentage, and an incomplete pie is created.
See also: pie, plot

plot(V_x, V_y)

Displays the two-dimensional linear plot connecting xy-points defined by the pair of elements from vectors V_x and V_y.

Output: A two-dimensional plot is displayed.

Argument options: plot(V_y) to plot the real values of V_y against their indices. That is, $V_x = [1\ 2\ 3\ \ldots]$. ✦ plot(V_{cmplx}), where V_{cmplx} is a vector of complex elements, to plot the equivalent of plot(real(V_{cmplx}), imag(V_{cmplx})). ✦ plot(V_x, V_y, str) to plot the line or points according to the color and type specification string str. For more information, see the entries for LineStyle and Color. ✦ plot(V_x, V_y, *line*(var, expr)) or plot(V_x, V_y, str, *line*(var, expr)) to specify additional properties to the line plotted. See the entry for line for more information. ✦ plot(V_{x1}, V_{y1}, ..., V_{xn}, V_{yn}), plot(V_{x1}, V_{y1}, str$_1$, ..., V_{xn}, V_{yn}, str$_n$), etc., to combine the n linear plots in one figure.

Additional information: Other properties of two-dimensional plots (e.g., x- and y-axes labels, grid specifications, etc.) are not set from within the plot command, but from within their own commands (e.g., xlabel, ylabel, grid). For more information, see the entries for those commands. ✦ Linear plots with other scaling conventions or coordinate types can be created using loglog, semilogx, semilogy, polar.

See also: line, grid, title, axes, xlabel, ylabel, plot3, comet, loglog, semilogx, semilogy, polar, subplot

plot3(V_x, V_y, V_z)

Displays the three-dimensional linear plot connecting xyz-points defined by the triplet of elements from vectors V_x, V_y, and V_z.

Output: A three-dimensional plot is displayed.

Argument options: plot3(V_x, V_y, V_z, str) to plot the line or points according to the color and type specification string str. For more information, see the entry for LineStyle in the *Graphics Control* chapter. ✦ plot3(V_x, V_y, V_z, *line*(var, expr)) or plot3(V_x, V_y, V_z, str, *line*(var, expr)) to specify additional properties to the line plotted. See the entry for line for more information. ✦ plot3(V_{x1}, V_{y1}, V_{z1}, ..., V_{xn}, V_{yn}, V_{zn}), plot3(V_{x1}, V_{y1}, V_{z1}, str$_1$, ..., V_{xn}, V_{yn}, V_{zn}, str$_n$), etc., to combine the n linear plots in one figure. ✦ plot3(M_x, M_y, M_z), where M_x, M_y, and M_z are matrices of equal dimensions, to combine in one figure the multiple linear plots generated from their corresponding columns.

Additional information: Other properties of three-dimensional plots (e.g., x-, y-, and z-axes labels, viewing angle, etc.) are not set from within the plot3 command, but from within their own commands (i.e., xlabel, ylabel, zlabel, view, etc.). For more information, see the entries for those commands.

See also: line, view, title, axes, xlabel, ylabel, zlabel, plot, surf, mesh

plotyy(V_{x1}, V_{y1}, V_{x2}, V_{y2})

Displays two two-dimensional linear plots connecting xy-points defined by the pairs of elements from the input vectors. V_{x1} and V_{y1} are labeled on the left y-axis, while V_{x2} and V_{y2} are labeled on the right y-axis.

Output: A two-dimensional plot is displayed.

Argument options: plotyy(V_{x1}, V_{y1}, V_{x2}, V_{y2}, str), where str is the name of a plotting function like *'loglog'* or *'stem'*, to use a function other than plot. ✦ plotyy(V_{x1}, V_{y1}, V_{x2}, V_{y2}, str$_1$, str$_2$) to use different functions for the left- and right-axis data.

See also: plot

[num$_x$, num$_y$] = pol2cart(num$_\theta$, num$_r$)
Converts the polar coordinate specified by θ and r values num$_\theta$ and num$_r$ into cartesian coordinates.
Output: Two numeric values representing x and y values are assigned to num$_x$ and num$_y$, respectively.
Argument options: [M$_x$, M$_y$] = cart2pol(M$_\theta$, M$_r$), where M$_\theta$ and M$_r$ are matrices of identical dimensions, to convert each corresponding pair of values. ♣ [M$_x$, M$_y$, M$_z$] = cart2pol(M$_\theta$, M$_r$, M$_z$) to transform values in cylindrical coordinates into three-dimensional cartesian coordinates.
Additional information: Use the plot or plot3 command to plot the resulting values.
See also: plot, plot3, cart2pol, cart2sph, sph2cart

polar(V$_\theta$, V$_\rho$)
Displays the two-dimensional polar plot of vector of θ angles V$_\theta$ versus vector of ρ angles V$_\rho$.
Output: A two-dimensional plot is displayed.
Argument options: feather(V$_\theta$, V$_\rho$, str) to use the line type specified by str. (See the listings for plot and line for more information.)
See also: plot, rose, compass, line

rose(V$_\theta$)
Displays a polar histogram of the elements of vector V$_\theta$.
Output: A two-dimensional polar plot is displayed.
Argument options: hist(V$_\theta$, n) to draw the n-bin polar histogram. The default is 20. ♣ hist(V$_\theta$, V$_x$) to plot the polar histogram of the elements in V$_\theta$ with respect to the bins specified in V$_x$. ♣ [V$_f$, V$_x$] = hist(V$_y$) to assign the frequency counts and bin specifications to V$_f$ and V$_x$, respectively. No plot is displayed, but the result can be shown using polar.
Additional information: The elements of V$_\theta$ must be angles measured in radians.
See also: polar, hist, bar, stairs

semilogx(V$_x$, V$_y$)
Displays the two-dimensional plot, with logarithmic scaling on the x-axis, of the values in vectors V$_x$ versus V$_y$.
Output: A two-dimensional plot is displayed.
Argument options: semilogx(V$_x$, V$_y$, str) to draw the semilog plot with line style specified by str. ♣ semilogx(V$_{x1}$, V$_{y1}$, str$_1$, ..., V$_{xn}$, V$_{yn}$, str$_n$) to draw the n semilog plots on the same set of axes.
Additional information: A base 10 logarithmic scale is used for the x-axis and a linear scale is used for the y-axis. ♣ To create a plot where both axes are scaled, use loglog. ♣ For more information on line styles and multiple plots, see the listing for plot.
See also: semilogy, loglog, plot

semilogy(V$_x$, V$_y$)
Displays the two-dimensional plot, with logarithmic scaling on the y-axis, of the values in vectors V$_x$ versus V$_y$.
Output: A two-dimensional plot is displayed.

Argument options: semilogy(V_x, V_y, str) to draw the semilog plot with line style specified by str. ✦ semilogy(V_{x1}, V_{y1}, str_1, ..., V_{xn}, V_{yn}, str_n) to draw the n semilog plots on the same set of axes.
Additional information: If there are any positive values in V_y, then all negative values in V_y are ignored. Otherwise, negative values are plotted. ✦ A base 10 logarithmic scale is used for the y-axis and a linear scale is used for the x-axis. ✦ To create a plot where both axes are scaled, use loglog. ✦ For more information on line styles and multiple plots, see the listing for plot.
See also: semilogx, loglog, plot

stairs(V)
Displays a stair graph where the height of the steps correspond to the value of the elements of vector V.
Output: A two-dimensional plot is displayed.
Argument options: stairs(V_x, V_y), where V_x and V_y are vectors of identical length, to plot the values of V_y at the corresponding horizontal locations in V_x. The values in V_x must be evenly spaced and ascending.
Additional information: Stair graphs are similar to bar graphs, except that the vertical lines dropping to the x-axis are excluded.
See also: plot, bar, stem, hist

Graphing Surfaces and Volumes

This chapter looks at plotting 3-D surfaces from data points or function values. It also examines plotting 3-D slices from volumetric data.

The MATLAB functions covered are:

- mesh, meshz, meshc, and waterfall plot 3-D *mesh* surfaces from z-matrix data.
- meshgrid computes z-matrix data from function values, $f(x,y)$.
- griddata computes z-matrix grid from data points, (x_i, y_i, z_i).
- surf, surfc, and surfl plot *shaded* surfaces from z-matrix data.
- contour, contourc, and contour3 plot contour lines from z-matrix data.
- pcolor and image plot shaded contours from z-matrix data.
- slice plots 3-D slices (as surfaces) from volumetric data (4-D).
- trimesh and trisurf plot surfaces from triangular grids.

See the *Graphing Properties* chapter for details on customizing graphs (e.g., logarithmic axes, legends, etc.).

`mesh, meshz, meshc, waterfall` - Plot 3-D *Mesh* Surface from a z-Matrix

Consider the following function which defines a surface:

$$z(x,y) = xe^{-x^2-y^2}.$$

To plot this function, first create an M-file, such as `nice.m`:

```
function z = nice(x, y)
    z = x .* exp(-x.^2 - y.^2);
```

Normally, you compute z-values from (x,y) points in a rectangular grid. Therefore, the next step is to define the series of x and y values (as vectors):

```
dx = 0.5;                   % x step size
dy = 1.0;                   % y step size

x = [-2 : dx : 2];          % series of x-values
y = [-2 : dy : 2];          % series of y-values
```

Then you invoke meshgrid to create a matrix, X, whose columns all contain the same x-values, and a matrix, Y, whose rows all contain the same y-values:

```
[X,Y] = meshgrid(x, y)
```

The result in this case is two 5 × 9 matrices, defining the 45 points, $(X_{i,j}, Y_{i,j})$, in the grid:

```
X =
    -2   -1.5   -1   -0.5    0   +0.5   +1   +1.5   +2
    -2   -1.5   -1   -0.5    0   +0.5   +1   +1.5   +2
    -2   -1.5   -1   -0.5    0   +0.5   +1   +1.5   +2
    -2   -1.5   -1   -0.5    0   +0.5   +1   +1.5   +2
    -2   -1.5   -1   -0.5    0   +0.5   +1   +1.5   +2
    -----------------------------------------****---->
                                              x=+1.5

Y =
    -2   -2   -2   -2   -2   -2   -2   -2   -2   |
    -1   -1   -1   -1   -1   -1   -1   -1   -1   |
     0    0    0    0    0    0    0    0    0   *   y=0
    +1   +1   +1   +1   +1   +1   +1   +1   +1   |
    +2   +2   +2   +2   +2   +2   +2   +2   +2   \|/
```

Now you can compute a matrix of z-values at these points:[46]

```
Z = nice(X, Y)
```

and the result (rounded to one decimal place) is as follows:

```
Z =
    0     0    0    0    0    0    0    0    0   |
    0   -.1  -.1  -.1    0  +.1  +.1  +.1    0   |
    0   -.2  -.4  -.4    0  +.4  +.4  +.2    0   *   y=0
    0   -.1  -.1  -.1    0  +.1  +.1  +.1    0   |
    0     0    0    0    0    0    0    0    0   \|/
    -----------------------------------------****---->
                                              x=+1.5
```

The dotted arrows and x and y equations in the above matrices are the authors' invention, not MATLAB output, and are meant to illustrate that z=+0.2 at the point (x=+1.5, y=0).

You can plot the $(X_{i,j}, Y_{i,j}, Z_{i,j})$ values as separate points using plot3; however, it is hard to visualize the surface, even when you add lines to connect the points to the x, y-plane at $z = -0.4$. See Figure 44 for the total result.

It is better to invoke mesh to connect the points to each other in a rectangular grid as shown in Figure 45.

The commands to produce this plot are:

---

[46] Alternately, you could just enter the matrix expression for z, Z = X .* exp(-X.^2 - Y.^2);, but in the long run it reduces effort (and errors) if you use M-files to define your functions.

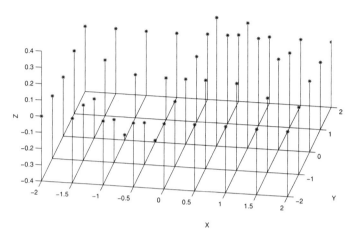

FIGURE 44. Plotting points on surface $z(x, y)$ using plot3

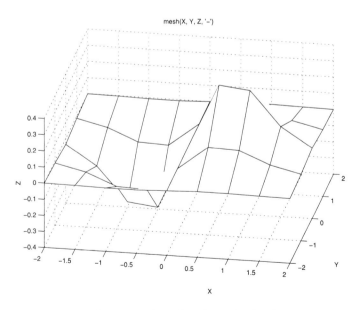

FIGURE 45. Plotting mesh surface through $z(x, y)$ using mesh

```
mesh(X, Y, Z, '-')

view([10,35])

set(gca, 'XTick', x)
set(gca, 'YTick', y)

xlabel('X')
ylabel('Y')
zlabel('Z')

title('mesh(X, Y, Z, ''-'')')

print -deps nice2.eps
```

Apart from the call to mesh, all the above functions are specialized commands that change the way the data is displayed. These commands are covered in detail in the *Graphics Properties* chapter.

meshc - Plot Contour Lines Below *Mesh* Surface

You can plot contour lines below the 3-D mesh surface using meshc, which takes the same arguments as mesh.

```
meshc(X, Y, Z, '-')
```

The result is shown in Figure 46 (left side).

meshz - Plot Vertical Lines around *Mesh* Surface

Using meshz, you can plot an "apron" of vertical lines around a mesh surface. meshz takes the same arguments as mesh except that you *cannot* specify a linetype for the mesh lines:

```
meshz(X, Y, Z)
```

The result is shown in Figure 46 (right side).

waterfall - Omit Mesh Lines in one Direction

You can omit the mesh lines in one direction, and so achieve a waterfall effect, by invoking waterfall instead of mesh:

```
waterfall(X, Y, Z)

view([10, 35])
```

The result is shown in Figure 47.

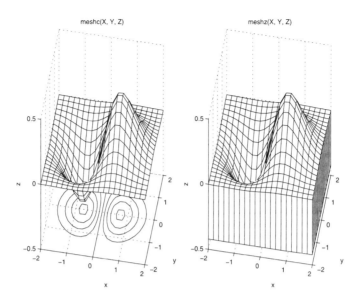

FIGURE 46. Plotting contours below mesh surfaces using meshc, and plotting vertical lines around surfaces using meshz.

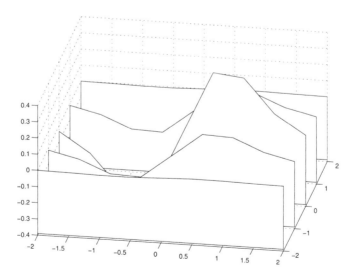

FIGURE 47. Omitting mesh lines in one direction using waterfall

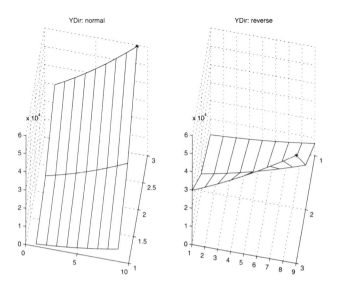

FIGURE 48. Plotting matrices using 'YDir', 'normal/reverse'

Tips for Visualizing Matrix Elements, $A_{i,j}$

If you omit X and Y in the invocation of mesh, it uses the row and column indices. For example, consider the matrix:

```
A = [ 11  12  13  14  15  16  17  18  19
      21  22  23  24  25  26  27  28  29
      31  32  33  34  35  36  37  38  39 ] .^ 3;
```

Now plot it using mesh:

```
mesh(A)
```

The result is shown in Figure 48 (left side).

The problem with this graph is that the labels on the y-axis (which correspond to row numbers 1 to 3), would look better plotted in *reverse* order, as shown on the right side of Figure 48.[47] To achieve this, you can issue the following *after* invoking mesh:

```
set(gca, 'YDir', 'reverse')    % Default: 'normal'
                               % ------------------
```

or the equivalent short form:

```
axis('ij')                     % Default: 'xy'
                               % -------------
```

Either of these commands will reverse the labeling on the y-axis, making it easier to visualize matrices.

---

[47]To help you compare the two graphs, we plotted a * at the location of the last element of A.

The following commands were issued to over-ride the default numbers that were chosen for the axes:

```
[m, n] = size(A);

set(gca, 'XLim', [1 n], 'YLim', [1 m])
set(gca, 'XTick', [1:n], 'YTick', [1:m])
```

where XLim specifies the data values at the start (1) and end (9) of the *x*-axis, and XTick specifies the location of the tick marks (1, 2, ..., 9) on the *x*-axis.[48]

In both graphs, we made the surface stand out more by adding grid lines to the three back-planes, using:

```
grid on          /or/          grid('on')
```

How mesh Treats Complex Numbers and NaN, with Applications

Consider the following function for a hemisphere:

$$z = \sqrt{1 - x^2 - y^2}$$

Create a matrix of *z* values corresponding to *x* and *y* varying from -1 to 1:

```
x = [-1 : 0.2 : 1];
y = [-1 : 0.2 : 1];

[X, Y] = meshgrid(x, y);

Z = sqrt(1 - X.^2 - Y.^2);
```

Notice that some of the elements of Z are complex; e.g., Z(1,1):

```
Z(1,1)
```

which is $\sqrt{-1}$:

```
ans =
       0 + 1.0000i
```

When you try to use mesh to plot the surface defined by Z:

```
mesh(X, Y, Z)
```

you get the error:

```
??? Error using ==> surface
Argument must be real.
```

To quickly visualize the pattern of complex numbers in Z, use the quiver function discussed in the previous chapter:

```
quiver(X, Y, real(Z), imag(Z), '-')

axis([-1.2 1.2   -1.2 1.2])
```

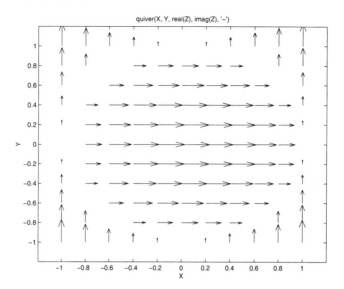

FIGURE 49. Visualizing the complex numbers in the matrix we wish to pass to mesh

The result is shown in Figure 49.

There are three ways to get around this "complex" problem:

1. Replace all complex elements of Z with the value 0.

    Because the complex numbers in this example are all purely imaginary (i.e., the real parts are 0), you can simply extract the real part of *all* elements of Z:

    ```
    Z = real(Z)
    ```

    However, a more general method (which would also zero out complex elements with *non-zero* real parts) is to first create a matrix which is all ones where Z is real, and all zeros where Z is complex:[49]

    ```
    keep = (imag(Z) == 0);
    ```

    and then multiply it (element-wise) by Z to zero out the complex elements:

    ```
    Z = Z .* keep;
    ```

    Now you can plot the resulting Z using mesh(X, Y, Z). You can actually do all of the above in one line:

    ```
    mesh( X,   Y,   Z.*(imag(Z) == 0) )
    ```

    The result is shown in Figure 50.

---

[48] Refer to the *Graphics Properties* chapter for more information on using set to modify axis attributes, etc.

[49] For a further explanation of this sort of technique, please refer to the *Programming in* MATLAB chapter.

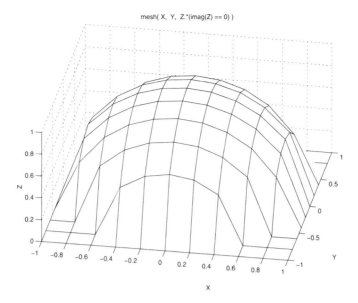

FIGURE 50. Mesh plot of matrix after "zeroing-out" complex elements

2. Replace all complex elements of Z with the value NaN, meaning "not a number". In this case, MATLAB does not draw mesh lines to those points at all.[50]

   This replacement can be accomplished by finding the indices of all elements of Z that have a non-zero imaginary part.

   ```
   I = find( imag(Z) ~= 0 );
   ```

   or simply:

   ```
   I = find( imag(Z) );
   ```

   Then store NaN in those elements

   ```
   Z(I) = NaN;
   ```

   and plot a mesh surface using:

   ```
   mesh(X, Y, Z)
   ```

   The result is shown in Figure 51 (left side).

   Notice that the surface is jagged along the lower edge, where $x^2 + y^2 = 1$.

   While we're at it, let's change the middle element of Z to NaN, and observe the effect on the mesh plot:

   ```
   Z(6,6) = NaN;
   mesh(X, Y, Z)
   ```

---

[50] NaN normally represents the result of an illegal operation on floating-point numbers such as $\frac{0}{0}$, $\frac{\inf}{\inf}$, or $0 \times \inf$. MATLAB also lets you use NaN to indicate *undefined* or *missing* matrix elements in calls to *some* of its functions, such as mesh.

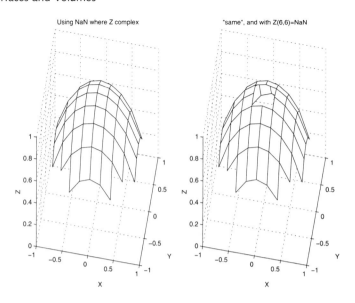

FIGURE 51. Mesh plot of matrix after replacing complex elements with NaN

The result is shown in Figure 51 (right side). Notice that the four panels adjacent to the point (x=0, y=0, z=1) are no longer visible. Also, you can see through to the underside of the surface.

Hence, setting elements of Z to NaN provides a means of making the surface transparent in places where that would be beneficial.

3. Plot $z(x, y)$ over a polar domain instead of a rectangular domain. This is covered in detail in the next section and eliminates the jaggedness problem *and* the need to eliminate complex elements of Z—because there won't be any.

Mesh Surfaces from Non-rectangular (e.g., Polar) Grids

Consider the function from the last section:

$$z = \sqrt{1 - x^2 - y^2}.$$

By converting (x, y) to polar coordinates (a, r), you can re-express z as:

$$z = \sqrt{1 - r^2} \quad \text{for all angles } a.$$

What you want to do is plot this function over a polar grid of (a, r) points. Because $r = 1$ will be one of the grid lines, this should eliminate the jaggedness encountered earlier.

First, define a *polar* grid of angle and radius values, and compute z at each (a, r) point:

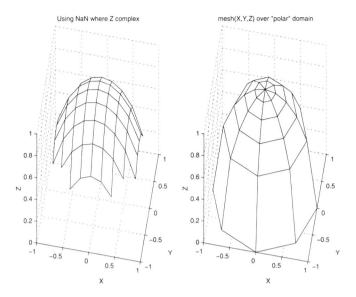

FIGURE 52. Mesh plot of matrix defined over *polar* grid

```
a = [0 : pi/4 : 2*pi];
r = [0 :  0.2 :  1  ];

[A, R] = meshgrid(a, r);

Z = sqrt(1 - R.^2);
```

Now convert the polar (*a*, *r*) coordinates to Cartesian (*x*, *y*) coordinates for mesh, which requires Cartesian coordinates:

```
[X, Y] = pol2cart(A, R);
```

Then plot the surface using mesh:

```
mesh(X, Y, Z)
```

The result is shown in Figure 52.

Notes

- In the above example, you plotted a matrix of *z*-values defined over a *polar* grid. The same techniques can be used to plot *z*-values over other such *curvi-linear* domains.

griddata - Compute *z*-Matrix from Data Points (x_i, y_i, z_i)

Often data is not available on a grid (recti-linear or otherwise), but rather is *irregularly* distributed in the *xy*-plane.

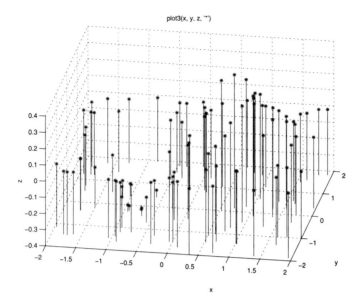

FIGURE 53. Plot of (x, y, z) points, where (x, y) points are irregularly distributed

```
x = rand(100,1);        % x between 0 and 1
x = 4*x - 2;            % x now between -2 and 2

y = 4*rand(100,1) - 2;  % y between -2 and 2
```

For example, create 100 randomly distributed (x, y) points:

and calculate z values from these points, using the nice function defined earlier:

```
z = nice(x, y);
```

Now plot these (x, y, z) points, and draw lines to the bottom plane of the graph using the following commands:

```
b = -0.4 * ones(100, 1);    % Bottom plane (z=-0.4)

plot3(x, y, z, '*')

hold on
plot3([x';
       x'],    [y';              ...
               y'],    [z';      ...
                        b'],    '-')
hold off

view([10, 35])
grid on
set(gca, 'ZLim', [-0.4  0.4])

xlabel('x');   ylabel('y');    zlabel('z')
title('plot3(x, y, z, ''*'')')

print -deps grid1.eps
```

The result is shown in Figure 53.

Notice that the (x, y) points do not lie on a regular grid. Therefore, you *cannot* use them to define matrices X, Y, and Z and plot a surface. What you want to do is interpolate, using this irregularly distributed data, values of z on a regular (x, y) grid. The MATLAB command for this is griddata:

```
dxi = 0.5;
dyi = 1.0;

xi = [-2 : dxi : 2];            % series of x-values
yi = [-2 : dyi : 2];            % series of y-values

[XI,YI] = meshgrid(xi, yi);

ZI = griddata(x, y, z, XI, YI);
```

This computes values of ZI at the points given in the matrices XI and YI. The result is:

```
ZI =
   +.1    0     0     0     0     0     0     0    0
    0   -.1   -.1   -.1    0   +.1   +.1   +.1    0
    0   -.2   -.4   -.4    0   +.4   +.4   +.2    0
    0   -.1   -.1   -.1    0   +.1   +.1   +.1    0
    0    0     0     0     0     0     0     0    0
```

This is close to the result of evaluating the function itself at the regular (x, y) points, to the one-digit accuracy shown.[51]

Now plot the result:

```
mesh(XI, YI, ZI, '-')
```

The result is shown in Figure 54, with the original (x, y) data points plotted on the bottom plane.

Notes

- See the section *Triangular Grid Plotting* for details on functions that allow you to plot directly from irregular data.

`surf`, `surfc`, `surfl` - Plot *Shaded* Surface from z-Matrix

surf is similar to mesh except that it plots a *shaded* surface from a z-matrix. By default, the shading is proportional to the z-values. Continue with the Z matrix from first section:

```
dx = 0.5;
dy = 1.0;

x = [-2 : dx : 2];              % series of x-values
```

---

[51] The obvious exception is ZI(1,1), which would be 0.0 if calculated directly from the nice function, instead of being interpolated from nearby (x, y, z) data points.

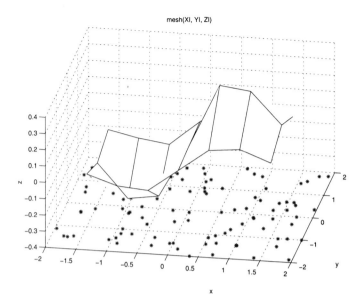

FIGURE 54. Mesh plot of interpolated (XI, YI, ZI) points on grid, above original irregularly distributed (x, y) points.

```
y = [-2 : dy : 2];          % series of y-values
[X,Y] = meshgrid(x, y);
Z = nice(X, Y);
```

You can plot it as a shaded surface using surf. First, set the colormap to gray:[52]

```
colormap(gray)
surf(X, Y, Z)
```

This produces the shaded surface shown in Figure 55 (left side).

The surface would look much better with more surface patches, so increase the Z-matrix from 5 × 9 to 21 × 21. The result is shown on the right side of Figure 55.

You can change the shading to represent some other quantity than Z (such as temperature). For example, make the shading be proportional to the sum of x and y at a point:

```
C = X + Y;
surf(X, Y, Z, C)
```

The result is shown in Figure 56 (on the left).

For comparison, make the shading proportional to the *absolute* value of z at a point:

---

[52]Refer to the *Graphics Properties* chapter for information on the colormap function.

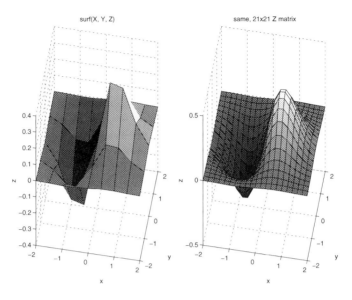

FIGURE 55. Surface plots of (x, y, z) points defined on: (1) a coarse grid, and (2) a finer grid.

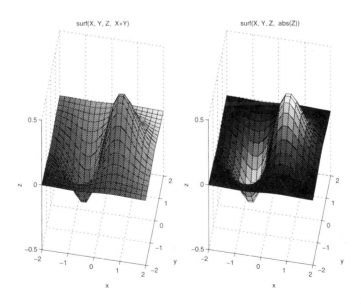

FIGURE 56. Surface plots of (x, y, z) points, varying the shading to be proportional to: (1) X + Y, and (2) abs(Z).

212 Surfaces and Volumes

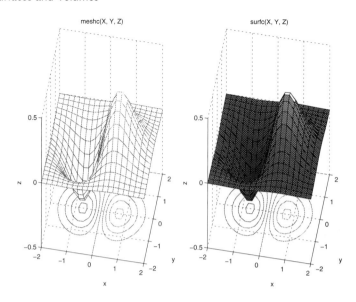

FIGURE 57. Plotting contours below shaded surface using surfc

```
C = abs(Z);
surf(X, Y, Z, C)
```

The result is shown in Figure 56 (on the right).

surfc - Plot Contour Lines below *Shaded* Surface

You can plot contour lines below a 3-D shaded surface using surfc, which takes the same arguments as surf:

```
surfc(X, Y, Z)
```

The result is shown in Figure 57.

surfl - Plot *Shaded* Surface with Lighting

We can add lighting to our 3-D shaded surface using surfl. It takes the same arguments as surf, *plus* an argument s to specify the direction of the light source for highlights on the surface:

```
surfl(X, Y, Z, s)
```

s can be specified in [x, y, z] coordinates or in spherical coordinates [azimuth, elevation]. The default value of s is $45°$ counter-clockwise from the current view — more or less over your right shoulder. In other words, if [AZ, EL] = view, then the value of s defaults to [AZ, EL] + [45, 0].

Let's plot our surface with two different light sources. In the lefthand plot, the light source (whose azimuth is indicated by the added arrow) is 45° *clockwise* from the viewpoint, and in the righthand plot, the light source is 45° counterclockwise (the default):

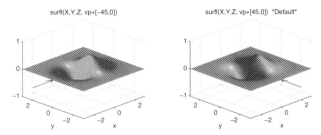

The program which produced these plots is as follows:

```
colormap(gray)

x = [-3 : 0.2 : 3];
y = [-3 : 0.2 : 3];

[X, Y] = meshgrid(x, y);
Z      = nice(X, Y);

for i=1:2
      subplot(2,2, i)

      vp = [-45, 30];

      if i==1,       surfl(X,Y,Z, vp+[-45,0]),
            title('surfl(X,Y,Z, vp+[-45,0])'),
            arrow([-3 -5], [0 0], .5),                 end

      if i==2,       surfl(X,Y,Z, vp+[45,0]),
            title('surfl(X,Y,Z, vp+[45,0])  *Default*'),
            arrow([0 0], [-3 -5], .5),                 end

      axis([-3 3  -3 3   -1 1])
      view(vp);

      xlabel x
      ylabel y

      shading interp
end

print -deps surfl1.eps
```

It makes use of the function in `arrow.m`:

```
function h = arrow(x, y, w)

%-----------------------------------------------

angle = atan2( y(2)-y(1),  x(2)-x(1) );

h(1) = line(x, y);

%-----------------------------------------------
```

```
        dx = w*cos(angle + pi/6);
        dy = w*sin(angle + pi/6);

        h(2) = line([x(1) x(1)+dx], ...
                    [y(1) y(1)+dy]);

    %-------------------------------------------------

        dx = w*cos(angle - pi/6);
        dy = w*sin(angle - pi/6);

        h(3) = line([x(1) x(1)+dx], ...
                    [y(1) y(1)+dy]);

        set(h, 'color', [0 0 0])
```

Notes

- Notice the use of the gray color map. It and bone, copper, and pink are good choices as surfl uses their uppermost colors (white, etc.) for highlights.

- Notice the use of shading interp, which causes the color to **vary** throughout each surface patch, based on linear interpolation from the colors in the corners.

  The default is shading faceted which has black mesh lines super-imposed on shading flat. shading flat, in turn, causes the color in each surface patch to be a **constant** color based on the colors in the corners.

- An optional fifth argument to surfl allows you to control the type of light being shone on the surface. Refer to the *Command Listing* section for further information.

- Refer to the *Animation* chapter for other methods of adding light sources to a plot.

contour, contourc, contour3 - Plot Contour Lines from z-Matrix

Consider the nice function used earlier:

$$z(x,y) = xe^{-x^2-y^2}$$

and the matrix of Z values computed from it:

```
Z =
      0    0    0    0    0    0    0    0    0
      0  -.1  -.1  -.1    0  +.1  +.1  +.1    0
      0  -.2  -.4  -.4    0  +.4  +.4  +.2    0
      0  -.1  -.1  -.1    0  +.1  +.1  +.1    0
      0    0    0    0    0    0    0    0    0
```

You can plot contour lines through this matrix using contour. For example, plot seven contour lines at evenly-spaced z-values:

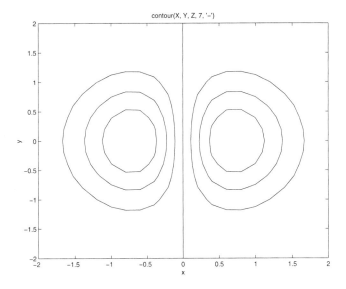

FIGURE 58. Plotting contours through surface defined by X, Y, and Z

```
contour(X, Y, Z, 7, '-')
```

The result is shown in Figure 58.

You can label the contour lines labeled with their z-values using clabel. For example:

```
C = contour(X, Y, Z, 7, '-');
clabel(C)
```

C is a matrix containing the coordinates of the contour lines for use by clabel. The result is shown in Figure 59 (left side). The labels are plotted at random locations along their respective contours.

Also, clabel will put the labels wherever *you* choose with your mouse pointer when you call:

```
clabel(C, 'manual');
```

A sample result is shown in Figure 59 (right side).

Finally, you can select the *specific z*-values for the contours. To do this with the above Z matrix, first use min and max to compute the range of z-values.

```
min(min(Z))
max(max(Z))
```

Which result in:[53]

```
ans =
     -0.3894
```

---

[53] Remember that the displayed copy of Z that we have has been rounded to one-digit accuracy. The actual Z has more digits.

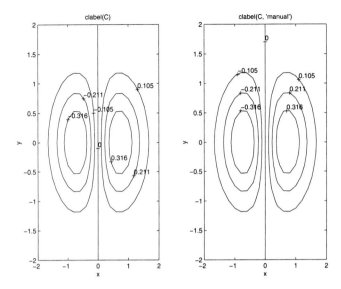

FIGURE 59. Contours labels

```
ans =
        0.3894
```

So, now let's plot contour lines for the nine integer values between −0.4 and +0.4:

```
zvalues = [-0.4 : 0.1 : 0.4];

C = contour(X, Y, Z, zvalues, '-');
clabel(C, 'manual');
```

The result is shown in Figure 60.

As you can see, there are no contour lines for the values of −0.4 and +0.4; the z-values never quite reach those levels.

The *Graphing Properties* chapter covers ways of changing the line styles of individual contour lines.

contour3 - Plot Contour Lines in 3-D

You can plot the contour lines in 3-D using contour3, which takes the same arguments as contour.

```
zvalues = [-0.4 : 0.1 : 0.4];

contour3(X, Y, Z, zvalues, '-')
```

The result is shown in Figure 61. clabel does not work in conjunction with contour3.

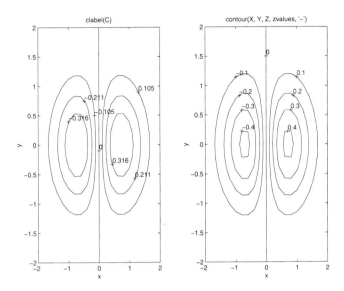

FIGURE 60. Contours lines at specific z-values (labelled)

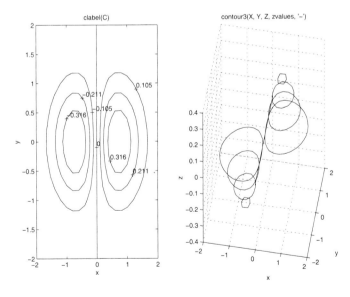

FIGURE 61. Contours lines at specific z-values, plotted in 3-D (unlabelled)

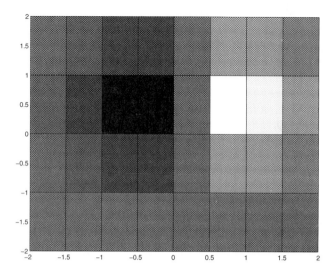

FIGURE 62. Plotting surfaces using pcolor

pcolor — Pseudocolor (Checkerboard) Plots

pcolor produces a surface plot viewed from above so that it resembles a checkerboard. For example plot the nice function using pcolor:

```
colormap(gray)

pcolor(X,Y,Z)

print -depsc pcolor1.eps
```

The result is shown in Figure 62.

image - Raster Image Plotting

pcolor is more suited to plotting data, whereas image is more suited to plotting rasterized images such as photographs.

image produces a plot that looks similar to that produced by pcolor, but differs in the following important ways:

1. With pcolor, the rectangular patches are plotted with their *vertices* at the z-values from the matrix. However, with image, the rectangular patches are plotted with their *centers* over the z-values from the matrix.

2. With pcolor, the z-values are scaled to span the range of the color map. With image, the z-values are not scaled - they are direct indices into the color map.

For example, plot the nice function using image. Notice that we had to scale up our Z values to map onto a larger range of the color map.

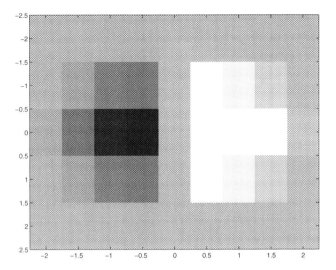

FIGURE 63. Plotting surfaces using image

```
colormap(gray)

image(x,y,Z*100+50)

print -depsc image1.eps
```

The result is shown in Figure 63.

Rather than using a colormap, you can specify the colors in the image **directly** by passing a $m \times n \times 3$ matrix of color values to image.

```
Zmin = min(min(Z));
Zmax = max(max(Z));
Z = (Z-Zmin)/(Zmax-Zmin);

[m,n] = size(Z);
C = zeros(m,n,3);

for i=1:m
    for j=1:n
        rgb = [1 1 1]*Z(i, j);
        C(i, j, 1) = rgb(1);
        C(i, j, 2) = rgb(2);
        C(i, j, 3) = rgb(3);
    end
end

image(C)
```

Here we have scaled the elements of Z to make them values between zero and one, and then stored them as the red, green, and blue color values in a color matrix, C, for plotting with image.

This produces a similar plot to that shown in Figure 63.

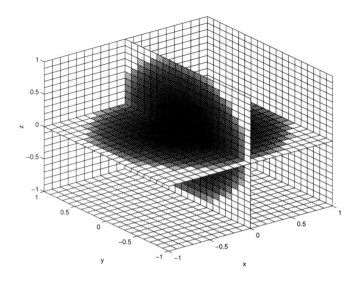

FIGURE 64. Plotting volumetric slices using slice

slice - Volumetric Plots

The slice function allows you to plot functions of three variables by plotting parallel shaded surfaces for you.

For example, consider the following function:

$$v = f(x, y, z) = \sqrt{1 - x^2 - y^2 - z^2}$$

which could represent temperature at points inside a hemisphere.

To visualize it, see Figure 64.

The commands to produce this plot are as follows:

```
x = [-1 : .1 : 1];
y = [-1 : .1 : 1];
z = [-1 : .1 : 1];

[X,Y,Z] = meshgrid(x,y,z);

V = sqrt(1. - X.^2 - Y.^2 - Z.^2);
V = real(V);

Sx = [0 1];
Sy = [1];
Sz = [-1 0];

slice(X, Y, Z, V, Sx, Sy, Sz)

xlabel x
ylabel y
zlabel z

colormap(1-gray)
print -depsc slice1.eps
```

Notes

- meshgrid is called to produce X, Y and Z matrices from the vectors of values in x, y, and z.
- Then values of V are computed over the grid of data points. (The imaginary part of V was removed as slice will not plot complex numbers.)
- slice is then called with:

    X, Y, Z the domain of the data.

    V the volumetric data to be plotted.

    Sx, Sy, Sz vectors containing the x, y, and z values at which we want planes to be plotted.

- Refer to the *Command Listing* for details on how to specify arbitrary cutting planes (i.e., not parallel to the planes of the axes).

Triangular Grid Plotting

MATLAB provides a means of plotting surfaces directly from *irregularly distributed* data points *without having to first interpolate to a regular grid*.

Let's return to our irregular data from the earlier section on griddata. As before, let's create 100 randomly distributed (x, y) points.

```
x = rand(100,1);        % x between 0 and 1
x = 4*x - 2;            % x now between -2 and 2

y = 4*rand(100,1) - 2;  % y between -2 and 2
```

and calculate z values from these points, using the nice function defined earlier:

```
z = nice(x, y);
```

These points are plotted in Figure 53.

Now let's create mesh and surface plots directly from our irregularly distributed data.

```
colormap(gray)

%-----------------------------------------------------------

tri         = delaunay(x, y);

subplot(2,2,1);         trimesh(tri, x, y, z)
                        title('trimesh(tri, x, y, z)')

subplot(2,2,2);         trisurf(tri, x, y, z)
                        title('trisurf(tri, x, y, z)')

%-----------------------------------------------------------

[XI, YI]    = meshgrid(-2:0.5:2, -2:0.5:2);
        ZI = griddata(x, y, z, XI, YI);
```

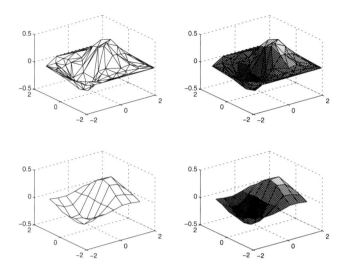

FIGURE 65. Mesh and surface plots from triangularized data (top) and interpolated data (bottom).

```
    subplot(2,2,3);               mesh(    XI, YI, ZI)
                            title('mesh(   XI, YI, ZI)')

    subplot(2,2,4);               surf(    XI, YI, ZI)
                            title('surf(   XI, YI, ZI)')

%-----------------------------------------------------------

print -deps trinice1.eps
```

We first invoked the delaunay function to compute triangles through the data. Then we used trimesh and trisurf to create the plots. For comparison, we used meshgrid, griddata, mesh, and surf to plot the same data, but interpolated to a rectangular grid.

The results are shown in Figure 65.

Figure 66 shows the Voronoi diagram of the triangularization of this data.

This was produced by the following commands:

```
    [vx, vy] =  voronoi(x, y, tri);
  plot(vx, vy, '-',      x, y, '.');

  axis([-2, 2, -2, 2]);

  print -deps trinice2.eps
```

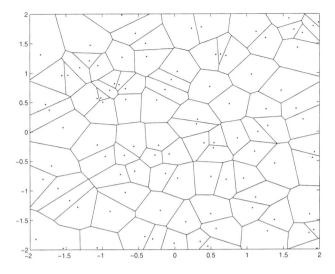

FIGURE 66. Voronoi diagram

Command Listing

bar3(M)
Displays a three-dimensional bar graph where the height of the "towers" corresponds to the value of the elements of matrix M.
Output: A three-dimensional plot is displayed.
Argument options: bar3(M, num) to specify a tower width of num. Values greater than 1 cause overlapping bars. ✤ bar3(V_y, V_z), where V_y and V_z are vectors of identical length, to plot the values of V_z at the corresponding horizontal locations in V_y.
✤ bar3(M, str) to supply an optional value to control the plot. Available values are 'detached', 'grouped', and 'stacked'. ✤ V_{hndl} = bar3(M) to return a vector of surface object handles.
Additional information: bar3h, bar, barh

bar3h(M)
Displays a horizontal three-dimensional bar graph where the length of the "towers" corresponds to the value of the elements of matrix M.
Output: A three-dimensional plot is displayed.
Argument options: bar3h(M, num) to specify a tower width of num. Values greater than 1 cause overlapping bars. ✤ bar3h(V_y, V_z), where V_y and V_z are vectors of identical length, to plot the values of V_z at the corresponding horizontal locations in V_y.
✤ bar3h(M, str) to supply an optional value to control the plot. Available values are 'detached', 'grouped', and 'stacked'. ✤ V_{hndl} = bar3h(M) to return a vector of surface object handles.
Additional information: bar3, bar, barh

[num$_{az}$, num$_{el}$, num$_r$] = cart2sph(num$_x$, num$_y$, num$_z$)

Converts the cartesian coordinate specified by x, y, and z values num$_x$, num$_y$, and num$_z$ into spherical coordinates.

Output: A numeric value representing azimuth (in radians) is assigned to num$_{az}$, a numeric value representing elevation (in radians) is assigned to num$_{el}$, and a numeric value representing radius is assigned to num$_r$.

Argument options: [M$_{az}$, M$_{el}$, M$_r$] = cart2sph(M$_x$, M$_y$, M$_z$), where M$_x$, M$_y$, and M$_z$ are matrices of identical dimensions, to convert each corresponding set of x, y, and z values.

See also: sphere, mesh, sph2cart, pol2cart, cart2pol

clabel(cplot)

Displays the contour plot cplot with each contour labeled with its value.

Output: A two-dimensional contour plot is displayed.

Argument options: clabel(cplot, V) to label only those contour levels present in the vector V. If any elements of V do not correspond to contours, they are ignored.
* clabel(cplot, 'manual') to allow manual contour labeling control with the mouse.

Additional information: Each individual contour line is label once for each invocation of clabel. Subsequent invocations without clearing the plot, cause extra labels to appear. * When using the 'manual' option, the closest contour to where the mouse was clicked is labelled. Individual contours can receive multiple labels. To end labelling, hit **Return**. * cplot has no option for returning vectors to plot with plot or any other such command.

See also: contour, contourc

contour(M$_z$)

Plots a contour plot for the surface defined by z-matrix M$_z$.

Output: A two-dimensional line plot is produced.

Argument options: contour(M$_z$, n) to specify that n contour *levels* be used.
* contour(M$_z$, V) to specify that contour *levels* corresponding to the values in vector V be used. * contour(V$_x$, V$_y$, M$_z$) to define the spacing on the x and y axes. Each x and y value specified in vectors V$_x$ and V$_y$ is equally spaced from its neighbors.
* M$_c$ = contour(M$_z$) to store the contour matrix in M$_c$. The command contourc also provides the same service.

Additional information: If the values defining the x and y axes are not monotonically increasing, but are irregularly spaced, then the contours are not drawn properly.
* If you do not specify the number or location of contour levels, contour chooses for itself. * To label the contour lines, use the clabel command. * To create a contour plot in three-dimensions, use contour3.

See also: clabel, contour3, contourc, contourf, surfc, meshc

contour3(M$_z$)

Plots a three-dimensional contour plot for the surface defined by z-matrix M$_z$.

Output: A three-dimensional line plot is produced.

Argument options: contour3(M$_z$, n) to specify that n contour *levels* be used.
* contour3(M$_z$, V) to specify that contour *levels* corresponding to the values in vector

V be used. ✦ contour3(V_x, V_y, M_z) to define, with the values in vectors V_x and V_y, the ranges of the x and y axes. ✦ M_c = contour3(M_z) to store the contour matrix in M_c.
Additional information: If the values defining the x and y axes are not monotonically increasing, but are irregularly spaced, then the contours are not drawn properly.
✦ If you do not specify the number or location of contour levels, contour3 chooses for itself. ✦ To create a contour plot in two-dimensions, use contour.
See also: contour, contourc, contourf, surfc, meshc

M_c = **contourc(M_z)**

Creates the plotting matrix necessary for a two-dimensional contour plot of the surface defined by z-matrix M_z.
Output: A plotting matrix is produced.
Argument options: M_c = contourc(M_z, n) to specify that n contour *levels* be used.
✦ M_c = contourc(M_z, V) to specify that contour *levels* corresponding to the values in vector V be used. ✦ M_c = contourc(V_x, V_y, M_z) to define the spacing on the x and y axes. Each x and y value specified in vectors V_x and V_y is equally spaced from its neighbors.
Additional information: If the values defining the x and y axes are not monotonically increasing, but are irregularly spaced, then the contours are not drawn properly.
✦ If you do not specify the number or location of contour levels, contourc chooses for itself. ✦ The command contour calls contourc. ✦ For more information on the structure of M_c, see the on-line help file for contourc.
See also: contour, contour3, contourf, meshc

contourf(M_z)

Plots a filled contour plot for the surface defined by z-matrix M_z.
Output: A two-dimensional patch plot is produced.
Argument options: contourf(M_z, n) to specify that n contour *levels* be used.
✦ contourf(M_z, V) to specify that contour *levels* corresponding to the values in vector V be used. ✦ contourf(V_x, V_y, M_z) to define the spacing on the x and y axes. Each x and y value specified in vectors V_x and V_y is equally spaced from its neighbors.
✦ M_c = contourf(M_z) to store the contour matrix in M_c. The command contourc also provides the same service.
Additional information: For more details, see the entry for contour.
See also: contour, contour3, contourc

V_i = **convhull(V_x, V_y)**

Computes indices to points on the convex hull of x- and y-vectors V_x and V_y.
Output: A vector of indices is assigned to V_i.
Argument options: V_i = convhull(V_x, V_y, M_{tri}), where M_{tri} is the results of a call to delaunay, to use that triangulation in the computation.
See also: delaunay, voronoi

[M_x, M_y, M_z] = **cylinder(V)**

Computes the coordinate matrices of a unit cylinder aligned with the z-axis and with radius defined by the equally-spaced elements of vector V.
Output: Three equally sized matrices are assigned to M_x, M_y, and M_z.

Argument options: [M_x, M_y, M_z] = cylinder to use the default radius vector of [1 1].
♦ [M_x, M_y, M_z] = cylinder(V, posint) to specify that n points be used around the circumference of the cylinder. Default is 20.
Additional information: Use either mesh, grid, or surf to display the cylinder.
See also: sphere, mesh, surf, grid

M_{tri} = delaunay(V_x, V_y)

Computes a set of triangles from x- and y-vectors V_x and V_y, such that no data points are contained in any triangle's circumcircle.
Output: An $n \times 3$ matrix of triangle indices is assigned to M_{tri}.
Argument options: M_{tri} = delaunay(V_x, V_y, *sorted*) to assume that the x- and y vectors have already been sorted and duplicate elements removed.
See also: dsearch, tsearch, convhull, voronoi

int = dsearch(V_x, V_y, M_{tri}, num$_x$, num$_y$)

Determines the index of the nearest point to (num$_x$, num$_y$ in the triangularization M_{tri} of V_x and V_y.
Output: An integer index is assigned to int.
See also: tsearch, voronoi, convhull, delaunay

num = diffuse(num$_x$, num$_y$, num$_z$, V_{ls})

Computes the reflectance of a diffuse surface with the normal vector with components num$_x$, num$_y$, and num$_z$, using a light source from the direction defined by vector V_{ls}.
Output: A numerical value between 0.0 and 1.0 is assigned to num.
Additional information: The reflection is the amount of light reflected from a surface, where 0.0 represents no reflection and 1.0 represents complete reflection.
♦ The lighting direction V_{ls} can be a three-element vector representing the x, y, and z coordinates of the light source, or two-element vectors specifying the azimuth and the elevation. ♦ This command is used repeatedly by the surfl command to create a three-dimensional shaded surface plot. ♦ For information on the algorithm used, see the on-line help file.
See also: surfl, specular, surfnorm

M_{zu} = griddata(V_x, V_y, V_z, M_{xu}, M_{yu})

Interpolates the z-values corresponding to uniform mesh grid specifiers M_{xu} and M_{yu}, from the non-uniform data vectors V_x, V_y, and V_z.
Output: A matrix of z-values is returned.
Argument options: [M_{xu}, M_{yu}, M_{zu}] = griddata(V_x, V_y, V_z, V_{xu}, V_{yu}) to automatically use meshgrid on V_{xu} and V_{yu}. The resulting mesh grid matrices are returned in M_{xu} and M_{yu}.
Additional information: Typically, M_{zu}, M_{xu}, and M_{yu} are used in a call to mesh or surf.
♦ An inverse distance method is used in the interpolation.
See also: meshgrid, mesh, surf, interp1, interp2

mesh(M_x, M_y, M_z)

Plots the hidden-line surface defined by M_z, a matrix of function values evaluated at mesh grid points defined by M_x and M_y.

Output: A three-dimensional plot is produced.
Argument options: mesh(V_x, V_y, M_z) to use vectors V_x and V_y to define the mesh grid. See the command listing for meshgrid for more information. * mesh(M_z) to use a mesh grid linked to the matrix positions of the elements in M_z. * mesh(M_x, M_y, M_z, M_c) to specify that the values in matrix M_c be scaled and used as pointers into the current color map for the figure. By default, the values in M_z are used.
Additional information: All of the matrices passed to mesh must be of equal dimensions. * Many attributes of the plot can be manipulated. For more information, see the *Graphics Controls* chapter.
See also: meshgrid, meshc, meshz, trimesh, waterfall, *colormap*, *axis*, *view*, surf, contour, image

meshc(M_x, M_y, M_z)

Plots the hidden-line surface together with the contour lines defined by M_z, a matrix of function values evaluated at mesh grid points defined by M_x and M_y.
Output: A three-dimensional plot is produced.
Argument options: For more information on alternate parameter sequences, see the entry for mesh.
Additional information: All of the matrices passed to meshc must be of equal dimensions. * The contour lines are projected onto the axes-plane beneath the surface plot. * If the values defining the mesh grid are not monotonically increasing, but are irregularly spaced, then the contours are not drawn properly. * Many attributes of the plot can be manipulated. For more information, see the *Graphics Controls* chapter.
See also: meshgrid, mesh, meshz, *colormap*, *axis*, *view*, surfc, contour, image

meshz(M_x, M_y, M_z)

Plots the hidden-line *curtain* surface defined by M_z, a matrix of function values evaluated at mesh grid points defined by M_x and M_y.
Output: A three-dimensional plot is produced. For more information on alternate parameter sequences, see the entry for mesh.
Additional information: All of the matrices passed to meshz must be of equal dimensions. * The *curtain* is a set of planes dropped from the edges of the surface to the axes-plane beneath the surface. * Many attributes of the plot can be manipulated. For more information, see the *Graphics Controls* chapter.
See also: meshgrid, mesh, meshc, *colormap*, *axis*, *view*, surf, contour, image

[M_x, M_y] = meshgrid(V_x, V_y)

Generates two matrices to be used in specification of three-dimensional plots from x and y vectors V_x and V_y.
Output: A matrix with every row equalling V_x is assigned to M_x and a matrix with every column equalling V_y is assigned to M_y.
Argument options: [M_x, M_y] = meshgrid(V) to use V as both the row and column vector.
Additional information: Once M_x and M_y are created, they are typically used in a matrix formula to create M_z. M_z is then used in the command mesh or surf.
See also: griddata, mesh, surf

pcolor(M_x, M_y, M_c)

Plots the pseudocolor (checkerboard) plot defined by M_c, a matrix of color values evaluated at mesh grid points defined by M_x and M_y.
Output: A two-dimensional plot is produced.
Argument options: surf(V_x, V_y, M_c) to use vectors V_x and V_y to define the mesh grid. See the command listing for meshgrid for more information. ✦ surf(M_c) to use a mesh grid linked to the matrix indices of the elements in M_c. ✦
Additional information: All of the matrices passed to pcolor must be of equal dimensions. ✦ The figure created here is a surface viewed from directly above (see surf).
See also: meshgrid, surf, surfc, *colormap*, mesh, image

quiver(M_x, M_y, M_{dx}, M_{dy})

Produces a two-dimensional quiver plot with arrows starting at positions in M_x and M_y, with direction and relative magnitude determined by M_{dx} and M_{dy}.
Output: A two-dimensional plot is produced.
Argument options: (V_x, V_y, M_{dx}, M_{dy}) to define the arrow points with vectors V_x and V_y. See meshgrid for more details. ✦ (M_{dx}, M_{dy}) to define the arrow positions according to the indices of the given matrices. ✦ (M_x, M_y, M_{dx}, M_{dy}, num_s) to scale the length of each individual arrow by a factor of num_s. ✦ (M_x, M_y, M_{dx}, M_{dy}, str_{ls}) to provide a line specification to be used for the arrows.
See also: quiver3, meshgrid, contour, plot

quiver3(M_x, M_y, M_z, M_{dx}, M_{dy}, M_{dz})

Produces a three-dimensional quiver plot with arrows starting at positions in M_x, M_y, and M_z, with direction and relative magnitude determined by M_{dx}, M_{dy}, and M_{dz}.
Output: A three-dimensional plot is produced.
Argument options: (M_z, M_{dx}, M_{dy}, M_{dz}) to plot arrows at equally-spaced positions defined by M_z. ✦ (M_x, M_y, M_z, M_{dx}, M_{dy}, M_{dz}, num_s) to scale the length of each individual arrow by a factor of num_s. ✦ (M_x, M_y, M_z, M_{dx}, M_{dy}, M_{dz}, str_{ls}) to provide a line specification to be used for the arrows.
See also: quiver, meshgrid, contour, plot3, scatter

ribbon(V_x, V_y)

Displays a three-dimensional ribbon plot where the two-dimensional line defined by V_x and V_y is "stretched" in the z-direction.
Output: A three-dimensional plot is displayed.
Argument options: ribbon(M_x, M_y) to plot one ribbon for each column of M_x and M_y. ✦ ribbon(V_x, V_y, num) to specify a width of num for each ribbon.
See also: plot

hndl = slice(M_x, M_y, M_z, M_v, V_x, V_y, V_z, num_x)

Creates the volumetric slice plot of the volume contained in num_x-row matrix M_v, at locations specified by matrices M_x, M_y, and M_z and with x-, y-, and z-planes (slices) at values specified in vectors V_x, V_y, and V_z.
Output: A handle to a graphic is assigned to hndl.

Argument options: hndl = slice(M_v, V_x, V_y, V_z, num_x) to use the indices of the volume data to specify the locations.
See also: surf, meshgrid

num = **specular**(num_x, num_y, num_z, V_{ls}, V_{vp})

Computes the reflectance of a specular (metallic) surface with the normal vector with components num_x, num_y, and num_z, using a light source from the direction defined by vector V_{ls} and a viewpoint defined by V_{vp}.
Output: A numerical value between 0.0 and 1.0 is assigned to num.
Additional information: The reflection is the amount of light reflected from a surface, where 0.0 represents no reflection and 1.0 represents complete reflection.
✦ Both lighting direction V_{ls} and viewpoint V_{vp} can be three-element vectors representing x, y, and z coordinates, or a two-element vector specifying azimuth and the elevation. ✦ This command is used repeatedly by the surfl command to create a three-dimensional shaded surface plot. ✦ For information on the algorithm used, see the on-line help file.
See also: surfl, diffuse, surfnorm

[num_x, num_y, num_z] = **sph2cart**(num_{az}, num_{el}, num_r)

Converts the spherical coordinate specified by azimuth, elevation, and radius values num_x, num_y, and num_z into cartesian coordinates.
Output: Three numeric values representing x, y, and z values are assigned to num_x, num_y, and num_z, respectively.
Argument options: [M_x, M_y, M_z] = sph2cart(M_{az}, M_{el}, M_r), where M_{az}, M_{el}, and M_z are matrices of identical dimensions, to convert each corresponding set of values.
See also: sphere, mesh, cart2sph, pol2cart, cart2pol

[M_x, M_y, M_z] = **sphere**(n)

Creates the coordinates matrices, each of size n+1 × n+1, for the unit sphere centered at the origin.
Output: Three n+1 × n+1 matrices are assigned to M_x, M_y, and M_z.
Argument options: sphere(n) to simply display the unit sphere using an n+1 × n+1 grid. ✦ sphere with no arguments to display the unit sphere using an 21 × 21 grid.
Additional information: Use either mesh, grid, or surf to display the sphere.
See also: cylinder, surf, grid, mesh

stem(V_y)

Displays a two-dimensional stem plot of the discrete data in vector V_y.
Output: A two-dimensional plot is displayed.
Argument options: stem(V_x, V_y) to plot the values in V_y at the x-values in V_x. The two vectors must have an equal number of elements. ✦ stem(V_x, V_y, str) to plot the stem lines according to the line specification str.
Additional information: A stem plot draws lines from the x-axis to the values in V_y, where a circle is attached at the end.
See also: stem3, bar, plot, stairs

stem3(M_z)

Displays a three-dimensional stem plot of the discrete data in matrix M_z.
Output: A three-dimensional plot is displayed.
Argument options: stem(M_x, M_y, M_z) to plot the values in M_z at the x- and y-values in M_x and M_y. The three matrices must have equal dimensions. ✦ stem(M_z, str) to plot the stem lines according to the line specification str.
Additional information: A three-dimensional stem plot draws lines from the x,y-plane to the values in M_z, where a circle is attached at the end.
See also: stem, bar3, plot3, stairs

surf(M_x, M_y, M_z)

Plots the shaded surface defined by M_z, a matrix of function values evaluated at mesh grid points defined by M_x and M_y.
Output: A three-dimensional plot is produced.
Argument options: surf(V_x, V_y, M_z) to use vectors V_x and V_y to define the mesh grid. See the command listing for meshgrid for more information. ✦ surf(M_z) to use a mesh grid linked to the matrix indices of the elements in M_z. ✦ surf(M_x, M_y, M_z, M_c) to specify that the values in matrix M_c be scaled and used as pointers into the current color map for the figure. By default, the values in M_z are used.
Additional information: All of the matrices passed to surf must be of equal dimensions. ✦ Many attributes of the plot can be manipulated. For more information, see the *Graphics Controls* chapter. ✦ For more information on how the surface is shaded, see the on-line help file.
See also: meshgrid, trimesh, surfc, surfl, *colormap*, *axis*, *view*, mesh, contour, pcolor, image

surfc(M_x, M_y, M_z)

Plots the shaded surface together with the contour lines defined by M_z, a matrix of function values evaluated at mesh grid points defined by M_x and M_y.
Output: A three-dimensional plot is produced.
Argument options: For more information on alternate parameter sequences, see the entry for surf.
Additional information: All of the matrices passed to surfc must be of equal dimensions. ✦ The contour lines are projected onto the axes-plane beneath the surface plot. ✦ If the values defining the mesh grid are not monotonically increasing, but are irregularly spaced, then the contours are not drawn properly. ✦ Many attributes of the plot can be manipulated. For more information, see the *Graphics Controls* chapter. ✦ For more information on how the surface is shaded, see the on-line help file for surf.
See also: meshgrid, surf, surfl, *colormap*, *axis*, *view*, mesh, contour, image

surfl(M_x, M_y, M_z, V_{ls})

Plots the lighted surface defined by M_z, a matrix of function values evaluated at mesh grid points defined by M_x and M_y, and the light source from the direction defined by vector V_{ls}.
Output: A three-dimensional plot is produced.
Argument options: surfl(M_x, M_y, M_z) to use the default direction for the light source of 45 degrees counterclockwise from the current view position. ✦ surfl(V_x, V_y, M_z, V_{ls})

to use vectors V_x and V_y to define the mesh grid. See the command listing for meshgrid for more information. ✦ surfl(M_z, V_{ls}) to use a mesh grid linked to the matrix indices of the elements in M_z.
Additional information: The lighting direction V_{ls} can be a three-element vector representing the x, y, and z coordinates of the light source, or a two-element vector specifying the azimuth and the elevation. ✦ All of the matrices passed to surfl must be of equal dimensions. ✦ The lighting scheme used is a combination of diffuse, specular, and ambient color models. ✦ Many attributes of the plot can be manipulated. For more information, see the *Graphics Controls* chapter.
See also: meshgrid, surf, surfc, diffuse, *colormap*, *axis*, *view*, mesh, contour, image

trimesh(M_{tri}, M_x, M_y, M_z)
Plots the hidden-line surface defined by M_z, a matrix of function values evaluated at mesh grid points defined by M_x and M_y, using the triangularization data in M_{tri}.
Output: A three-dimensional plot is produced.
Argument options: trimesh(M_{tri}, M_x, M_y, M_z, M_c) to specify that the values in matrix M_c be scaled and used as pointers into the current color map for the figure. By default, the values in M_z are used.
Additional information: M_{tri} is an $m \times 3$ matrix defining the faces of the triangles.
✦ All of the other matrices passed to mesh must be of equal dimensions. ✦ Many attributes of the plot can be manipulated. For more information, see the *Graphics Controls* chapter.
See also: mesh, trisurf, meshgrid, *colormap*, *axis*, *view*, surf, contour, image

trisurf(M_{tri}, M_x, M_y, M_z)
Plots the shaded surface defined by M_z, a matrix of function values evaluated at mesh grid points defined by M_x and M_y, using the triangularization data in M_{tri}.
Output: A three-dimensional plot is produced.
Argument options: trisurf(M_{tri}, M_x, M_y, M_z, M_c) to specify that the values in matrix M_c be scaled and used as pointers into the current color map for the figure. By default, the values in M_z are used.
Additional information: M_{tri} is an $m \times 3$ matrix defining the faces of the triangles.
✦ All of the other matrices passed to surf must be of equal dimensions. ✦ Many attributes of the plot can be manipulated. For more information, see the *Graphics Controls* chapter. ✦ For more information on how the surface is shaded, see the on-line help file.
See also: surf, meshgrid, trisurf, surfc, surfl, *colormap*, *axis*, *view*, mesh, contour, pcolor, image

int = tsearch(V_x, V_y, M_{tri}, num_x, num_y)
Determines the index of the enclosing delaunay triangle for point (num_x, num_y in the triangularization M_{tri} of V_x and V_y.
Output: An integer index is assigned to int.
See also: dsearch, voronoi, convhull, delaunay

voronoi(V_x, V_y)
Plots the Voronoi diagram of for the points in x- and y-vectors V_x and V_y.

Output: Not applicable.
Argument options: voronoi(V_x, V_y, M_{tri}), where M_{tri} is the results of a call to delaunay, to use that triangulation in the computation. [$V_{x,out}$, $V_{y,out}$] = voronoi(V_x, V_y) to assign the vertices of the diagram to $V_{x,out}$ and $V_{y,out}$.
See also: dsearch, tsearch, convhull, delaunay

waterfall(M_x, M_y, M_z)

Plots the hidden-line surface with the *column* lines missing defined by M_z, a matrix of function values evaluated at mesh grid points defined by M_x and M_y.
Output: A three-dimensional plot is produced.
Argument options: For more information about valid parameter sequences for waterfall, see the entry for mesh.
Additional information: To plot the mesh without the *row* lines, call waterfall(M_x', M_y', M_z').
See also: mesh

Animation

MATLAB allows you to create animations of mathematical functions over time. However, you can *also* animate models of physical entities (*characters*) such as people, planets, and particles.

In this chapter we will create an animation involving two *characters*: (1) a "cherry-picker truck", and (2) a street lamp.

The example will illustrate the following main features of animation in MATLAB:

- patch, surf - create and color characters.
- light, set(gca) - specify lighting sources.
- set(gca) - specify camera position and view.
- set(gcf) - specify scene rendering quality.
- moviein, getframe - capture frames of movie.
- movie - play movie.
- qtwrite, bmpwrite - save movie.

In addition, we have created two *high-level* functions to make *characters* travel along a specified route over time: [54]

- route, travel

We have also created a whole library of *low-level* functions for use in creating and animating hierarchical *characters* (the *Animation Toolbox*). These can also be used to animate MATLAB's camera and lights.

The functions in the *Animation Toolbox* include:

- cube, normal - complement MATLAB's cylinder and sphere graphics primitives.
- adopt, disown, clone - create hierarchical characters.
- family - handle of a part and of its descendents.

---

[54] These functions are available from `http://www.pracapp.com/matlab/`

- nudge, scale, turn - transform characters.
- front, aim, at - aim front of character at another.
- pivot, about, along, place - position character's pivot.
- left, center, right, align - align character on pivot.
- setcolor, getcolor - set and get color of a part.
- video - save animation (.mov, .avi or .bmp).

The MathWorks and third-party companies have developed high-level tools for animations. For example:

- see several animated demos in the MATLAB demo
- see penddemo in Simulink (inverted pendulum animation)
- see *Working Model* by Knowledge Revolution on the MathWorks web site
- see *VRLink* from Terasoft on the MathWorks website

Setting the Stage

We will create an animation involving two characters: (1) a "cherry-picker truck", and (2) a street lamp. But first we must "set the stage":

```
%-------------------------------------------------
% setting the stage:
%-------------------------------------------------

hold on

view(3)
viewmenu

HI = 30;
    axis([ -HI/2,  HI/2,   -HI/2,  HI,   -HI,  HI])

axis vis3d

zoom fill
```

We turn hold on for the entire animation, so that new plotting commands keep adding to the current axes.

We selected the default 3-D view using view(3). Then we created a View menu item in the figure window (using viewmenu [55]) so that we can quickly select various views of the axes (top, bottom, left, right, front, and back) during "rehearsal".

---

[55] viewmenu is part of the *Animation Toolbox*.

We defined the bounds of our axes using axis(v). Then we used the axis vis3d option to specify that the axes proportions not change when the view is changed.

Finally we used the zoom fill option to scale the objects so that they do not need to be resized whenever the view is changed.

Loading Film

Next, we specify the size and rendering quality of our animation frames.

```
%-------------------------------------------------
% loading film:
%-------------------------------------------------

set(gcf, 'Renderer', 'ZBuffer')

set(gcf, 'Units',    'Pixels')
set(gcf, 'Position', [0 0 320 240])

set(gca, 'Units',    'Normalized')
set(gca, 'Position', [0.1 0.1  0.8 0.8])
```

Best results are obtained using the 'ZBuffer' rendering technique (rather than the default of 'painters').

We set our figure size to whatever we like (320 by 240 pixels in this example). Another common choice for multimedia applications is 640 by 480 pixels.

Finally, we position the axes to entirely fill up the figure.

Lights, Camera

Here we light our scene, and move the camera to its initial position.

```
%-------------------------------------------------
% lights, camera:
%-------------------------------------------------

set(gca, 'AmbientLightColor', [0.5  0.5  0.5])

    sun = light;
set(sun, 'Position', [-1 -1 1], 'Style', 'Infinite')

    camera = gca;
set(camera, 'CameraPosition', [1 -1  1 ])
set(camera, 'CameraTarget',   [0  0  0.5])
```

First we set the ambient lighting to medium gray.

Then we place the sun in the sky, as a far-away light. We will define another light (a light bulb, up-close) later in this example.

Finally, we position the camera, and aim it at a target in the scene. (Notice that the camera is actually defined through axis properties.)

Characters, Props

We define the three characters in our scene: joe, truck, and lamp.

```
%----------------------------------------------------
% actors, props:
%----------------------------------------------------

joe   = person;        torso = joe.torso;

truck = cherry;        body  = truck.body;
                       foot  = truck.foot;

lamp  = lamppost;      base  = lamp.base;
                       bulb  = lamp.bulb;
```

They are instances of characters defined by the person, cherry, and lamppost functions in the *Animation Toolbox*.

Each of these characters is composed of several parts which are stored in a structure. For example, Joe's torso is stored in joe.torso. The other parts of joe are subparts of joe.torso. They include all the parts of the body that connect to the torso:

- joe.neck,
- joe.head (subpart of joe.neck),
- joe.arm_left,
- joe.arm_right,
- joe.leg_left,
- joe.leg_right.

These functions are described in detail in the section, *Defining Characters*, and should serve as a model for defining your own characters.

Make-up

We have created a setcolor function to provide a consistent means of coloring different types of objects. Here we use it to change the color of parts of our characters.

```
%----------------------------------------------------
% make-up:
%----------------------------------------------------

setcolor( family(torso), 'b' )

setcolor( family(body),  'g' )
setcolor( family(foot),  'y' )

setcolor( family(base),  'r' )
setcolor( family(bulb),  'y' )
```

```
shading  interp
lighting phong
material shiny
```

setcolor only works on the specified part(s). We have used our family function so that the children (and all descendents) of the part are also colored.

We have also specified shading, lighting, and material properties for all the objects in the scene using MATLAB's functions for doing so:

- shading interp gives the smoothest shading.
- lighting phong gives the best lighting effects.
- material shiny gives the best reflections.

See the *Graphics Properties* chapter for details on these properties.

"Places, Please!"

[We recommend that you skim this section for now, but **re-read** it after you have finished reading this chapter, especially once you've read the section: *Defining Characters*).]

We now use various functions in the *Animation Toolbox* to change the position and orientation of the characters.

```
%-------------------------------------------------
% places and poses, please!
%-------------------------------------------------

place(torso, at(foot))
nudge(torso, 'y',  1)
nudge(torso, 'z', -1)

adopt(foot, torso)          %- move as a unit

aim(camera, at(torso));     %- Joe IS the star!

%-------------------------------------------------

x = [10    5    0    0    0];
y = [20   15   10    5    0];
z = [ 0    0    0    0    0];
t = [ 0  0.5   1  1.5    3];

road = route(x, y, z, t);

travel(body, road, 0);

%-------------------------------------------------

place(base, [-2 0 0]);
turn( base, 'z', 90)

hidden(bulb)                %- The plot thickens!
```

First we placed joe (his torso and all subparts) in the truck's picker (known as foot), and then nudged joe into place.

Then we used adopt to make joe part of the truck's picker, so that joe moves when the picker moves.

Then we aimed the camera at joe.

We then used route to define the path (in road) that the truck is to travel. Then we used travel to move the truck to the beginning (t=0) of the road. (Note that the front of the truck will automatically face down the road.)

Finally, we placed the base of the lamp beside the road, and turned the lamp so that it overhangs the road. Then we used hidden to hide the bulb as a means of turning it off.

"Action!"

Using a for-loop, we advance through the frames of the animation, making slight changes in each frame.

```
%--------------------------------------------------
% action!
%--------------------------------------------------

secs = t(end) - t(1);
   fps = 15;                    % frames per second

n = secs * fps;
M = moviein(n, gcf);

for i=1:n
       time = t(1) + (i-1)/(n-1) * secs;

       travel(body,    road,     time)

       turn(truck.thigh, 'z',  -90/n)

       turn(truck.thigh, 'y',   45/n)
       turn(truck.shin,  'y',  -90/n)
       turn(truck.foot,  'y',   45/n)

       if i > n/2,   visible(bulb),   end

       aim(camera, at(torso))
       drawnow

       M(:,i) = getframe;
end

movie(M, 2, fps);              %- Review twice
```

We computed the number of frames in n. Each frame is stored in a column of a "movie" matrix, M, which we initialized using MATLAB's moviein function.

Inside the for-loop we computed the current time. Then we made the truck travel to the corresponding position on the road.

In each frame we turned the picker $\frac{1}{n}^{th}$ of -90 degrees about the z axis, so that by the end of *n* frames the picker had turned a total of 90 degrees.

Notice that in the middle of the movie we turned on the lamp. Also notice that in every frame we adjusted the camera so that it was always aimed at joe no matter where he was in the scene.

At the end of each loop, we updated the screen using drawnow, and we captured the current frame using MATLAB's getframe function.

Finally, we played back the whole movie using MATLAB's movie function.

"It's a Print!"

We wish to have the option of playing back the movie from other movie players outside of MATLAB. Thus we finish our program with:

```
%-------------------------------------------------
% Write out video for playback outside Matlab:
%-------------------------------------------------

video(M, 'truck', fps)

hold off
```

The video function is provided as part of the *Animation Toolbox*. It determines what operating system you are using, and writes out an appropriate movie file:

Macintosh:	truck.mov
Windows:	truck.avi
Other:	truck01.bmp, ... truck45.bmp

Note: If we only wish to play back the movie from within MATLAB, we can simply save it using:

```
save M fps truck.mat
```

and later we can load and play back the movie using:

```
load truck.mat
movie(M, fps)
```

Defining Characters

The *Animation Toolbox* allows you to easily create arbitrarily complex characters built from other hierarchical characters and/or MATLAB's graphics objects. Let's use our simplest character, the lamp-post, to see how characters are defined in general. Here is a diagram of the lamp-post:

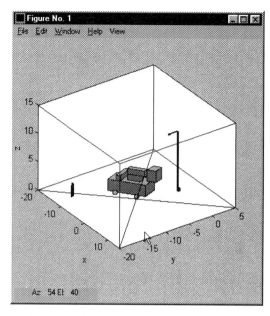

Each time you call the lamppost function it plots a new lamp-post. The function also returns the handles of the lamp-post's components for your use in a structure.

```
function lamp = lamppost

%-----------------------------------------------------
% Create and size individual components:
%-----------------------------------------------------
         base    = normal(cube);
         post    = normal(cube);
         support = normal(cube);
         housing = normal(cube);

         bulb    = light;

scale( base, 'x' ,    0.5)
scale( base, 'y' ,    0.5)
scale( base, 'z',             0.5)

scales(post,       [0.2,  0.2, 10.0])
scales(support,    [0.1,  2.0,  0.1])
scales(housing,    [0.3,  0.3,  0.1])

set(  bulb, 'Position', [0, 0, 0])
set(  bulb, 'Style',    'Local')
set(  bulb, 'Color',    'y')         %- y=yellow

%-----------------------------------------------------
% Assemble pieces:
%-----------------------------------------------------
center(base,    'x' )
center(base,    'y' )

center(post,    'x' )
center(post,    'y' )
nudge( post,    'z',                   0.5)
```

```
            center(support, 'x'  )
            nudge( support,   'y',     0.1/2 - 2)
            nudge( support,    'z',                              0.5 + 10)

            center(housing, 'x'  )
            nudge( housing,   'y' ,   0.1/2 - 2)
            nudge( housing,    'z',                              0.5 + 10)

            center(bulb,      'x'  )
            nudge( bulb,       'y' ,   0.1/2 - 2 - 0.3/2)
            nudge( bulb,        'z',                             0.5 + 10)

            %-------------------------------------------------------
            % Fasten together pieces.  Call this the front view:
            %-------------------------------------------------------

            adopt(base, post)
            adopt(        post, support)
            adopt(              support, housing)
            adopt(                     housing, bulb);

            facing(base)

            %-------------------------------------------------------

            lamp.base     = base;
            lamp.post     =    post;
            lamp.support  =       support;
            lamp.housing  =          housing;
            lamp.bulb     =             bulb;
```

First the function defines the five separate components of the lamp-post. Four components are based on MATLAB's cube function, and the fifth component is based on MATLAB's light function. [56]

Then the scale or scales functions are used to scale the sides of the cube to the appropriate lengths.

Then we use MATLAB's set function to position the light, make it a *local* light source, and change its color to yellow.

Then the pieces are assembled (i.e., moved into place relative to each other).

Once the pieces are assembled, they are "fastened" by using adopt to specify the hierarchical relationships.

The facing function specifies that the **front** of the lamp-post is now facing *forward* (the negative y-axis). Now, if we were to position the lamp-post on a path, using travel (as we did with the truck), travel would automatically rotate the lamp-post to face forwards along the specified path.

Finally, the lamppost function returns a structure consisting of the handles of the five parts of the lamp-post. Notice that the base is the top-most part in the hierarchy.

---

[56] The normal function scales an object to be one unit wide in all three directions. It also defines a pivot point at the center of the object about which the object can be scaled and rotated.

242 Animation

Therefore, when we subsequently move the base in our main program, the other parts move as well.

The functions that create joe and the truck are very similar to the lamppost function, and are given below.

Creating Joe

Joe was created using the person function in the *Animation Toolbox*:

```
function p = person

       head = normal(cube);
scale(head,      'x' , 0.2)
scale(head,      'y' , 0.2)
scale(head,      'z', 0.2)

       neck = normal(cube);
scale(neck,      'x' , 0.15)
scale(neck,      'y' , 0.15)
scale(neck,      'z', 0.15)

       torso = normal(cube);
scale(torso,     'x' , 0.5)
scale(torso,     'y' , 0.2)

       arm_left = normal(cube);
scale(arm_left, 'x' , 0.1)
scale(arm_left, 'y' , 0.1)
       arm_right = clone(arm_left);

       leg_left = normal(cube);
scale(leg_left, 'x' , 0.15)
scale(leg_left, 'y' , 0.15)
       leg_right = clone(leg_left);

arms = [arm_left, arm_right];
legs = [leg_left, leg_right];

%- - - - - - - - - - - - - - - - - - - - - - - - - - - - - - - -

center([head, neck, torso, arms, legs], 'x')
center([head, neck, torso, arms, legs], 'y')

nudge(torso,        'z',   1)
nudge(neck,         'z',   1 + 1)
nudge(head,         'z',   1 + 1 + 0.15)

nudge(arm_left,     'x' ,   0.3)
nudge(arm_right,    'x' ,  -0.3)
nudge(arms,         'z',   1)

nudge(leg_left,     'x' ,   0.2)
nudge(leg_right,    'x' ,  -0.2)

pivot([arms, legs], 'z', 1)

%- - - - - - - - - - - - - - - - - - - - - - - - - - - - - - - -

adopt(torso, neck)
adopt(     neck, head)
```

```
        adopt(torso, arms)
        adopt(torso, legs)

        facing(torso)

%------------------------------------------------

    p.torso      = torso;
    p.neck       =          neck;
    p.head       =                      head;

    p.arm_left   =          arm_left;
    p.arm_right  =          arm_right;

    p.leg_left   =          leg_left;
    p.leg_right  =          leg_right;
```

Creating the Truck

The cherry-picker truck was created using the cherry function in the *Animation Toolbox*:

```
function truck = cherry

            body = normal(cube);
    scales(body, [9    3      2    ])
    nudges(body, [0   0.5    0.5])

            cab  = normal(cube);
    scales(cab,  [1    3      1    ])
    nudges(cab,  [2   0.5    2.5])

                            [X, Y, Z] = cylinder;
            front =          surf(X, Y, Z);
    turn(   front, 'x',  90)
    normal(front );
    scale(  front, 'y',   4)       % wheel base
    nudge(  front, 'x',   1.5)

            back  = clone(front);
    nudge(  back,  'x',   5)
    adopt(  body, [cab, front, back])

%-------------------------------------------------

            thigh                   = normal(cube);
                    shin            = normal(cube);
                            heal    = normal(cube);
                                foot = normal(cube);

    adopt(  thigh,shin)
    adopt(          shin,heal)
    adopt(                  heal,foot)

    scale(  thigh,                   'x',  3)
    scale(          shin,            'x',  4)
    scales(                         foot,      [2 2 2])

    nudges(                         foot,      [1  -0.5  -0.5])
    nudge(                 [heal    ], 'x',  4                )
```

```
nudge(      [shin          ], 'z',          1   )
nudges([thigh              ],    [4   1.5   2.5])

pivot( thigh,               'x',  3)

%------------------------------------------------

adopt( body, thigh)

turn( body, 'z', 90)
facing(body)

truck.body  = body;
truck.cab   =        cab;
truck.front =        front;
truck.back  =        back;
truck.thigh =        thigh;
truck.shin  =              shin;
truck.heal  =                    heal;
truck.foot  =                           foot;
```

Relative versus Absolute Co-ordinates

Most of the time it is more convenient to use relative coordinates when defining characters (e.g., nudge, turn).

However, if you so desire you can use place to move an object to an absolute location, and you can use aim to turn an object to face an absolute location:

Relative	Absolute
nudge, nudges	place
turn	aim

Examples of both place and aim are found in the *"Action!"* section.

Path Animation

To make a character move to specific points on a path over time we *could* use the place (or nudge) function. However, the travel function makes this task simpler.

In our example, we first defined the path the truck was to travel along. We did this using the route function which returns the coefficients of a smooth *spline* curve passing through the points.

If you would prefer a straight-line interpolation through the points, then use routes instead of route. In this case, think of your road as consisting of a series of (straight-line) *routes*, instead of one (smooth) *route*.

Function Reference

Refer to help animate in the *Animation Toolbox*, which you can download from our web site (at `http://www.pracapp.com/matlab/`)

V = getframe
Creates a single movie frame by taking a snapshot of the current figure.
Output: A column vector representing one movie frame is assigned to V.
Argument options: getframe(hndl) to create the movie frame from the object with handle hndl. ✤ getframe(hndl, [num_l, num_b, num_w, num_h]) to get a rectangular area of figure hndl, defined by the left, bottom, width, and height, num_l, num_b, num_w, num_h, respectively. The values are in the current units of hndl.
Additional information: Typically, getframe is used in a loop to create a complete movie animation. For example for j = 1:10, M(:, j) = getframe, end.
See also: movie, moviein

movie(M)
Displays the movie frames contained in matrix M.
Output: A movie window is created.
Argument options: (M, n) to run the movie n times. A negative value causes each subsequent display to be run in the opposite direction. ✤ (M, V) to specify the ordering of frames with V. The first element is the number of times and the remaining elements are the ordering of the individual frames. ✤ (M, n, num_{fps}) to play the movie at num_{fps} frames per second, where the hardware allows it. (default is 12). ✤ (hndl, M) to play the movie in the figure or axes object pointed to by hndl. ✤ (hndl, M, n, num_{fps}, [num_x, num_y]) to play the movie in hndl at relative location [num_x, num_y].
Additional information: Typically, getframe is used in a loop to create the movie frames, M.
See also: getframe, moviein

M = moviein(n)
Creates a matrix large enough to hold a movie containing n frames.
Output: A matrix is assigned to M.
Additional information: The size of M also depends on the size of the current figure window.
See also: getframe, movie

Graphics Properties

In the preceding three chapters we established that each graphic object (figure, axes, text, line, patch, surface, image) has a unique identifier called a *handle*.

We used these handles in the set and get commands to change and determine properties of the objects.

For example, in the chapter on *Graphing Points and Curves*, we changed the scaling of the y-axis from linear to logarithmic:

```
set(gca, 'YScale', 'log')
```

The gca function used above returns the handle of the current set of axes.[57] Similarly, the gcf function (discussed below) returns the handle of the current figure.

This chapter presents:

1. each object's properties and possible values,

2. examples showing typical customizations to objects, and

3. other typical operations on graphical objects (e.g., deleting).

The MATLAB functions covered are:

- get and set

- gcf, gca, and gco

- text, plot, fill, surface, image

- figure, axes, delete, cla, and clf

- propedit, uisetcolor, uisetfont

---

[57] "gca" is an acronym for **get current axes**.

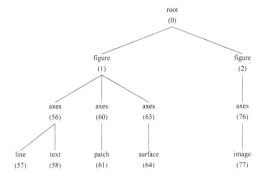

FIGURE 67. Hierarchy of graphics objects, using quads.m as an example. (Same as Figure 73.)

Object Properties and Values

The following eight tables (Tables 1 to 8) show *all* properties for *every* graphic object (except for the *root* window, covered later). Here are some general notes about the tables and the properties they display:

1. Popular properties are marked "*". Note that axis properties are the most popular.

2. The tables show **default** property values, except in some cases where a *null* default value is replaced with a value from the following example. (Figure 67 is based on that example and is discussed in detail in a later section.)

   For alternative property values, see the *Command Listing* section.

3. Property and value names are not case-sensitive.

4. Property and value names can be abbreviated. For example, the 'Visible' property can be abbreviated to 'V' for all objects, *except* axes and text.

   These two objects both also have a 'VerticalAlignment' property, so two-letter (or longer) abbreviations are necessary. We do not recommend using abbreviations in programs, as they make the programs difficult to read.

5. As several objects share many of the same properties, the tables should help you tell at a glance which properties go with which objects, and which do not.

6. A dash (-) in the table means the property is *not* supported for that type of object.

248 Graphics Properties

TABLE 1. Properties of Graphic Elements in Example

Function: (Handle)	0 (0)	gcf (1)	gca (56)	text (58)	plot (57)	fill (61)	surf (64)	image (77)
'Type'	root	figure	axes	text	line	patch	surface	image
'Color'	-	[0 0 0]	none	[0 0 0]	[1 1 0]	-	-	-
'Position'	-	[x y w h]	[x y w h]	[x y z]	-	-	-	-
'Units'	'pixels'	'pixels'	'normalized'	'data'	-	-	-	-
'Visible'	'on'	'on'	'on'	'on'	'on'	'on'	'on'	'on'
'EraseMode'	-	-	-	'normal'	'normal'	'normal'	'normal'	'normal'
'Children'	[1,2]	[56,60,63]	[57,58]	[]	[]	[]	[]	[]
'Parent'	[]	[0]	[1]	56	56	60	63	76
'Tag'	"	"	"	"	"	"	"	"
'Selected'	"	"	"	"	"	"	"	"
'SelectionHighlight'	"	"	"	"	"	"	"	"
'HandleVisibility'	'on'	'on'	'on'	'off'	'on'	'on'	'on'	'on'
'ButtonDownFcn'	"	"	"	"	"	"	"	"
'CreateFcn'	"	"	"	"	"	"	"	"
'DeleteFcn'	"	"	"	"	"	"	"	"
'BusyAction'	"	"	"	"	"	"	"	"
'Clipping'	'on'	'on'	'on'	'off'	'on'	'on'	'on'	'on'
'Interruptible'	'on'	'on'	'on'	'on'	'on'	'on'	'on'	'on'
'UserData'	[]	[]	[]	[]	[]	[]	[]	[]

TABLE 2. Properties of Graphic Elements in Example, *continued*

Function: (Handle)	0 (0)	gcf (1)	gca (56)	text (58)	plot (57)	fill (61)	surf (64)	image (77)
'CallbackObject'	[]	-	-	-	-	-	-	-
'ErrorMessage'	''	-	-	-	-	-	-	-
'ShowHiddenHandles'	'off'	-	-	-	-	-	-	-
'CurrentFigure'	1	-	-	-	-	-	-	-
'CurrentAxes'	-	56	-	-	-	-	-	-
'CurrentObject'	-	[]	-	-	-	-	-	-
'CurrentMenu'	-	1	-	-	-	-	-	-
'CurrentCharacter'	-	''	-	-	-	-	-	-
'CurrentPoint'	-	[0 0]	[(2 by 3)]	-	-	-	-	-

TABLE 3. Properties of Graphic Elements in Example, *continued*

Function:	0	gcf	gca	text	plot	fill	surf	image
'Colormap'	-	[(64 by 3)]	-	-	-	-	-	-
'FixedColors'	-	[ ]	-	-	-	-	-	-
'MinColormap'	-	[64]	-	-	-	-	-	-
'ShareColors'	-	'yes'	-	-	-	-	-	-
'Renderer'	-	'painters'	-	-	-	-	-	-
'Dithermap'	-	''	-	-	-	-	-	-
'MenuBar'	-	'figure'	-	-	-	-	-	-
'Name'	-	''	-	-	-	-	-	-
'NumberTitle'	-	'on'	-	-	-	-	-	-
'Pointer'	-	'arrow'	-	-	-	-	-	-
'Resize'	-	'on'	-	-	-	-	-	-
'SelectionType'	-	'normal'	-	-	-	-	-	-

TABLE 4. Properties of Graphic Elements in Example, *continued*

Function:	0	gcf	gca	text	plot	fill	surf	image
'KeyPressFcn'	–	''	–	–	–	–	–	–
'WindowButtonDownFcn'	–	''	–	–	–	–	–	–
'WindowButtonMotionFcn'	–	''	–	–	–	–	–	–
'WindowButtonUpFcn'	–	''	–	–	–	–	–	–
'ResizeFcn'	–	''	–	–	–	–	–	–
'CloseRequestFcn'	–	''	–	–	–	–	–	–
*								
'InvertHardcopy'	–	'on'	–	–	–	–	–	–
'PaperPosition'	–	[0.25 2.5 8 6]	–	–	–	–	–	–
'PaperType'	–	'usletter'	–	–	–	–	–	–
'PaperOrientation'	–	'portrait'	–	–	–	–	–	–
'PaperUnits'	–	'inches'	–	–	–	–	–	–
'PaperSize'	–	[8.5 11]	–	–	–	–	–	–
'NextPlot'	–	'add'	–	–	–	–	–	–
'BackingStore'	–	'on'	–	–	–	–	–	–

TABLE 5. Properties of Graphic Elements in Example, *continued*

Function:	0	gcf	gca	text	plot	fill	surf	image
'Box'	-	-	'off'	-	-	-	-	-
'DrawMode'	-	-	'normal'	-	-	-	-	-
'NextPlot'	-	-	'replace'	-	-	-	-	-
'Layer'	-	-	'bottom'	-	-	-	-	-
'PlotBoxAspectRatio'	-	-	[1 1 1]	-	-	-	-	-
'DataAspectRatio'	-	-	[1 1 1]	-	-	-	-	-
'View'	-	-	[0 90]	-	-	-	-	-
'Projection'	-	-	'orthographic'	-	-	-	-	-
'Xform'	-	-	[(4 by 4)]	-	-	-	-	-
'ColorOrder'	-	-	[(6 by 3)]	-	-	-	-	-
'CLim'	-	-	[0 1]	-	-	-	-	-
'CLimMode'	-	-	'auto'	-	-	-	-	-
'X|Y|ZColor'	-	-	[1 1 1]	-	-	-	-	-
'AmbientLightColor'	-	-	[1 1 1]	-	-	-	-	-
'LineStyleOrder'	-	-	'-'	-	-	-	-	-
'LineWidth'	-	-	0.5	-	0.5	0.5	0.5	-
'GridLineStyle'	-	-	':'	-	-	-	-	-
'X|Y|ZGrid'	-	-	'off'	-	-	-	-	-

*

TABLE 6. Properties of Graphic Elements in Example, *continued*

Function:	0	gcf	gca	text	plot	fill	surf	image
'X|Y|ZLim'	-	-	[0 1]	-	-	-	-	-
'X|Y|ZLimMode'	-	-	'auto'	-	-	-	-	-
'X|Y|ZScale'	-	-	'linear'	-	-	-	-	-
'X|Y|ZDir'	-	-	'normal'	-	-	-	-	-
'TickLength'	-	-	'in'	-	-	-	-	-
'X|Y|ZTick'	-	-	[0.01 0.025]	-	-	-	-	-
'X|Y|ZTickMode'	-	-	[0 0.5 1]	-	-	-	-	-
	-	-	'auto'	-	-	-	-	-
'X|Y|ZTickLabels'	-	-	['0'; '0.5'; '1']	-	-	-	-	-
'X|Y|ZTickLabelMode'	-	-	'auto'	-	-	-	-	-
'XAxisLocation'	-	-	'bottom'	-	-	-	-	-
'YAxisLocation'	-	-	'left'	-	-	-	-	-

TABLE 7. Properties of Graphic Elements in Example, *continued*

Function:	0	gcf	gca	text	plot	fill	surf	image
'Title'	–	–	59	–	–	–	–	–
'XLabel'	–	–	66	–	–	–	–	–
'YLabel'	–	–	67	–	–	–	–	–
'ZLabel'	–	–	68	–	–	–	–	–
'FontName'	–	–	'Helvetica'	'Helvetica'	–	–	–	–
'FontSize'	–	–	12	12	–	–	–	–
'FontUnits'	–	–	'points'	'points'	–	–	–	–
'FontAngle'	–	–	'normal'	'normal'	–	–	–	–
'FontWeight'	–	–	'normal'	'normal'	–	–	–	–
'String'	–	–	–	'Hello'	–	–	–	–
'Interpreter'	–	–	–	'tex'	–	–	–	–
'HorizontalAlignment'	–	–	–	'left'	–	–	–	–
'VerticalAlignment'	–	–	–	'middle'	–	–	–	–
'Rotation'	–	–	–	0	–	–	–	–
'Extent'	–	–	–	[x y w h]	–	–	–	–
*								

TABLE 8. Properties of Graphic Elements in Example, *continued*

Function:	0	gcf	gca	text	plot	fill	surf	image
'LineStyle'	-	-	-	-	''	'-'	'-'	-
'Marker'	-	-	-	-	'none'	'none'	'none'	-
'MarkerSize'	-	-	-	-	6	[]	6	-
'MarkerEdgeColor'	-	-	-	-	'auto'	'auto'	'auto'	-
'MarkerFaceColor'	-	-	-	-	'none'	'none'	'none'	-
'XData'	-	-	-	-	[1 2 3 4]	[1 2 3 ..]	[(1 by 3)]	[1 3]
'YData'	-	-	-	-	[1 3 2 4]	[1 3 2 ..]	[(3 by 1)]	[1 3]
'ZData'	-	-	-	-	[]	[]	[(3 by 3)]	-
'MeshStyle'	-	-	-	-	-	-	'both'	-
'CData'	-	-	-	-	-	[]	[]	-
'CDataMapping'	-	-	-	-	-	'scaled'	'scaled'	'direct'
'Faces'	-	-	-	-	-	[]	-	-
'Vertices'	-	-	-	-	-	[]	-	-
'FaceVertexCData'	-	-	-	-	-	[]	-	-
'EdgeColor'	-	-	-	-	-	[0 0 0]	[0 0 0]	-
'FaceColor'	-	-	-	-	-	[r g b]	'flat'	-
'EdgeLighting'	-	-	-	-	-	'none'	'none'	-
'FaceLighting'	-	-	-	-	-	'flat'	'flat'	-
'AmbientStrength'	-	-	-	-	-	[]	[]	-
'DiffuseStrength'	-	-	-	-	-	[]	[]	-
'SpecularStrength'	-	-	-	-	-	[]	[]	-
'SpecularExponent'	-	-	-	-	-	[]	[]	-
'SpecularColorReflectance'	-	-	-	-	-	[]	[]	-

Alternate Commands for Changing Property Values

Matlab provides convenient commands to set many specific object properties. For example, instead of:

```
set(gca, 'XGrid', 'on')
set(gca, 'YGrid', 'on')
```

you can issue:

```
grid on
```

or:

```
grid('on')
```

The set and non-set alternatives are shown in Table 9. Our recommended choice is indicated with an arrow.

We recommend the set form in some cases, and the non-set form in others, depending on readability and ease of use.

Properties Common to Several Objects

Most properties are unique to one or two objects. However, Table 1 is devoted to properties that *most* objects have in common. The three main properties are listed first:

1. 'Color'
2. 'Position'
3. 'Units'

Color Property

The four objects that have a Color property are: figure, axes, text, and line.

figure The figure color is the color of the whole window. Different figures can have different colors.

> Some users prefer medium gray as the figure color. This can be achieved on a *figure-at-a-time basis* using:
>
> ```
> set(gcf, 'Color', [0.5 0.5 0.5])
> ```
>
> To make this the default color for *all future* figure windows you can use the following form of set:
>
> ```
> set(0, 'DefaultFigureColor', [0.5 0.5 0.5])
> ```

TABLE 9. Alternate Commands for Changing Property Values

subplot(m,n,p)	←	axes([xlo, ylo, wid, ht])
axis([xlo xhi, ...	→	set(gca, 'XLim', [xlo xhi])
ylo yhi, ...	→	set(gca, 'YLim', [ylo yhi])
zlo zhi])	→	set(gca, 'ZLim', [zlo zhi])
axis('ij')	→	set(gca, 'YDir', 'reverse')
axis('xy')	→	set(gca, 'YDir', 'normal')
		AR = get(gca, 'AspectRatio')
axis('equal')	←	set(gca, 'AspectRatio', [AR(1), 1./1.])
axis('square')	←	set(gca, 'AspectRatio', [1./1., AR(2)])
axis('normal')	←	set(gca, 'AspectRatio', [NaN , NaN])
	←	set(gca, 'XGrid', 'on')
grid('on')	←	set(gca, 'YGrid', 'on')
	←	set(gca, 'ZGrid', 'on')
view([az, el])	←	set(gca, 'View', [az, el])
view(2)	←	set(gca, 'View', [0., 90.])
view(3)	←	set(gca, 'View', [-37.5, 30.])
colormap(map)	←	set(gcf, 'Colormap', map)
caxis([lo, hi])	←	set(gca, 'CLim', [lo, hi])
clf	←	j = get(gcf, 'children')
	←	delete(j)
cla	←	j = get(gca, 'children')
	←	delete(j)

We will explain this command in detail in the section *Changing Default Property Values*.

axes The axes color is 'none' by default.[58] In other words it has no color of its own, but lets the figure color show through from underneath.

You can change this; for example, light gray looks good as the axes color:

```
set(gca, 'Color', [0.8  0.8  0.8])
```

See Figure 68 for a light gray axes area in the middle of a medium gray figure. It was produced as follows:

```
plot([1 2 3 4], ...
     [1 3 2 4], 'w-')

medium = [0.5  0.5  0.5];
light  = [0.8  0.8  0.8];
```

---

[58]'None' may be specified as the color only with axes, not with any other objects.

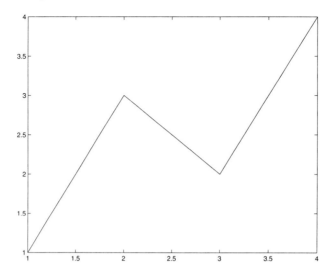

FIGURE 68. Light gray axes area with medium-gray figure in the background

```
set(gcf, 'Color', medium)    % Figure
set(gca, 'Color', light)     % Axes

print -deps figaxes.eps
```

text The default color for text is black. There are two ways of changing the color to be used for a text string. One is to specify the color in the text command itself:

```
text(2, 2, 'Stable Region', 'Color', 'r')
```

or:

```
text(2, 2, 'Stable Region', 'Color', [1 0 0])
```

The other way is to get the handle of the text object, and use set to change the color. The handle of the text object is an optional output argument of the text command:

```
h = text(2, 2, 'Stable Region');

set(h, 'Color', 'r')
```

or:

```
set(h, 'Color', [1 0 0])
```

The handle method is preferred because: (1) it's easier to read two short commands than one long command, and (2) you may have other properties of the text string that you wish to set later using the handle.

plot The default color for the first set of points in a given call to plot is blue ([0 0 1]). Subsequent lines in the same call to plot:

```
plot(x1, y1, '-',  ...
     x2, y2, '--', ...
     x3, y3, ':')
```

are plotted in a sequence of colors, according to the ColorOrder property of the **axes**. Ascertain the sequence as follows:

```
get(gca, 'ColorOrder')
```

The result is: [59]

```
ans =

         0        0   1.0000    <--- blue    = 'b'
         0   0.5000        0    <--- green   = 'g'
    1.0000        0        0    <--- red     = 'r'
         0   0.7500   0.7500    <--- cyan    = 'c'
    0.7500        0   0.7500    <--- magenta = 'm'
    0.7500   0.7500        0    <--- yellow  = 'y'
    0.2500   0.2500   0.2500    <--- grey
```

Once the end of this sequence is reached, it is repeated as often as necessary. However, you can override this default color selection. One way is as part of the plot command:

```
plot([1 2 3], [1 4 9], 'Color', 'r')
```

or:

```
plot([1 2 3], [1 4 9], 'Color', [1 0 0])
```

The other way is by using set:

```
h = plot([1 2 3], [1 4 9])

set(h, 'Color', 'r')
```

or:

```
set(h, 'Color', [1 0 0])
```

Note: remember that another way to have multiple lines on a set of axes is to call plot multiple times. Make sure you issue hold on before the second call to plot, however, so that the first plot is not discarded. For example:

```
plot([1 2 3], [-1 -4 -9], 'Color', 'r')

hold on
    plot([1 2 3], [0 0 0], 'Color', 'g')
    plot([1 2 3], [1 4 9], 'Color', 'b')
hold off
```

surface The following two functions are used to specify the colors that correspond to data values in surface plots:

1. colormap - specifies a series of colors to which surface data values are mapped.
2. caxis - specifies data values that correspond to the first and last color map entries. (Default: minimum and maximum values of surface data.)

---

[59] We have embellished the output by adding the color names, and inserting a blank line in the output.

The default color map for a figure window is hsv, with 64 distinct color entries. If the color map has been changed, you can reset it using either:

```
colormap(hsv)
```

or:

```
colormap(hsv(64))
```

In the following, we have requested only 16 distinct colors using:

```
colormap(hsv(16))
```

The ninth color (cyan) has entries of:

```
>> map = hsv(16);
>>
>> map(9, 1:3)

ans =
     0     1     1
```

Available color maps include bone, cool, copper, flag, gray, hot, hsv, jet, pink, prism, spring, summer, autumn, winter, colorcube, and lines.

You can create your own color map. For example, we created our own in bluered.m, and we can invoke it using:

```
colormap(bluered(16))
```

This is what the code for bluered.m looks like:

```
function map = bluered(rows)

if nargin < 1
    colormap    = get(gcf, 'colormap');
    [rows, cols] = size(colormap);
end

red   = linspace(0, 1, rows);    % /_
green = linspace(0, 0, rows);    % |_
blue  = linspace(1, 0, rows);    %  _\

map = [red', green', blue'];
```

and here is what bluered(8) produces:

```
ans =
         0         0    1.0000
    0.1429         0    0.8571
    0.2857         0    0.7143
    0.4286         0    0.5714
    0.5714         0    0.4286
    0.7143         0    0.2857
    0.8571         0    0.1429
    1.0000         0         0
```

As you can see, red increases from 0 to 1, green stays at 0, and blue decreases from 1 to 0.

Consider the following side-by-side surface plots. Notice that in *each* case the minimum and maximum surface values are mapped to the first and last colors in the color map. As a result, the two surfaces have inconsistent coloring, with respect to each other:

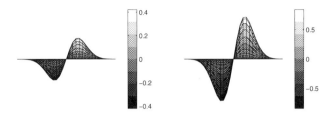

To force both plots to use the same color scaling, we use the caxis command in *both* plots, to override the surface minimum and maximum values with our own values:

```
caxis([-1 1])
```

The result is that z-value -1 is mapped to the first color in the color map, and z-value 1 is mapped to the last color in the color map, in *both* plots:

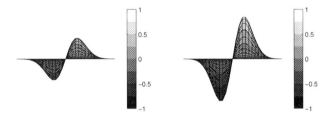

The complete program to produce the above plots is as follows:

```
colormap(gray(8))

x = [-3 : 0.2 : 3];
y = [-3 : 0.2 : 3];

[X, Y] = meshgrid(x, y);

for i=1:2
    subplot(2,2, i)

    Z = i * nice(X, Y);

    surf(X, Y, Z, Z);

    caxis([-1 1])      % <=========

    axis([-3 3   -3 3   -1 1])
    view([0, 0])

    axis off
    colorbar('vert')
end

print -deps twonice2.eps
```

image The data in image plots indexes directly into the color map. For example:

```
A = magic(5)

x = [-2  2];
y = [ 1  5];
image(x, y, A)

colormap(1 - gray(25))
colorbar('vert')

print -depsc myimage.eps
```

produces the following matrix:

```
A =
    17    24     1     8    15
    23     5     7    14    16
     4     6    13    20    22
    10    12    19    21     3
    11    18    25     2     9
```

and the following image plot:

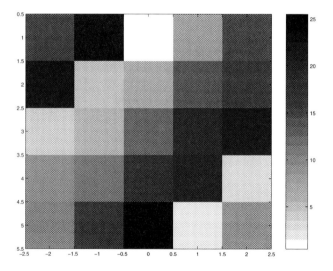

Notice that we inverted the gray color map (1 - gray(25)) so that larger indices produce darker colors.

Notes

- image can also be directly passed an $m \times n \times 3$ matrix of RBG values, thereby bypassing the color map.

- Under Microsoft Windows and on the Macintosh, you can invoke the function uisetcolor to pop up a window containing a pallette of the available colors.

TABLE 10. Default units for objects

Object	pixels	inches	points	centimeters	normalized	data
root	pixels					x
figure	pixels					x
axes					normalized	x
text						data
plot	x	x	x	x	x	data
fill	x	x	x	x	x	data
surf	x	x	x	x	x	data

Units and Position Properties

You can change the unit of measure used to position and size the following four objects [60]

1. the "root" window,

2. figures,

3. axes and

4. text.

The available units are:

- 'pixels',

- 'inches', 'points', 'centimeters',

- 'normalized', and 'data'.

The default and allowable units for each object are shown in Table 10. Notice that 'data' units are not available until you get down to text objects and below in the table.

These units are used mostly by the 'Position' property. The 'Position' property specifies the **location** and **size** of an object *inside* of the object enclosing it.

Specifically (in reverse order):

- The location of *text* is specified within the bounds of **axes**. [61]

- The location and size of *axes* are specified within the bounds of **figure windows**.

---

[60] The other objects do not allow you to change their units: plot, fill, and surf *only* work in 'data' units.

[61] Actually, text objects can extend beyond the bounds of the axes if their Clipping property is turned off (default).

- The location and size of *figure windows* are specified within the bounds of the **"root" window**.

This "hierarchy" is indicated by the indentation in Table 10.

For example, let's position the current figure (gcf) in the center of the root window (0), and make it half the size of the root window. First, we ascertain the size of the root window using:

```
screensize = get(0, 'ScreenSize');

r_left   = screensize(1);    %    1
r_bottom = screensize(2);    %    1

r_width  = screensize(3);    %  1024
r_height = screensize(4);    %   768
```

The root window's 'ScreenSize' property is similar to the 'Position' property of figures and axes in that it is a vector with four numbers:

- The **x** and **y** coordinates of the lower-left corner. Here they are (1,1).

- The **width** and **height** of the object. In this example they are 1024 and 768 pixels.

To center the figure window on the screen, and make it half the size of the screen, specify the figure's 'Position' property as follows, working in pixels, the default units for figures:

```
left   = r_left   + r_width/4;
bottom = r_bottom + r_height/4;

width  = r_width/2;
height = r_height/2;

position = [left, bottom, width, height];

set(gcf, 'Position', position)
```

Now if we issue:

```
pos = get(gcf, 'Position')
```

it will report:

```
pos =

     256   192   512   384
```

This is illustrated in Figure 69.

Change the figure's 'Units' property to 'Normalized', and **get** the position again:

```
set(gcf, 'Units', 'Normalized')

pos = get(gcf, 'Position')
```

It will report:

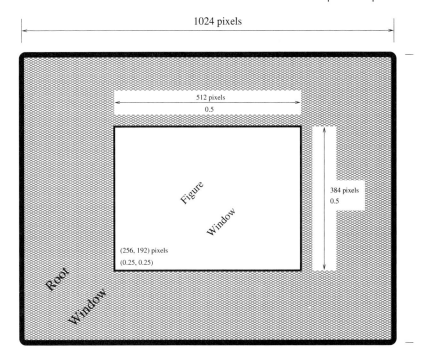

FIGURE 69. Positioning figure in the middle of the screen using pixels and normalized units

```
pos =
      0.2500    0.2500    0.5000    0.5000
```

Normalized units work on a scale of 0 to 1 in both *x* and *y* directions. They are usually easier to use when positioning figures in the root window, and when positioning axes (and GUI controls [62]) in the figure window.

For example, in place of the eleven lines above to position the figure, issue:

```
set(gcf, 'Units', 'Normalized')
set(gcf, 'Position', ...
         [0.25,  0.25,  0.5,  0.5 ])
```

The commands are more compact here because we do not need to know the actual size of the bounding area (i.e., the screen) when working in normalized units.

Controlling Axis Placement - Three Ways

You can control the position of a set of axes in a figure window in three ways:

1. subplot(m,n,p) command - *easiest*,
2. subplot(m,n,p) tricks,

---

[62] Refer to the chapter *Graphical User-Interface Functions* for GUI controls.

3. set(gca,'Position',...) command - *most flexible*.

The following MATLAB program (`quads.m`) illustrates all three:

```
figure
    subplot(2,2, 1)
        plot([1 2 3 4], ...
             [1 3 2 4], '-')
        text(2.5, 2.5, 'Decline')
        title('Fig.1, Top Left')
    subplot(2,2, 2)
        fill([1 2 3 4   4 1], ...
             [1 3 2 4   0 0], [.8 .8 .8])
        title('Fig.1, Top Right')
    subplot(2,1, 2)
        surf([ 0    0.5    0 ;   ...
              0.5    1    0.5;   ...
               0    0.5    0 ]);
        title('Fig.1, Bottom')
    print -deps quads1.eps

figure
    axes
        set(gca, 'Position', ...
                  [0.1    0.1    0.8    0.8])
%                 left bottom width height
        image(64*[ 0    0.5    0 ;    ...
                  0.5    1    0.5;    ...
                   0    0.5    0 ]);
        axis('ij')          %% Not to be confused
                            %%    with "axes".
        title('Fig.2')
    print -deps quads2.eps
```

It produces two separate figure windows, shown in Figures 70 and 71.

Notes

subplot(m,n,p) command The first two plots use subplot, dividing the figure window into m=2 rows and n=2 columns of axes:

```
subplot(2,2, 1)
   ...
subplot(2,2, 2)
   ...
```

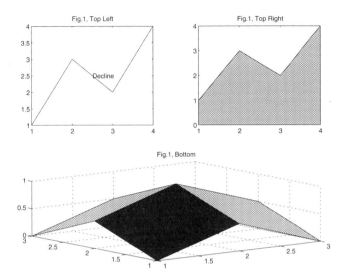

FIGURE 70. First figure with three sets of axes created using the subplot command

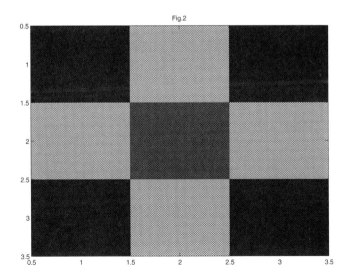

FIGURE 71. Second figure with one set of axes created (and positioned) using the axes command

Then they plot in positions p=1 and p=2 of these m×n=4 plots:

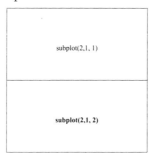

subplot(m,n,p) tricks For the third graph in this figure, we tell subplot that we are working with m=2 rows and n=1 column of axes (two plots, one above the other):

```
subplot(2,1, 2)
...
```

Then it plots in position p=2 of these:

This subplot trick (varying m and n between calls), can help create custom page layouts with minimal effort.

set(gca,'Position',...) In the second figure we have one set of axes. It was created explicitly using the axes command. [63]

Then we specified a new position for the axes to over-ride the default position:

```
axes
        set(gca, 'Position', ...
                 [0.1    0.1    0.8    0.8])
    %            left bottom width height
```

Recall that the default value of the 'Units' property for axes is 'Normalized', so it is quite straight-forward to position the axes.

By means of another example, to lay out a 2 × 2 tableau of axes, you could issue the following:

---

[63] Up until now, we have usually created axes implicitly via calls to functions such as plot.

```
axes('Position', [.1 .6  .3 .3])
    text(0.5, 0.5, '1')

axes('Position', [.6 .6  .3 .3])
    text(0.5, 0.5, '2')

axes('Position', [.1 .1  .3 .3])
    text(0.5, 0.5, '3')

axes('Position', [.6 .1  .3 .3])
    text(0.5, 0.5, '4')
```

(Notice that the 'Position' property is included as part of the axes commands.)

Alternatively, we could create all the axes at once, saving the handle of each set of axes. Then further down in the MATLAB program we could specify which of the pre-created axes to plot on:

```
a1 = axes('Position', [.1 .6  .3 .3])
a2 = axes('Position', [.6 .6  .3 .3])
a3 = axes('Position', [.1 .1  .3 .3])
a4 = axes('Position', [.6 .1  .3 .3])

axes(a1)
    text(0.5, 0.5, '1')

axes(a2)
    text(0.5, 0.5, '2')

axes(a3)
    text(0.5, 0.5, '3')

axes(a4)
    text(0.5, 0.5, '4')

print -deps tableau2.eps
```

In either case, the result is the same set of four plots, as shown in Figure 72.

Positioning Text and Other Objects

text The 'Position' property for text is simply the point at which it is drawn, using three-dimensional coordinates. For example:

```
h = text(1, 2, 'string')
get(h, 'Position')
```

It returns a three-element vector with the z-value set to 0:

```
ans =
     1     2     0
```

See the section *Text Properties* for other properties which affect the actual location of and the space occupied by text objects.

plot, surf These other objects do not have a 'Position' property, but rather have 'XData', 'YData', and 'ZData' properties to specify the position of points that comprise them.

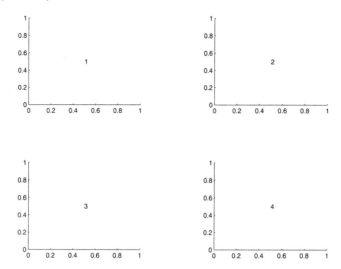

FIGURE 72. Four plots in one figure. axes was called directly to specify the position, rather than using subplot.

Hierarchy of Objects

Note that the preceding example consisted of:

1. one figure containing three sets of axes, and
2. another figure containing one set of axes.

and each set of axes in turn contains: text, lines, surfaces, etc.

This suggests there is a hierarchy of graphics objects, and indeed even the lines of the program above have been indented to show this:

1. **root window**,
2. **figures** in *root window*,
3. **axes** in *figures*, and
4. **text, lines, etc.** in *axes*.

Figure 73 shows the four-level hierarchy of graphics objects in Matlab, using the previous program as an example. The figure also shows the handles of the various graphics objects (rounded to one decimal place).

We obtained the handles running a general-purpose M-file written to display all objects and their children. The output for this example is:

```
root        0.0
  figure      1.0
    axes         56.0
```

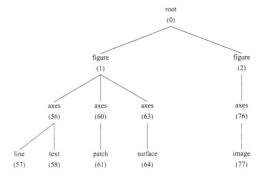

(Not shown are each axes's other children: title, xlabel, ylabel)

FIGURE 73. Hierarchy of graphics objects, using `quads.m` as an example. (Handles in parentheses are rounded to one decimal place.)

```
      (title)     59.0
      (xlabel)    66.0
      (ylabel)    67.0
      line        57.0
      text        58.0
   axes         60.0
      (title)     62.0
      (xlabel)    69.0
      (ylabel)    70.0
      patch       61.0
   axes         63.0
      (title)     65.0
      (xlabel)    72.0
      (ylabel)    73.0
      surface     64.0
figure      2.0
   axes         76.0
      (title)     78.0
      (xlabel)    79.0
      (ylabel)    80.0
      image       77.0
```

The function that produced this output is called showkids, and is shown below. To invoke it, simply type: showkids.

```
function kids

%-------------------------------------------------

root_ = 0;
type  = get(root_, 'Type');

              fprintf(1, '%0s %-10s %.1f\n', ...
                      '', type, root_);

kids_root_ = sort( get(root_, 'Children') );
     [m,n] = size(kids_root_);

%-------------------------------------------------

for i=1:m*n
```

272 Graphics Properties

```
        figure_ = kids_root_(i);
        type    = get(figure_, 'Type');

                fprintf(1, '%2s %-10s %.1f\n', ...
                        '', type, figure_);

        kids_figure_ = sort( get(figure_, 'Children') );
             [m,n] = size(kids_figure_);

%       ---------------------------------------

        for j=1:m*n

            axes_ = kids_figure_(j);
            type  = get(axes_, 'Type');

                fprintf(1, '%4s %-10s %.1f\n', ...
                        '', type, axes_);

            kids_axes_ = sort( get(axes_, 'Children') );

            if strcmp(type, 'axes')     %%% Could be uimenu
                                        %%%  or uicontrol.
                title_  = get(axes_, 'Title');
                xlabel_ = get(axes_, 'XLabel');
                ylabel_ = get(axes_, 'YLabel');

                fprintf(1, '%6s %-10s %.1f\n', ...
                        '', '(title)', title_);
                fprintf(1, '%6s %-10s %.1f\n', ...
                        '', '(xlabel)', xlabel_);
                fprintf(1, '%6s %-10s %.1f\n', ...
                        '', '(ylabel)', ylabel_);
            end

            [m,n] = size(kids_axes_);

%           -----------------------------------

            for k=1:m*n

                kid_ = kids_axes_(k);
                type = get(kid_, 'Type');

                fprintf(1, '%6s %-10s %.1f\n', ...
                        '', type, kid_);

            end

%           -----------------------------------

        end

%       ---------------------------------------
    end
%---------------------------------------------
```

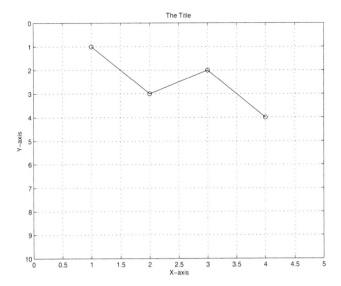

FIGURE 74. Plot customized using only non-set commands

Commonly-Used Properties

Axes Properties

The following program plots some data, and adds labelling, grid lines, etc., all using the usual non-set commands:

```
plot([1 2 3 4], [1 3 2 4], 'o')

hold on
plot([1 2 3 4], [1 3 2 4], '-')
hold off

% - - - - - - - - - - - - - - -

title('The Title')
xlabel('X-axis')
ylabel('Y-axis')

% - - - - - - - - - - - - - - -

grid on
axis([0, 5,   0, 10])
axis('ij')

print -deps axispop1.eps
```

The output is shown in Figure 74. Notice that the numbering on the *y*-axis is reversed, thanks to the axis('ij') command.

Now consider the equivalent program that *only* uses set commands to change property values. Below the comments are *additional* and *useful* set commands that have *no* Matlab command equivalents:

```
    plot_1_ = plot([1 2 3 4], [1 3 2 4]);
    set(plot_1_, 'LineStyle', 'o')

    hold on
    plot_2_ = plot([1 2 3 4], [1 3 2 4]);
    set(plot_2_, 'LineStyle', '-')
    hold off

%- - - - - - - - - - - - - - - -

    title_ = get(gca, 'Title');
    set(title_, 'String', 'The Title')

    xlabel_ = get(gca, 'XLabel');
    set(xlabel_, 'String', 'X-axis')

    ylabel_ = get(gca, 'YLabel');
    set(ylabel_, 'String', 'Y-axis')

%- - - - - - - - - - - - - - - -

    set(gca, 'XGrid', 'on')
    set(gca, 'YGrid', 'on')

    set(gca, 'XLim', [0  5])
    set(gca, 'YLim', [0 10])

    set(gca, 'YDir', 'reverse')

%------------------------------%
% The following "set"s have NO  %
% equivalent Matlab commands:   %
%------------------------------%

    set(plot_1_, 'MarkerSize', 12);

    set(plot_2_, 'LineWidth', 4.0)

%- - - - - - - - - - - - - - - -

    set(title_,  'FontAngle', 'Italic')
    set(xlabel_, 'FontAngle', 'Italic')
    set(ylabel_, 'FontAngle', 'Italic')

%- - - - - - - - - - - - - - - -

    set(gca, 'GridLineStyle', '--')
    set(gca, 'TickDir', 'out')
    set(gca, 'LineWidth', 4.0)

    set(gca, 'XTick',         [1; ...
                               2; ...
                               3; ...
                               4])
    set(gca, 'XTickLabels', [' One '; ...
                             ' Two '; ...
                             'Three'; ...
                             ' Four']   )

    set(gca, 'FontName',   'Times')
    set(gca, 'FontWeight', 'Bold')

    print -deps axispop2.eps
```

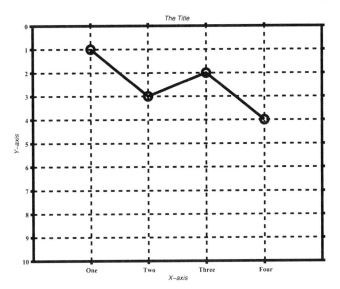

FIGURE 75. Plot customized using *only* set commands

The output is shown in Figure 75.

Notes

- You *can* specify properties directly in the plot command (and many others), and so avoid using the set command. However, this often makes for long and hard-to-read commands, and so we recommend using separate set commands for each property value you specify.

    For example, we *could* have made the marker be 'o', *and* changed the 'LineWidth' to 4.0 on our second plot by issuing:

    ```
    plot([1 2 3 4], [1 3 2 4], 'o', ...
         'LineWidth', 4.0)
    ```

    or by explicitly specifying the 'Marker' property name in front of the 'o':

    ```
    plot([1 2 3 4], [1 3 2 4], ...
         'Marker', 'o',        ...
         'LineWidth', 4.0)
    ```

    Similarly we could have specified *italics* directly for our title and axis labels using:

    ```
    title('Title',   'FontAngle', 'Italic')
    xlabel('X-axis', 'FontAngle', 'Italic')
    ylabel('Y-axis', 'FontAngle', 'Italic')
    ```

- Notice that the command:

    ```
    title_ = get(gca, 'Title')
    ```

does **not** return the title character string. Instead it returns the *handle* of a **text** object containing the title. This is also true for the 'XLabel' and 'YLabel' axis properties.

Notice that an underscore (_) was added to the end of the variable name title_ to remind ourselves that it is a handle. [64]

- The last group of set commands has no equivalent Matlab commands. Three are particularly useful:

  1. The 'GridLineStyle' property can be handy. Acceptable values are the same as for plot: '-' (*solid*), '--' (*dashed*), ':' (*dotted*), and '-.' (*dash-dot*). The default is dotted.
  2. The 'XTick' property allows you to place labelled tick marks wherever you like. For example, we chose not to place tick marks at the axis extremities.
  3. The 'XTickLabels' property allows you to override the numbers that would normally be plotted as tick mark labels with your own strings.

Text Properties

There are several properties unique to text objects, plus others in common only with axes objects:

- 'String',
- 'FontName', 'FontSize', 'FontAngle', 'FontWeight',
- 'HorizontalAlignment', 'VerticalAlignment',
- 'Rotation', and 'Extent' (read-only).

'String' is the character string. You can change the 'String' property if you like:

```
h = text(x, y, 'Critical Point')
set(h, 'String', 'Increasing')
```

'FontName' 'FontName' is the typeface to be used. The four most popular fonts are usually available: Courier (mono-spaced), Times, Helvetica (default), and Symbol (greek and math characters). [65]

Samples of each are shown in Figure 76. For example, to plot a string in 'Times', you could issue:

---

[64] The underscore does somewhat resemble the handle on a frying pan!

[65] The number of fonts available depends on the underlying graphics system for the screen and printer. Under Microsoft Windows and on the Macintosh, you can invoke the function uisetfont to pop up a window containing a scrolling list of available fonts.

FIGURE 76. Four most popular fonts (`Courier`, Times, Helvetica, Symbol)

```
text(1, 2, 'Increase', ...
         'FontName', 'Times')
```

or:

```
h = text(1, 2, 'Increase');
set(h, 'FontName', 'Times')
```

FontSize is the height of the characters, measured in points. The default is 12 points.[66] Here are samples of various sizes:

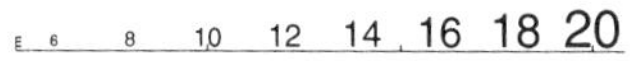

FontAngle is the angle of the individual letters. The default is 'normal'. The other two choices are 'italic' and 'oblique'.[67] Here are samples of each:

FontWeight is the weight of the characters. The default is 'normal' for no bolding. The other choices are 'light', 'demi' and 'bold'. Here are samples of each:

'HorizontalAlignment' specifies whether the text string is to be left-justified, right-justified or centered at the x position of the text string. (The default is 'Left'.)

'VerticalAlignment' specifies whether the y position of the text string represents the: top, *cap*, middle, baseline or bottom. (The default is 'Middle'.)

---

[66] One point is approximately $\frac{1}{72}$ of an inch.
[67] See the 'Rotation' property below for rotating the entire character string.

The various combinations of 'HorizontalAlignment' and 'VerticalAlignment' are shown below. The crosses represent the location of the (x, y) position of the string:

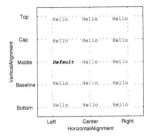

'**Rotation**' is the angle of rotation (in degrees) of the text about its (x, y) position. The default is 0 degrees.

'**Extent**' (read-only) is similar to the 'Position' property for figures and axes. It is a vector with four elements indicating the lower left hand corner of an imaginary box bounding the string, as well as the box's width and height:

```
[left, bottom,  width, height]
```

You can only get the value of this property, you can't directly set it. [68]

Contour Line Properties

One method of changing line style of individual contour lines is simply to make multiple calls to contour.

The second method makes use of the fact that you can invoke contour as follows:

```
dx = 0.5;                    % x step size
dy = 1.0;                    % y step size

x = [-2 : dx : 2];           % series of x-values
y = [-2 : dy : 2];           % series of y-values

[X,Y] = meshgrid(x, y)
Z = nice(X, Y)

[C, h] = contour(X, Y, Z, 7)
```

where h is a vector of "handles" to the individual contour lines. (C is a matrix containing the corresponding (x, y, z) values for the contour lines.)

This is handy if you want to change how certain contours are plotted, e.g., you might want negative contours to be dashed (or red, etc.).

Here is a procedure to accomplish this: [69]

---

[68] The 'Extent' property's value is affected by the 'Position', 'HorizontalAlignment' and 'VerticalAlignment' properties of the text string.

[69] Notice that, except for the three lines of Matlab commands between the dashed lines, this function has all the necessary code to parse the contour line data from matrix C for *any* application, not just ours here. For details on the format of the C matrix, issue: help contourc.

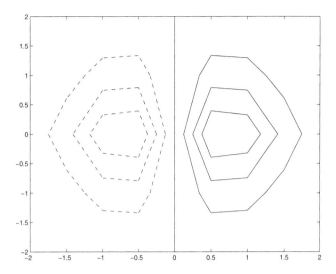

FIGURE 77. Plot showing results of changing 'LineStyle' of individual *negative* contour lines to "dashed"

```
function negcon(C, h, linestyle)

[m,n] = size(C);

j = 1;
nlines = 0;

while 1

     if j >= n,    break,    end

          level = C(1, j);
          pairs = C(2, j);
          nlines = nlines + 1;

          x = C(1, j+1:j+pairs);
          y = C(2, j+1:j+pairs);

%         -------------------------------
          if level < 0.
               set(h(nlines), 'LineStyle', linestyle)
          end
%         -------------------------------

     j = j + (1 + pairs);

end
```

Then we can invoke it as follows:

```
negcon(C, h, '--')

print -deps negdemo.eps
```

The results are shown in Figure 77.

Note: where possible, use simpler forms of the contour function. For example, in this case we could achieve similar results using the following (assuming we know what specific contour levels to use in advance):

```
contour(X, Y, Z, [0., .1, .2, .3], '-')
hold on
contour(X, Y, Z, [    -.1, -.2, -.3], '--')
hold off
```

Hardcopy Properties

When you issue the print command,[70] the current figure is copied to the specified hardcopy device.

By default, this hardcopy will have the following properties:

PaperSize The default for this property is **8.5 × 11 inches** (*x*-size × *y*-size).

PaperPosition The default for this property is such that the figure is centered on the page with 0.25 inch left/right margins and 2.5 inch top/bottom margins. In other words, it is the same as would be produced by the command:

```
set(gcf, 'PaperPosition', ...
         [0.25,  2.5,   8.0,   6.0])
  %      Left  Bottom Width Height
```

Consequently, the figure is scaled to fit in a print area of **8 × 6 inches** (*x*-size × *y*-size), as shown in Figure 78.

Changing the PaperPosition Property

Changing the 'PaperPosition' property is straightforward. For example, to specify 1 inch margins all around, except for a 1.5 inch margin on the left side, issue:

```
set(gcf, 'PaperPosition', ...
         [1.5,  1.0,  (8.5-1.5-1), (11-1-1)])
  %      Left Bottom    Width        Height
```

Changing the PaperSize Property

Changing the PaperSize property is less straightforward, because you cannot change it directly. Rather, you must change one or more of the following three *other* properties which influence the PaperSize property's value. These are:

---

[70] Or when you select the Print menu item under the File menu with Microsoft Windows or on the Macintosh.

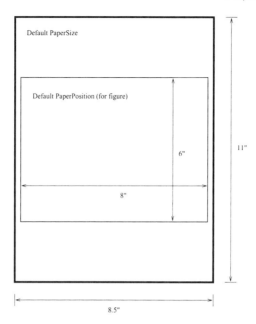

FIGURE 78. Default 'PaperPosition' is an 8 × 6 inch print area centered on page

1. PaperUnits

2. PaperOrientation

3. PaperType

This relationship is illustrated in Figure 79. Now let's look at each of these three properties in detail:

PaperUnits The two available values of PaperUnits are inches and centimeters, with inches being the default.

If you change the value of PaperUnits that you want to work in, then the PaperSize (as well as the 'PaperPosition') is reported in those new units. For example, assuming a 'PaperSize' of 8.5 × 11 inches:

```
>> set(gcf, 'PaperUnits', 'centimeters')
>> get(gcf, 'PaperSize')

ans =

   21.5736    27.9188
```

PaperOrientation The two available values for PaperOrientation are Portrait and Landscape, with Portrait as the default.

If you change the value of PaperOrientation that you want to work in, then the PaperSize will be reported in reverse order. For example, assuming a PaperSize of 8.5 × 11 inches (x-size × y-size):

282 Graphics Properties

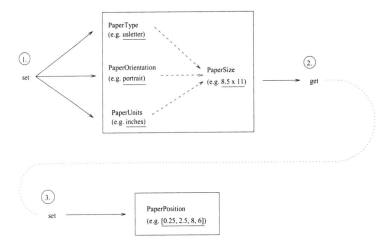

FIGURE 79. The value of a figure's 'PaperSize' property is influenced by changes to one or more of its 'PaperUnits', 'PaperOrientation' and 'PaperType' properties

```
>> set(gcf, 'PaperOrientation', 'landscape')
>> get(gcf, 'PaperSize')

ans =

    11.0000    8.5000
```

Note: Changing the PaperOrientation has no effect on the **value** of the 'PaperPosition', as shown in Figure 80.

Therefore, you may want to change the 'PaperPosition' to a value such that the figure area would appear to have been rotated as well. This is easily done:

```
pos = get(gcf, 'PaperPosition')

set(gcf, 'PaperPosition', ...
         [pos(2), pos(1), ...
          pos(4), pos(3)])
```

PaperType The available values for PaperType are: usletter, uslegal, a3, a4letter, a5, b4, and tabloid.

For your reference, Table 11 shows the size of each type of paper in inches and centimeters.

For example, let's tell MATLAB that we would like our figure to be in landscape and have one inch margins all around when it is printed.

```
set(gcf, 'PaperType',        'usletter')     % 8.5 x 11
set(gcf, 'PaperOrientation', 'landscape')    %  11 x 8.5

size   = get(gcf, 'PaperSize');
width  = size(1);                             %  11
height = size(2);                             %        8.5

set(gcf, 'PaperPosition', [      1,        1, ...
                           width-2, height-2]);
```

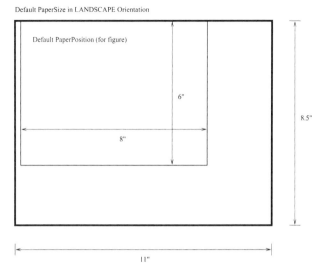

FIGURE 80. The value of a figure's 'PaperPosition' property is *not* affected by changing between portrait and landscape mode

TABLE 11. 'PaperType's and their sizes (in Portrait orientation)

PaperType	inches	centimeters
usletter	8.50 × 11.00	21.57 × 27.92
uslegal	11.00 × 14.00	27.92 × 35.53
a3	11.69 × 16.53	29.68 × 41.95
a4letter	8.27 × 11.69	20.98 × 29.68
a5	5.85 × 8.26	14.84 × 20.97
b4	9.85 × 13.92	24.99 × 35.32
tabloid	11.00 × 17.00	27.92 × 43.15

Changing the InvertHardcopy Property

Although figures have white lines on a dark background, the hardcopy has dark lines on a white background.

This is due to the figure's 'InvertHardCopy' property which by default is set to on. Lines and text are affected, but not surface and patch shading.

In certain cases it is beneficial to turn 'off' this property, for example when producing output for a Postscript film recorder.

Handle Manipulation

This final section covers:

1. Determining *all* handles
2. Determining *all* properties/values
3. Changing **default** property values
4. Hiding and deleting objects

Determining *all* Handles

figures, axes Most of the time you create figures and axes sequentially, and don't go back to change previous ones.

Hence you can usually get by with the functions gcf and gca for getting the handles of the *current* figure and axes, respectively.

However, if you think you may go back and change the properties of either a figure or a set of axes, then you should save their handles. One way is to issue:

```
plot(...)

fig1 = gcf;
ax1  = gca;
```

Alternatively, you can explicitly create the figure and axes, and then save the handles:

```
       fig1 = figure;

       ax1  = axes;
% or:  ax1  = subplot(...);

       plot(...)
```

text, plot, etc. All high-level plotting functions optionally return the handle of the object created.

```
h = text(...)

h = plot(...)
h = fill(...)

h = surf(...)
h = image(...)
```

It is common to save the handle in a MATLAB variable for subsequent use, for although you can specify options directly, such as:

```
text(x, y, string, 'FontName', 'Times', ...
                   'FontAngle', 'Italic')
```

it is often easier to read and modify the program if you split up lines like this:

```
h = text(x, y, string)
set(h, 'FontName', 'Times')
set(h, 'FontAngle', 'Italic')
```

But ... Suppose you created a graph using interactive commands, and didn't save the handle of an object you want to change?

In this case click on the object, and then issue the command:

```
h = gco
```

where gco stands for *get current object*. It includes figures and axes as well as text, plots, etc.

Note: You should check that you selected the right object before proceeding by checking its 'Type':

```
get(h, 'Type')
```

Notice that the 'Type' of three of the objects is not the same as name of the function we usually use to create them. These are plot, fill, and surf. Their corresponding types are:

High-Level Function:	Type:
plot	'line'
fill	'patch'
surf	'surface'

Line, patch and surface are also the names of low-level functions in MATLAB. They *also* produce the same type of objects, but you generally do not call them directly. Instead you rely on their higher-level counterparts: plot, fill, and surf.

Determining *all* Properties/Values

A handy feature of the get command is that if you pass it only a handle (without a property name)

```
h = text(0.5, 0.5, 'Liquid Phase')

get(h)
```

it displays all the property names and values currently in effect for that object:

```
Color = [0 0 0]
EraseMode = normal
Editing = off
Extent = [0.497696 0.467836 0.177419 0.0526316]
FontAngle = normal
FontName = Helvetica
FontSize = [10]
FontUnits = points
FontWeight = normal
HorizontalAlignment = left
Position = [0.5 0.5 0]
Rotation = [0]
String = Liquid Phase
Units = data
Interpreter = tex
VerticalAlignment = middle

ButtonDownFcn =
Children = []
Clipping = off
CreateFcn =
DeleteFcn =
BusyAction = queue
HandleVisibility = on
Interruptible = on
Parent = [2.00037]
Selected = off
SelectionHighlight = on
Tag =
Type = text
UserData = []
Visible = on
```

Similarly if you issue set with only a handle:

```
set(h)
```

it displays all the properties that you can set. Where there is a list of options, it displays them as well, with the default value inside brace brackets:

```
Color
EraseMode: [ {normal} | background | xor | none ]
Editing: [ on | off ]
FontAngle: [ {normal} | italic | oblique ]
FontName
FontSize
FontUnits: [ inches | centimeters | normalized | {points} | pixels ]
FontWeight: [ light | {normal} | demi | bold ]
HorizontalAlignment: [ {left} | center | right ]
Position
Rotation
String
Units: [ inches | centimeters | normalized | points | pixels | {data} ]
Interpreter: [ {tex} | none ]
VerticalAlignment: [ top | cap | {middle} | baseline | bottom ]
```

```
ButtonDownFcn
Children
Clipping: [ {on} | off ]
CreateFcn
DeleteFcn
BusyAction: [ {queue} | cancel ]
HandleVisibility: [ {on} | callback | off ]
Interruptible: [ {on} | off ]
Parent
Selected: [ on | off ]
SelectionHighlight: [ {on} | off ]
Tag
UserData
Visible: [ {on} | off ]
```

If you provide an output argument when using get or set as above, the results are returned in a structure.

Changing **Default** Property Values

You may find that you are applying the same customizations to objects over and over. A common example is specifying the 'FontName' property for text objects:

```
text(x, y, string, 'FontName', 'Times')
```

To change the default font to 'Times' for all subsequent text objects on *that set of axes*, issue:

```
set(gca, 'DefaultTextFontname', 'Times')
         ^^^
```

If you want 'Times' to be the default for *all* subsequent text objects on *all* axes in *that figure window*, issue:

```
set(gcf, 'DefaultTextFontname', 'Times')
         ^^^
```

Finally, if you want 'Times' to be the default for all subsequent text objects in *all figure windows*, issue:

```
set(0, 'DefaultTextFontname', 'Times')
       ^
```

The following question arises:

> What if you have set the 'DefaultTextFontname' in the root window (0), the current figure (gcf), and the current axes (gca) to different values? *Which font gets used?*

The answer is: whichever object (root window, figure, axes) is above and closest in the object hierarchy to the text objects (see Figure 81). Text objects always come under axes, therefore the gca's 'DefaultTextFontname' setting is used.

288   Graphics Properties

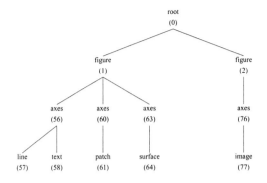

(Not shown are each axes's other children: title, xlabel, ylabel)

FIGURE 81. Hierarchy of graphics objects, using `quads.m` as an example. (Same as Figure 73.)

Another Example

Another common customization is changing the default color of figure windows. For example, to make them all medium-gray, issue:

```
set(0,'DefaultFigureColor', [.5 .5 .5])
```

Note: You must set default properties (such as 'DefaultFigureColor') in an object that (in the hierarchy) is above the new objects to be created.

For example, if you try to set the default in an object on the same level:

```
set(gcf,'DefaultFigureColor', [.5 .5 .5])
```

you will get an error:

```
??? Error using ==> set
Invalid default property for figure
property: 'DefaultFigureColor'.
```

General Method for Changing Default Values

Now compare changing the default value of a property with directly setting a value. For example, let's change the 'MarkerSize' property of a line produced by a call to plot:

```
h = plot(x, y, '-')
set(h, 'Markersize', 12)
```

But, to change the default 'MarkerSize' for subsequent plots, issue one of the following (again depending on the scope needed):

```
set( 0 , 'DefaultLineMarkersize', 12)
```
or:
```
set(gcf, 'DefaultLineMarkersize', 12)
```
or:
```
set(gca, 'DefaultLineMarkersize', 12)
```

Remember, always set the 'Default...' property in an "ancestor" of the objects you will subsequently be creating. Here 0, gcf and gca are all ancestors.

The name of the property is simply the usual property name (e.g. 'Markersize') preceded by the word 'Default' and then the 'Type' of the object(s) you want to affect ('Figure', 'Axes', 'Text', 'Line', 'Patch', 'Surface', 'Image').

Restoring Properties to Default values

Suppose you have explicitly specified a font on a text string:

```
h = text(x, y, string, 'FontName', 'Courier')
```

and then you decide to set it back to the default font. You can do so by specifying 'Default' as the value of the 'FontName' property:

```
set(h, 'FontName', 'Default')
```

Likewise, you can restore all other properties to their default value by specifying the value as the string 'Default'.[71]

Restoring Properties to Factory Default Values

You can also restore a property to its so-called "factory" value, that is, the default value of the property before it was (possibly) overridden.

For example, returning to our first example, we could change the default 'FontName' for text from the factory setting of 'Helvetica' to 'Times', and plot a string:

```
set(0, 'DefaultTextFontname', 'Times')
h = text(x, y, string)
```

and then restore the string to the factory-default font:

```
set(h, 'FontName', 'Factory')
```

How to Remove default Property Values

Consider again the last example:

```
set(0, 'DefaultTextFontname', 'Times')
```

To undo this Default value, simply issue the same command, but specify the value as 'Remove':

```
set(0, 'DefaultTextFontname', 'Remove')
```

---

[71] If you want to reset *all* properties (except 'Position') of an object to their default values, use the reset function. For example: reset(h).

Hiding and Deleting objects

Hiding To hide an object (and any objects below it in its hierarchy), simply set its 'Visible' property value to 'off'. It remains invisible until this property is set back 'on'.

Deleting To delete an object (and any objects below it in its hierarchy), use the delete function [72] and pass it the handle of the object. For example, an oft-used command is:

```
delete(gcf)
```

which deletes the current figure.

Clearing To "clear" an object is to delete all of the objects below it, but not the object itself.

clf clears the current figure, and cla clears the current axes.

Command Listing

AmbientLightColor [num_r, num_g, num_b]
Specifies surrounding light color.
Additional information: A light object must be present for this property to take effect.
See also: set, get, axes

AmbientStrength num
Specifies the strength of the axes ambient light on a patch object.
See also: set, get, AmbientLightColor, patch, surface

AutomaticFileUpdates *on/off*
Specifies whether automatic file updates are to be performed.
See also: set, get, root

hndl = axes($name_1$, $expr_1$, ..., $name_n$, $expr_n$)
Creates an axes object with the n properties specified by $name_1$ through $name_n$ to the values $expr_1$ through $expr_n$, respectively.
Output: A handle to an axes object is assigned to hndl.
Argument options: axes(hndl) to set the current axes to hndl. ♣ hndl = axes to create the axes with all default property values and assign it to handle hndl.
Additional information: Axes objects are the children of figure objects and the parents of specific graphics objects such as line, text, surface, patch, and image objects.

---

[72] The delete function in MATLAB is also used to delete files.

♣ The available properties include *AmbientLightColor*, *Box*, *BusyAction*, *ButtonDownFcn*, *CameraPosition*, *CameraTarget*, *CameraTargetMode*, *CameraUpVector*, *CameraUpVectorMode*, *CameraViewAngle*, *CameraViewAngleMode*, *Children*, *CLim*, *CLimMode*, *Clipping*, *Color*, *ColorOrder*, *CreateFcn*, *DataAspectRatio*, *DataAspectRatioMode*, *DeleteFcn*, *DrawMode*, *FontAngle*, *FontName*, *FontSize*, *FontUnits*, *FontWeight*, *GridLineStyle*, *HandleVisibility*, *Interruptible*, *Layer*, *LineStyleOrder*, *LineWidth*, *NextPlot*, *Parent*, *PlotBoxAspectRatio*, *PlotBoxAspectRatioMode*, *Projection*, *Position*, *Selected*, *SelectionHighlight*, *Tag*, *TickLength*, *TickDir*, *TickDirMode*, *Title*, *Type*, *Units*, *UserData*, *View*, *Visible*, *XAxisLocation*, *XColor*, *XDir*, *XGrid*, *XLabel*, *XLim*, *XLimMode*, *XScale*, *XTick*, *XTickLabel*, *XTickLabelMode*, *XTickMode*, *YAxisLocation*, *YColor*, *YDir*, *YGrid*, *YLabel*, *YLim*, *YLimMode*, *YScale*, *YTick*, *YTickLabels*, *YTickLabelMode*, *YTickMode*, *ZColor*, *ZDir*, *ZGrid*, *ZLabel*, *ZLim*, *ZLimMode*, *ZScale*, *ZTick*, *ZTickLabels*, *ZTickLabelMode*, and *ZTickMode*, See the individual entries for these properties for information on valid and default values. ♣ The values of individual properties can be set or retrieved using set and get, respectively. As well, the axis command provides a more intuitive interface to setting many axes properties. ♣ Until an axes object is applied to a figure (typically by being included in a plot or plot3 command), nothing is displayed.
See also: plot, plot3, subplot, axis, set, get, line, text, image, patch, surface, figure, gca, cla

axis([x_{min}, x_{max}, y_{min}, y_{max}])

Sets the minimum and maximum x- and y-axis values, i.e., the values of properties XLim and YLim, of the current axes.
Output: Not applicable.
Argument options: axis([x_{min}, x_{max}, y_{min}, y_{max}, z_{min}, z_{max}]) to set the x-, y-, and z-axis values of the current axes. ♣ axis('auto') to set the axes scaling mode to automatic for the x-, y-, and z-axis. See XLimMode for more information. ♣ axis(axis) to freeze the scaling for the current axes, forcing subsequent plots to use that scaling if hold is on. ♣ V = axis to assign a four- or six-element vector, containing the scaling values of the current axes, to V. ♣ axis('square') to set the current axes region to be square. This is equivalent to manipulating the Position property. ♣ axis('equal') to specify that the scaling factors and the tic mark increments are equal for each axis. ♣ axis('ij') or axis('xy') to draw the graph in *matrix* coordinate form or *Cartesian* coordinate form, respectively. This is equivalent to altering the property YDir. ♣ axis('on') or axis('off') to turn all axis labeling and tick marks on or off. ♣ axis(str$_1$, str$_2$, str$_3$) to set labeling properties for the x-, y-, and z-axis. str$_1$ sets the tick label mode, and can be 'auto' or 'manual'. str$_2$ sets whether axis labeling and tick marks appear or not, and can be 'on' or 'off'. str$_3$ sets coordinate form is used, and can be 'ij' or 'xy'. ♣ [str$_1$, str$_2$, str$_3$] = axis('state') to return the current state of the tick label mode, the tick mark status, and the coordinate mode.
Additional information: The axis command is only useful when there is a current axes displayed. If you are trying to build a figure from scratch, without displaying it along the way, use the axes command and its associated property variables.
See also: plot, plot3, axes, XLim, XLimMode, YDir, Position

BackFaceLighting *reverselit/unlit/lit*

Specifies how faces are lit when pointing away from the camera.
See also: set, get, patch, surface

BackingStore *on/off*

Specifies whether figure windows get saved in pixel buffer when hidden.
See also: set, get, figure

grid *on*

Adds a box to the current axes.
Output: Not applicable.
Argument options: grid *off* to remove the box from the current axes. This is the default.
✽ grid to toggle the state of the box in the current axes.
See also: title, hidden, text, axes, grid, plot

Box *on/off*

Specifies whether the object is to be enclosed by a box (in two dimensions) or a cube (in three dimensions).
See also: set, get, box, axes

BusyAction *queue/cancel*

Specifies what should happen to events that could interrupt a callback function.
See also: set, get, axes, figure, image, light, line, patch, root, surface, text

clrmap$_{new}$ = **brighten**(num, clrmap)

Brightens or darkens the color map clrmap depending on the value of num.
Output: If $0 <$ num < 1, a brightened color map is assigned to clrmap$_{new}$. If $-1 <$ num < 0, a darkened color map is assigned to clrmap$_{new}$.
Argument options: brighten(num) to alter the current color map. No value is returned and the change is made immediately. ✽ clrmap = brighten(num) to assign to clrmap$_{new}$ the color map created when the current color map is altered.
Additional information: The further the value num is from 0, the greater the effect of brighten.
See also: colormap, rgbplot

ButtonDownFnc *str*

Specifies a string representing a MATLAB expression, value, or function that gets evaluated when the mouse button is pressed down in a graphics object.
Additional information: When the button is pressed, str gets passed to the function eval.
See also: set, get, eval, KeyPressFnc, WindowButtonDownFnc, CallBack, Interruptible, axes, figure, image, light, line, patch, root, surface, text

CallbackObject *hndl*

Contains the handle of the callback object that is currently executing.
Additional information: This property is *read-only*.
See also: get, root

CameraPosition [num$_x$, num$_y$, num$_z$]

Specifies the point from which the axes are viewed.

See also: set, get, CameraPositionMode, CameraTarget, CameraUpVector, CameraViewAngle, axes

CameraPositionMode *auto/manual*
Specifies whether the camera position is adjusted manually or automatically.
See also: set, get, CameraPosition, axes

CameraTarget [num_x, num_y, num_z]
Specifies the point in the axes which the camera is viewing.
See also: set, get, CameraTargetMode, CameraPosition, CameraUpVector, CameraViewAngle, axes

CameraTargetMode *auto/manual*
Specifies whether the camera target is adjusted manually or automatically.
See also: set, get, CameraTarget, axes

CameraUpVector [num_x, num_y, num_z]
Specifies the rotation vector of the camera.
See also: set, get, CameraUpVectorMode CameraPosition, CameraTarget, CameraViewAngle, axes

CameraUpVectorMode *auto/manual*
Specifies whether the camera "up vector" is adjusted manually or automatically.
See also: set, get, CameraUpVector, axes

CameraViewAngle [num_x, num_y, num_z]
Specifies the *field of view* for the camera.
See also: set, get, CameraViewAngleMode, CameraPosition, CameraTarget, CameraUpVector, axes

CameraViewAngleMode [num_x, num_y, num_z]
Specifies whether the camera view angle is adjusted manually or automatically.
See also: set, get, CameraViewAngle, axes

caxis([num_{min}, num_{max}])
Sets the minimum and maximum scaling values for the color axis, i.e., the value of property CLim, of the current axes.
Output: Not applicable.
Argument options: caxis('*auto*') to set the color axis scaling mode to automatic. See CLimMode for more information. ♣ V = caxis to assign a two-element vector, containing the scaling values of the current color axis, to V. ♣ caxis(*caxis*) to specify, when hold is turned on, that the current scaling limits be used for subsequent plots.
Additional information: The caxis command is only useful when there is a current axes displayed. If you are trying to build a figure from scratch, without displaying it along the way, use the axes command and its associated property variables. ♣ See the on-line help file for examples of using caxis.
See also: plot, plot3, axes, CLim, CLimMode, colormap

CData M
Specifies the color matrix that maps elements to a color map.
Additional information: If any elements of M are non-integer, they are rounded with floor. ✽ If any elements of M are out of the range of the color map, they are displayed as transparent. ✽ In the patch command, M is a vector mapping colors to the vertices of the patch. ✽ In the image command, M is a matrix mapping single colors to the elements of the matrix image. ✽ In the surface command, M is a matrix mapping single colors into a texture map for the surface. When using an image as a texture map the size of M and the surface data need not match. See FaceColor for details.
See also: set, get, image, surface, patch, colormap, CDataMapping, FaceColor

CDataMapping *scaled/direct*
Specifies whether indexed colors are interpreted in *scaled* or direct mode.
See also: set, get, image, surface, patch, colormap, CData, FaceColor

Children V$_{hndl}$
Specifies that the handles in vector V$_{hndl}$ are children of the current object.
Additional information: Typically, the handles in V$_{hndl}$ get set automatically when subproperties are defined within another property command such as axes or figure. Alternatively, children of an object can be specified using set.
See also: set, get, Parent, axes, figure, image, light, line, patch, root, surface, text

cla
Clears the current axis by deleting all objects that are its children.
Output: Not applicable.
Argument options: cla *reset* to not only delete the axes, but to reset all plotting values to their defaults, except for the Position property.
Additional information: This command has no effect on what is currently in memory.
✽ To clear the current plot display window, use clf.
See also: clf, reset, hold, clc

clf
Clears the current plot display window by deleting all axes from the current figure.
Output: Not applicable.
Argument options: clf *reset* to not only delete the axes, but to reset all other plotting values to their defaults.
Additional information: This command has no effect on what is currently in memory.
✽ To clear the current axes, use cla.
See also: cla, reset, hold, clc

CLim [num$_{min}$, num$_{max}$]
Specifies that related data values are mapped to the relevant color map from the minimum value of num$_{min}$ to the maximum value of num$_{max}$.
Argument options: CLim([*-inf*, num$_{max}$] or CLim([num$_{min}$, *inf*] to set the maximum or minimum value, respectively, and let the other extreme be automatically scaled.

Additional information: By default, the values of num_{min} and num_{max} are set the first and last entries in the color map, respectively. ✤ If specific values are set to num_{min} and num_{max}, the CLimMode property is affected.
See also: set, get, axes, CLimMode, Color, ColorOrder

CLimMode *auto/manual*
Specifies whether values in CLim should be used to calculate the color mapping limits.
Additional information: If specific values are set in CLim, the CLimMode property is automatically set to *manual*.
See also: set, get, axes, CLim, Color, ColorOrder

Clipping *on/off*
Specifies whether any portion of an object that is outside of the axes rectangle is still displayed.
See also: set, get, Position, axes, figure, image, light, line, patch, root, surface, text

close
Closes the current window.
Output: Not applicable.
Argument options: close hndl to close the window with handle hndl.
Additional information: No confirmation is requested when close is used, so exercise caution.
See also: gcf, figure, CloseRequestFcn, delete

CloseRequestFcn *str*
Specifies a string representing a MATLAB expression, value, or function that gets evaluated when the close command is used on a figure.
Additional information: When the button is pressed, str gets passed to the function eval.
See also: set, get, eval, CallBack, Interruptible, figure

Color [num_r, num_g, num_b]
Specifies that the current object has a color defined by RGB (red, green, blue) values num_r, num_g, and num_b, respectively.
Argument options: Color str to specify one of the preset colors. Some possible values of str are *yellow*, *y*, *magenta*, *m*, *cyan*, *c*, *red*, *r*, *green*, *g*, *blue*, *b*, *white*, *w*, *black*, and *k*.
Additional information: The values of num_r, num_g, and num_b must all be in the range [0, 1]. ✤ For an axes, Color specifies the background color of the axes rectangle, has a default value of the parent figure's background color (which is also set with Color), and can be specified as transparent with *none* (default).
See also: set, get, axes, figure, light, line, text, colormap, colorbar

colorbar()
Adds a vertical colorbar to the right of the current axis.
Output: If no current axis is defined, then an empty one is created.
Argument options: colorbar(*'horiz'*) to add a horizontal colorbar to the current axis.
✤ colorbar(*'vert'*) to add a vertical colorbar to the current axis. ✤ colormap(hndl$_{axis}$)

to add a colorbar to the axes defined by handle hndl$_{axis}$. ✣ hndl = colormap(hndl$_{axis}$) to return the handle of the colorbar associated with the axis defined by hndl$_{axis}$.
Additional information: Colorbars display the range of colors and their associated values for the current colormap. ✣ Calling colorbar with no arguments adds a new colorbar only if none exist for the current axis; otherwise the colorbar is just updated. ✣ Using the *'horiz'* and *'vert'* options, any number of colorbars can be added to the same axes.
See also: colormap, ColorMap

colordef *black*
Set the default axes background color to black.
Output: Not applicable.
Argument options: colordef *white* to set the default to white. ✣ colordef *none* to set the color to the MATLAB 4 default value. ✣ colordef(hndl, str), where str is one of *'black'*, *'white'*, or *'none'*, to change the default for a figure specified with hndl. ✣ hndl = colordef(*'new'*, str), where str is one of *'black'*, *'white'*, or *'none'*, to create a new figure with the specified default.
Additional information: When the background color is changed, the colors of other elements in the figure are adjusted to make the plot look good. ✣ For complete control over colors, use the handles to the figure.
See also: Color, whitebg, white, colormap, graymon, figure

colormap(clrmap)
Sets the current color map to clrmap.
Output: Not applicable.
Argument options: colormap(*'default'*) or colormap(hsv) to set the current color map to the default *hue-saturation-value* model. ✣ colormap(option(posint)) to set the current color map to a preset optional value. Valid options include *hot*, *cool*, *gray*, *autumn*, etc. The value posint defines how many individual shades you want. Default is 64.
✣ clrmap = colormap to assign the current colormap to the variable clrmap.
Additional information: All color maps must be matrices with exactly 3 columns and a number of rows equal to the number of colors in the map. Each row must have exactly three elements between 0 and 1, inclusive, which represent the intensity of red, green, and blue, respectively.
See also: surf, image, pcolor, brighten, hsv, caxis, rgbplot, colorbar

Colormap clrmap
Specifies that the current figure has a color map defined by the three-columned matrix clrmap.
Additional information: See the entry for colormap for details about the elements of clrmap. ✣ Colormap specifies the colors used in surface, image, and patch objects within the figure.
See also: set, get, figure, colormap, colorbar

ColorOrder clrmap
Specifies a default color ordering for lines whose colors are not specified.

Additional information: The colors in clrmap are cycled through from top to bottom.
♣ By default, clrmap contains the predefined colors *blue*, *green*, *red*, *cyan*, *magenta*, *yellow*, and *grey*, in that order. ♣ The ordering in ColorOrder is also used to choose pen colors when outputting a file to a plotter.
See also: set, get, axes, line

clrmap = **contrast**(M)
Creates a gray scale colormap from image M (i.e., a matrix of pointers into the current colormap), such that the various intensity levels are evenly distributed.
Output: A colormap is assigned to clrmap.
Argument options: clrmap = contrast(M, n) to specify that the resulting colormap has n rows.
See also: brighten, gray, colormap

CreateFcn str
Specifies a string representing a MATLAB expression, value, or function that gets evaluated when an object is created.
Additional information: When the button is pressed, str gets passed to the function eval.
See also: set, get, eval, CallBack, Interruptible, axes, figure, image, light, line, patch, root, surface, text

CurrentAxes hndl
Specifies that the axes object with handle hndl is the figure's current axes.
Additional information: If a figure has any axes as children, there will be a current axes. ♣ To retrieve the current axes of a figure, use get or gca. If no current axes exists, get creates one.
See also: set, get, figure, gca

CurrentCharacter char
Specifies the character char is the last key from the keyboard pressed while in a figure window.
Additional information: You cannot implicitly set the CurrentCharacter property.
See also: get, figure, CurrentPoint, CurrentMenu

CurrentFigure hndl
Specifies the current figure window.
Argument options: get(0, 'CurrentFigure') to create a current figure when none is specified.
Additional information: CurrentFigure is a property of the *root object*, which has a handle of 0.
See also: set, get, gcf, root, figure

CurrentObject hndl
Specifies that the object with handle hndl is under the current point.
Additional information: The topmost object in the stacking order is returned.
See also: set, get, figure, CurrentPoint

CurrentPoint [num$_x$, num$_y$]

Specifies that the location of the last mouse press (or release) in the current figure or axes was at [num$_x$, num$_y$].

Additional information: The distances are measured from the bottom-left corner of the figure window or axes area. The units used in the measurement are set by the Units property.

See also: set, get, figure, axes, Position, CurrentObject

DataAspectRatio [num$_x$, num$_y$, num$_z$]

Specifies the relative scaling of the three axes.

See also: set, get, DataAspectRatioMode, axes

DataAspectRatioMode *auto/manual*

Specifies whether the scaling of axes is automatic or manual.

See also: set, get, DataAspectRatio, axes

datetick(str, int)

Specifies that the str-axis tickmarks be labelled with the date string format int.

Output: Not applicable.

Additional information: str must be one of 'x', 'y', or 'z'. Default is 'x'. ✦ The data for the specified axes must be in serial date form. See the datenum command for more details. ✦ See the datestr command for more details on int.

See also: *datestr, datenum*, XTick, YTick, ZTick

DeleteFcn str

Specifies a string representing a MATLAB expression, value, or function that gets evaluated when an object is deleted.

Additional information: When the button is pressed, str gets passed to the function eval.

See also: set, get, eval, CallBack, Interruptible, axes, figure, image, light, line, patch, root, surface, text

Diary *on/off*

Specifies whether MATLAB keeps a record of input and output statements.

Additional information: The name of the file in which the diary is set is specified with the DiaryFile property.

See also: DiaryFile, diary, set, get, root

DiaryFile filename

Specifies that when MATLAB keeps a dairy, it writes to file filename.

Additional information: Turn recording on and off with the diary command or the Diary property.

See also: Diary, diary, set, get, root

DiffuseStrength num

Specifies the amount of reflection of diffuse light(s).

Additional information: num must be between 0 and 1.
See also: set, get, light, patch, surface

Dithermap *clrmap*
Specifies that the current figure has a dither map defined by the three-columned matrix clrmap.
Additional information: Dither maps are used to specify true-color data on pseudo-color displays.
See also: set, get, DithermapMode, figure, colormap

DithermapMode *auto/manual*
Specifies whether dither map creation is automatic or manual.
See also: set, get, Dithermap, figure

DrawMode *normal/fast*
Specifies the speed of three-dimensional rendering.
Additional information: While the *normal* option gives a truer picture, the *fast* option results in quicker drawing.
See also: set, get, axes

drawnow
Flushes the event queue and forces the completion of any pending drawing.
Output: Not applicable.
Argument options: drawnow(*'discard'*) to do the opposite action, i.e., remove all pending event from the queue.
Additional information: If there are any updates to the screen to be made, they are done. ✤ This command is only necessary when reading in an M-file, because typically the screen is not updated until the last statement is executed.
See also: pause, getframe

Echo *on/off*
Specifies whether MATLAB displays every line of script files as they execute.
See also: echo, set, get, root

EdgeColor [num_r, num_g, num_b]
Specifies that the current object has edge color defined by RGB (red, green, blue) values num_r, num_g, and num_b, respectively.
Argument options: EdgeColor *flat* to specify edges be colored with the average of the patch's color axis data. ✤ EdgeColor *interp* to specify the edge color be interpolated from the vertex values in CData. ✤ EdgeColor *none* to specify that edges not be drawn.
Additional information: By default, the edge color is black.
See also: set, get, surface, patch, FaceColor, Color, colormap

EdgeLighting *none/flat/gourand/phong*
Specifies the algorithm used to calculate the effect of light(s) on the edges of an object.
See also: set, get, surface, patch, EdgeColor, light

Editing on/off
Specifies whether a text object can be edited.
See also: set, get, text

EraseMode normal/background/xor/none
Specifies how an object is to be drawn, with regards to the screen color beneath it, when drawing or erasing.
Additional information: There are various advantages and disadvantages to each erase mode. See the on-line help file for more details.
See also: set, get, image, line, patch, surface, text, Visible

ErrorMessage str
Returns the last error message produced by MATLAB.
See also: get, root

Extent [num_l, num_b, num_w, num_h]
Specifies the bounding rectangle of an object with left and bottom coordinates of num_l and num_b, and width and height of num_w and num_h.
Additional information: The specifications are in the units specified by the Units property of the object.
See also: set, get, text, Units

FaceColor [num_r, num_g, num_b]
Specifies that the current object has face color defined by RGB (red, green, blue) values num_r, num_g, and num_b, respectively.
Argument options: FaceColor *flat* to specify face(s) be colored a single value taken from CData. ✽ FaceColor *interp* to specify the face color(s) be interpolated from the values in CData. ✽ FaceColor *texturemap* to specify the faces be colored according to the image contained in CData. The size of the matrix in CData does not have to correspond to the data for the surface—an appropriate mapping is done. FaceColor must be set before CData is changed. ✽ FaceColor *none* to specify that the face(s) not be filled with any color.
Additional information: For special face color mappings, you might want to set the EdgeColor property to *none* to make these lines disappear.
See also: set, get, surface, patch, EdgeColor, FaceLighting, CData, Color, colormap

FaceLighting none/flat/gourand/phong
Specifies the algorithm used to calculate the effect of light(s) on the faces of an object.
See also: set, get, surface, patch, FaceColor, light

Faces M
Specifies which vertices in the Vertices property are connected.
Additional information: Each column of M denotes a face and the rows specify the connectivity. See the on-line help for an explanation of M.
See also: set, get, Vertices, FaceVertexCData, patch

FaceVertexCData M
Specifies the color of faces defined using the Faces and Vertices properties.
Additional information: See the on-line help for an explanation of M for different types of coloring.
See also: set, get, Vertices, Faces, patch

hndl = **figure**(name$_1$, expr$_1$, ..., name$_n$, expr$_n$)
Creates a figure object with the n properties specified by name$_1$ through name$_n$ to the values expr$_1$ through expr$_n$, respectively.
Output: A handle to a figure object is assigned to hndl and a graph window is opened.
Argument options: figure(hndl) to set the current figure to hndl. ✦ hndl = figure to create and open the figure with all default property values and assign it to handle hndl.
Additional information: Figure objects are the children of the root object (i.e., 0) and the parents of specific graphics objects such as axes, uimenus, and uicontrols objects. ✦ The available properties include *BackingStore*, *BusyAction*, *ButtonDownFcn*, *Children*, *Clipping*, *CloseRequestFcn*, *Color*, *Colormap*, *CreateFcn*, *CurrentAxes*, *CurrentCharacter*, *CurrentObject*, *CurrentPoint*, *DeleteFcn*, *Dithermap*, *DithermapMode*, *FixedColors*, *HandleVisibility*, *IntegerHandle*, *Interruptible*, *InvertHardCopy*, *KeyPressFcn*, *MenuBar*, *MinColormap*, *Name*, *NextPlot*, *NumberTitle*, *PaperOrientation*, *PaperPosition*, *PaperPositionMode*, *PaperSize*, *PaperType*, *PaperUnits*, *Parent*, *Pointer*, *PointerShapeCData*, *PointerShapeHotSpot*, *Position*, *Renderer*, *RendererMode*, *Resize*, *ResizeFcn*, *Selected*, *SelectionHighlight*, *SelectionType*, *ShareColors*, *Tag*, *Type*, *Units*, *UserData*, *Visible*, *WindowButtonDownFnc*, *WindowButtonMotionFnc*, *WindowButtonUpFnc*, and *WindowStyle*. See the individual entries for these properties for information on valid values. ✦ The values of individual properties can be set or retrieved using set and get, respectively.
See also: plot, plot3, subplot, axis, set, get, line, text, image, patch, surface, figure, gcf, clf

FixedColors clrmap
Specifies that the current figure has fixed colors defined by the three-columned matrix clrmap.
Additional information: Originally, this color map contains only the colors black and white. Any subsequent colors that are specifically declared for the figure are added to the list, and any that are removed from the figure are removed from the list.
See also: get, figure, Colormap

FontAngle *normal/italic/oblique*
Specifies the angle of the font to be used with a text-based object.
Additional information: Different types of text-based objects have different default font angles.
See also: set, get, axes, text, uicontrol, FontSize, FontWeight, FontName, FontUnits

FontName str
Specifies the name of the font to be used with a text-based object.
Additional information: Different systems have different fonts available. ✦ Different types of text-based objects have different default fonts.
See also: set, get, axes, text, uicontrol, FontSize, FontWeight, FontAngle, FontUnits

FontSize n
Specifies the font size to be used with a text-based object.
Additional information: The specifications are in units set by FontUnits. ♣ Different types of text-based objects have different default font sizes.
See also: set, get, text, uicontrol, FontName, FontWeight, FontAngle, FontUnits

FontUnits *points/normalized/inches/centimeters/pixels*
Specifies which unit are used to measure text-based object.
See also: set, get, axes, text, uicontrol, FontSize, FontWeight, FontName, FontAngle

FontWeight *light/normal/demi/bold*
Specifies the weight of the font to be used with a text-based object.
Additional information: Different types of text-based objects have different default weights.
See also: set, get, text, uicontrol, FontSize, FontWeight, FontName, FontUnits

Format *short/shortE/long/longE/bank/hex/+/rat*
Specifies the default output format property for numerical values.
Additional information: See the format command for more details. ♣ Format is a property of the *root object*, which has a handle of 0.
See also: format, FormatSpacing, set, get, root

FormatSpacing *compact/loose*
Specifies whether MATLAB eliminates excess line feeds when displaying values.
Additional information: The format command can also be used to set spacing. ♣ FormatSpacing is a property of the *root object*, which has a handle of 0.
See also: format, Format, set, get, root

hndl = gca
Returns the handle to the current set of axes.
Output: A handle is assigned to hndl.
Additional information: Changing the current figure changes the current axes. ♣ Use the command axes to change the current axes. ♣ If no axes exist, one is automatically created.
See also: axes, subplot, gcf, gco, get, cla, hold

hndl = gcf
Returns the handle to the current figure.
Output: A handle is assigned to hndl.
Additional information: Each figure has a set of axes, which can be accessed with gca, associated with it. ♣ Use the command figure to change the current figure.
See also: figure, subplot, gca, gco, get, clf, delete, close

hndl = gco
Returns the handle to the current object in the current figure.
Output: A handle is assigned to hndl.

Argument options: hndl = gco(hndl$_{fig}$) to find the current object in the figure with handle hndl$_{fig}$.
Additional information: The current object in a figure is the last object clicked on with the mouse.
See also: figure, gca, gcf

get(hndl)
Lists all properties associated with the object with handle hndl.
Output: A list of equations with strings on the left-hand side is displayed. No value is assigned.
Argument options: V = get(hndl, str), where str is the name of one of the properties held by object hndl, to assign the value of that property to V. In most cases, a this result is a vector. ♣ get(hndl, 'Default') to list the default values for components of the object hndl. ♣ V = get(0, str), where str is a special string starting with Default and ending with a property name, such as DefaultEdgeColor, to assign the default value of that property to vector V.
Additional information: get is particularly useful when examining objects with many components or properties, such as figures. ♣ get is not recursive, it only examines subobjects to a depth of *one* level. ♣ If a matrix is too large to display conveniently, get returns the string *(too many rows)*. ♣ Use the command set to set object properties.
See also: set, gca, gcf

[V$_x$, V$_y$] = ginput(n)
Allows the user to input n points from the current figure by using a mouse or the arrow keys.
Output: Two column vectors representing the x and y values of each point selected are assigned to V$_x$ and V$_y$.
Argument options: [V$_x$, V$_y$] = ginput continues to gather points until the *Return* key is struck. ♣ [V$_x$, V$_y$, V$_b$] = ginput(n) or [V$_x$, V$_y$, V$_b$] = ginput to assign a vector of corresponding mouse/button values to V$_b$. If a keyboard button was pressed, the ASCII value of that key is returned. If a mouse button was pressed, its number, from the left, is returned.
Additional information: To specify a point, use the mouse or arrows keys to position the cursor and then press either a key or a mouse button. ♣ This command is particularly useful in allowing intuitive interaction with figures. ♣ This command is meant to be used with two-dimensional plots.
See also: plot, gtext

graymon
Toggles the default properties between values appropriate for gray-scale and color monitors.
Output: Not applicable.
Additional information: graymon cannot be used to affect individual windows, only the entire session. ♣ If used in the middle of a session, previously created windows are *not* affected.
See also: contrast, whitebg, colormap, figure

grid *on*
Adds grid lines to the current axes.
Output: Not applicable.
Argument options: grid *off* to remove grid lines from the current axes. This is the default. ✦ grid to toggle the state of grid lines in the current axes.
Additional information: The grid lines are always drawn out from the current axis tickmark locations.
See also: title, hidden, text, axes, box, plot

GridLineStyle *str*
Specifies that the current axes has a grid line style defined format string str.
Additional information: Valid options for str include - for a solid line (default value), -- for a dashed line, : for a dotted line, and -. for a dash-dot line, and none. ✦ To specify whether grid lines are drawn at all, use *grid*, *XGrid*, or YGrid.
See also: set, get, LineStyle, grid, XGrid, YGrid, axes

gtext(*str*)
Allows the user to specify placement of the string str on the current two-dimensional figure.
Output: Not applicable.
Additional information: To specify a placement, use the mouse or arrows keys to position the cursor and then press either a key or a mouse button. ✦ This command is particularly useful in labelling special points on a plot.
See also: text, ginput

HandleVisibility *on/off/callback*
Specifies whether an object's handles are visible to the command-line user.
Additional information: The callback option allows handles to be visible to callback functions, but not from command-line functions.
See also: set, get, ShowHiddenHandles, axes, figure, image, light, line, patch, root, surface, text

hidden *off*
Turns off hidden line removal for mesh plots.
Output: Not applicable.
Argument options: hidden *on* to turn on hidden line removal. This is the default. ✦ hidden to toggle the state of hidden line removal in the current mesh plot.
Additional information: Hidden line removal simply removes any lines that lie behind another part of the mesh from the current viewing angle.
See also: shading, grid, mesh

hold *off*
Turns off the holding of axes properties over subsequent plots.
Output: Not applicable.
Argument options: hold *on* to turn on holding of axes properties over subsequent plots. This is the default. ✦ hold to toggle the state of holding of axes properties.

Additional information: If hold is *off*, then all axes properties are reset for each new plot drawn.
See also: ishold, axis, cla

HorizontalAlignment *left/center/right*
Specifies the horizontal justification of a text object with respect to the text location point.
See also: set, get, text, Position, VerticalAlignment

colormap = **hsv**(posint)
Creates a color map formed by cycling hue-saturation-values.
Output: A posint ×3 color map matrix is returned.
Argument options: hsv to return a color map with a default 64 rows.
Additional information: The color map starts off at red, moves through yellow, green, cyan, blue, and magenta, then ends with red again. ✤ Each row has exactly three elements between 0 and 1, inclusive, which represent the intensity of red, green, and blue, respectively.
See also: colormap, prism, white, contrast, hsv2rgb, rgb2hsv

colormap$_{rgb}$ = **hsv2rgb**(colormap$_{hsv}$)
Converts a color map defined by hue-saturation-value values to one defined by red-green-blue values.
Output: A color map is returned.
Additional information: The three columns of the input matrix represent hue, saturation, and value, respectively. The three columns of the output represent red, green, and blue values, respectively. ✤ All values in both types of color maps are between 0 and 1, inclusive. ✤ For more information on the meaning of different values, see the help page for hsv2rgb.
See also: colormap, hsv, brighten, rgb2hsv

hndl = **image**(name$_1$, expr$_1$, ..., name$_n$, expr$_n$)
Creates a two-dimensional image object with the n properties specified by name$_1$ through name$_n$ to the values expr$_1$ through expr$_n$, respectively.
Output: A handle to an image object is assigned to hndl.
Argument options: image(M) to create and display an image file contained in matrix M. ✤ hndl = image(M) or hndl = image(V$_x$, V$_y$, M) to assign an image of M to the handle hndl.
Additional information: An image object is a direct mapping of the elements of a matrix to squares filled with the corresponding colors from a figure's color map.
✤ If any elements of M are non-integer, they are rounded with floor. ✤ If any elements of M are out of the range of the color map, they are displayed as transparent.
✤ The available properties include *BusyAction*, *ButtonDownFcn*, *CData*, *CDataMapping*, *Children*, *Clipping*, *CreateFcn*, *DeleteFcn*, *EraseMode*, *HandleVisibility*, *Interruptible*, *Parent*, *Selected*, *SelectionHighlight*, *Tag*, *Type*, *UserData*, *Visible*, *XData*, and *YData*. See the individual entries for these properties for information on valid values. ✤ The values of individual properties can be set or retrieved using set and get, respectively.
See also: plot, pcolor, axes, colormap, imagesc, floor

imagesc(M)

Displays a two dimensional image, contained in matrix M, whose colors are scaled to use the entire current color map.
Output: A two-dimensional plot is displayed.
Argument options: imagesc(M, [num_{low}, num_{high}]) to restrict the scaling between the color values num_{low} and num_{high}.
Additional information: Using the imagesc command affects the UserData property of the figure.
See also: image, colormap, colorbar

IntegerHandle *on/off*

Specifies whether object handles should be integers (*on*) or real numbers (*off*).
See also: set, get, figure

Interpreter *tex/none*

Specifies whether special character combinations are treated as TEX code or not.
See also: set, get, String, text

Interruptible *on/off*

Specifies whether a callback action for a graphics object can be interrupted.
See also: set, get, ButtonDownFnc, BusyAction, CreateFcn, DeleteFcn

InvertHardCopy *on/off*

Specifies whether black areas in figures be printed as black, and white as white.
Additional information: Most MATLAB figures have a larger proportion of black areas to white areas. When printed as is, these figures can cause serious strain on your printer. ✤ When figures containing colors are printed, the colors are mapped to black or white, depending on which looks better.
See also: title, hidden, figure

num = ishold

Determines whether hold is on.
Output: If hold is on, 1 is assigned to num. If hold is off, 0 is assigned to num.
Additional information: If hold is *off*, then all axes properties are reset for each new plot drawn.
See also: ishold, axis, cla

KeyPressFnc *str*

Specifies a string representing a MATLAB expression, value, or function that gets evaluated when a keyboard key is pressed down when a window has focus.
Additional information: When a key is pressed, str gets passed to the function eval.
See also: set, get, eval, ButtonDownFnc, CallBack, figure

Layer *bottom/top*

Specifies whether axis lines and tickmarks get drawn above or below children of Axes objects.
See also: set, get, axes

legend(str$_{l1}$, ..., str$_{ln}$)
Creates a legend on the current plot using str$_{l1}$ through str$_{ln}$ as the labelling strings.
Output: Not applicable.
Argument options: legend(str$_{t1}$, str$_{l1}$, ..., str$_{tn}$, str$_{ln}$) specify that line style str$_{li}$ be used in conjunction with label str$_{li}$. ✤ legend(hndl, str$_{l1}$, ..., str$_{ln}$) to assign the legend to the plot with handle hndl. ✤ legend(M$_{str}$) to use a string matrix. ✤ legend(V$_{hndl}$, M$_{str}$) to use a vector containing handles to line structures. ✤ legend(str$_{l1}$, ..., str$_{ln}$, int) to control where the legend gets placed. Less than int points can be covered by the legend. If −1 is provided, then the legend is forced to the outside of the plot. ✤ legend(off) to remove a legend from the current plot.
Additional information: Legends can be moved around on the plot with the mouse. See the on-line help file for further directions.
See also: plot

hndl = light(name$_1$, expr$_1$, ..., name$_n$, expr$_n$)
Creates a light object with the n properties specified by name$_1$ through name$_n$ to the values expr$_1$ through expr$_n$, respectively.
Output: A handle to a light object is assigned to hndl.
Additional information: The available properties include *BusyAction*, *ButtonDownFcn*, *Children*, *Clipping*, *Color*, *CreateFcn*, *DeleteFcn*, *HandleVisibility*, *Interruptible*, *Parent*, *Position*, *Selected*, *SelectionHighlight*, *Style*, *Tag*, *Type*, *UserData*, *Visible*. See the individual entries for these properties for information on valid values. ✤ The values of individual properties can be set or retrieved using set and get, respectively. ✤ The effect of a light object is modified by properties of other objects, such as *AmbientLight*, *SpecularColorReflectance*, *EdgeLighting*.
See also: plot, plot3, text

hndl = line(name$_1$, expr$_1$, ..., name$_n$, expr$_n$)
Creates a line object with the n properties specified by name$_1$ through name$_n$ to the values expr$_1$ through expr$_n$, respectively.
Output: A handle to a line object is assigned to hndl.
Argument options: line(V$_x$, V$_y$) to add the line corresponding to vectors V$_x$ and V$_y$ to the current two-dimensional figure. ✤ line(V$_x$, V$_y$, V$_z$) to add the line corresponding to vectors V$_x$, V$_y$, and V$_z$ to the current three-dimensional figure. ✤ hndl = line(V$_x$, V$_y$) or hndl = line(V$_x$, V$_y$, V$_z$) to assign a line to the handle hndl.
Additional information: The available properties include *BusyAction*, *ButtonDownFcn*, *Children*, *Clipping*, *Color*, *CreateFcn*, *DeleteFcn*, *EraseMode*, *HandleVisibility*, *Interruptible*, *LineStyle*, *LineWidth*, *Marker*, *MarkerSize*, *MarkerEdgeColor*, *MarkerFaceColor*, *Parent*, *Selected*, *SelectionHighlight*, *Tag*, *Type*, *UserData*, *Visible*, *XData*, *YData*, and *ZData*. See the individual entries for these properties for information on valid values. ✤ The values of individual properties can be set or retrieved using set and get, respectively.
✤ Until a line object is applied to a figure (typically by being included in a plot or plot3 command), nothing is displayed. ✤ For more information on how to specify line coordinates through the use of vectors and matrices, see the entries for plot and plot3.
See also: plot, plot3, text

LineStyle str
Specifies that the current object has a line style defined by string str.
Additional information: Valid options for str include - for a solid line (default value), -- for a dashed line, : for a dotted line, -. for a dash-dot line, none.
See also: set, get, line, patch, surface, Marker, LineWidth, MarkerSize

LineStyleOrder [str_1; ...; str_n]
Specifies a default ordering for lines styles used in an axes object.
Additional information: Each string must have exactly the same number of characters. This can be achieved using trailing blanks. ♣ The default line style ordering is to use all solid lines.
See also: set, get, axes, LineStyle

LineWidth num
Specifies that the current object has a line width of num points.
See also: set, get, axes, line, patch, surface, LineStyle

Marker str
Specifies that the current object has a marker defined by string str.
Additional information: Valid options for str include +, o for circles, *, ., x, square, diamond, ˆ for up-pointing triangle, v for down-pointing triangle, > for right-pointing triangle, < left up-pointing triangle, ˆ for up-pointing triangle, pentagram, hexagram, and none.
See also: set, get, line, patch, surface, MarkerSize

MarkerEdgeColor [num_r, num_g, num_b]
Specifies the edge color for a marker.
Additional information: The options auto (same as color property) and none (invisible) can also be specified.
See also: set, get, line, patch, surface, Color, Marker, MarkerFaceColor

MarkerFaceColor [num_r, num_g, num_b]
Specifies the face color for a marker.
Additional information: The options auto (same as color property) and none (transparent) can also be specified.
See also: set, get, line, patch, surface, Color, Marker, MarkerEdgeColor

MarkerSize num
Specifies the size of *markers* used in graphics objects.
Additional information: The default value is 6 points, each point equaling 1/72 of an inch.
See also: set, get, line, patch, surface, Marker

MenuBar none/figure
Specifies whether the menu bars that typically appear at the top of figure windows are suppressed.

MeshStyle both/row/column
Specifies which lines to draw on meshed object.
See also: set, get, surface, LineStyle, EdgeColor

MinColormap num
Specifies the minimum number of colors used to store a colormap.
See also: set, get, figure, colormap, Colormap

Name str
Sets a figure's title (in the title bar of a figure window), to str.
Additional information: The default string is the empty string. ♣ Unless the NumberTitle property is set to *off*, all figure window titles automatically begin with *Figure No. ###*, where ### is the figure's handle number.
See also: set, get, figure, NumberTitle

hndl = newplot
Determines, given the state of certain graphics properties, the handle to be used for a new plot axes.
Output: An axes handle is assigned to hndl.
Additional information: This command is meant to be used before commencing to use low-level graphical object creation commands. ♣ Depending on the value of NextPlot for the current figure and axes objects, a figure is either created or cleared and an axes is either created or cleared.
See also: axes, figure, hold, ishold

NextPlot add/replace/replacechildren
Specifies whether the next plot is added to the same set of axes or the same figure.
Additional information: This property is typically queried by higher-level commands such as plot, plot3, mesh, etc. ♣ The value of hold also affects NextPlot.
See also: set, get, hold, axes, figure

NormalMode auto/manual
Specifies whether vertex normals are automatically or manually generated.
See also: set, get, VertexNormals, patch, surface

NumberTitle on/off
Specifies whether the string *Figure No. ###*, where ### represents the figure number, is automatically prepended to the figure window title.
See also: get, set, figure, Name

orient landscape
Sets the orientation of printouts paper to landscape mode.
Output: Not applicable.

Argument options: orient *portrait* to set paper orientation to portrait mode. This is the default. ❋ orient *tall* to remove the automatic margins and print over the entire page.
Additional information: When more is enabled, the space key displays the next page, the Return key displays the next line, and the q key terminates display of the output. ❋ Paging of output works similarly for results of computations and displays of information and/or help files.
See also: print, PaperOrientation

PaperOrientation *portrait/landscape*
Specifies whether printed figures are to be printed in portrait or landscape mode.
See also: get, set, figure, PaperPosition, PaperSize, PaperType, PaperUnits, InvertHardCopy

PaperPosition [num_l, num_b, num_w, num_h]
Sets the location of a figure on the printed page.
Additional information: The value num_l determines the distance from the left side of the paper to the figure. The value of num_b determines the distance from the bottom of the page to the figure. ❋ The values num_w and num_h determine the width and height of the figure on the page. ❋ The distances represented depend on the value of PaperUnits.
See also: get, set, figure, PaperOrientation, PaperSize, PaperType, PaperUnits, InvertHardCopy

PaperPositionMode *auto/manual*
Specifies whether the PaperPosition property is used.
Additional information: If *auto* is specified, the printout is the same size as on the screen, centered on the page.
See also: set, get, PaperPosition

PaperSize [num_w, num_h]
Represents the width and height of the current PaperType in PaperUnits.
Additional information: You cannot implicitly set the values of num_w and num_h; they are defined by the values of PaperType and PaperUnits.
See also: get, figure, PaperOrientation, PaperPosition, PaperType, PaperUnits, InvertHardCopy

PaperType *usletter/uslegal/a3/a4/a5/b4/tabloid*
Specifies the type of paper being printed on.
See also: get, set, figure, PaperOrientation, PaperPosition, PaperSize, PaperUnits, InvertHardCopy

PaperUnits *normalized/inches/centimeters/points*
Sets the printing paper measurement units.
See also: get, set, figure, PaperPosition, PaperSize, PaperUnits

Parent hndl
Specifies that the handle hndl is the parent of the current object.

Additional information: You cannot implicitly set the parent of an object.
See also: get, line, image, surface, patch, axes, root, figure, text, Children

hndl = **patch**(name$_1$, expr$_1$, ..., name$_n$, expr$_n$)

Creates a filled polygon object with the n properties specified by name$_1$ through name$_n$ to the values expr$_1$ through expr$_n$, respectively.
Output: A handle to a patch object is assigned to hndl.
Argument options: patch(V$_x$, V$_y$, str) to add to the current two-dimensional figure the polygon whose vertices are defined by corresponding elements of vectors V$_x$ and V$_y$ filled with the color specified by string str. See the entry for Color for a listing of valid color strings. ✤ patch(V$_x$, V$_y$, n) to fill the inside of the polygon with the color corresponding to the nth. row in the current color map. ✤ patch(V$_x$, V$_y$, V$_c$), where V$_c$ is a vector of scalars of equal length to V$_x$ and V$_y$, to specify that the vertices are set to the corresponding color values and the polygon is filled with an interpolation of those colors. ✤ patch(V$_x$, V$_y$, V$_z$, str) to add the filled polygon corresponding to vectors V$_x$, V$_y$, and V$_z$ to the current three-dimensional figure. ✤ patch(M$_x$, M$_y$, M$_c$) or patch(M$_x$, M$_y$, M$_z$, M$_c$) to create a patch for each column of the vertex matrices. For more information on the effect of M$_c$, see the on-line help page. ✤ hndl = patch(V$_x$, V$_y$, str) or hndl = patch(V$_x$, V$_y$, V$_z$, str) to assign a patch to the handle hndl.
Additional information: If you are creating a figure with only patches, use fill or fill3. ✤ Polygons can be concave or self-intersecting. ✤ Unclosed polygons are automatically closed. ✤ The available properties include *AmbientStrength*, *BackFaceLighting*, *BusyAction*, *ButtonDownFcn*, *CData*, *CDataMapping*, *Children*, *Clipping*, *CreateFcn*, *DeleteFcn*, *DiffuseStrength*, *FaceVertexCData*, *EdgeColor*, *EdgeLighting*, *EraseMode*, *FaceColor*, *FaceLighting*, *Faces*, *HandleVisibility*, *Interruptible*, *LineStyle*, *LineWidth*, *Marker*, *MarkerEdgeColor*, *MarkerFaceColor*, *MarkerSize*, *NormalMode*, *Parent*, *Selected*, *SelectionHighlight*, *SpecularStrength*, *SpecularExponent*, *SpecularColorReflectance*, *Tag*, *Type*, *UserData*, *VertexNormals*, *Vertices*, *Visible*, *XData*, *YData*, and *ZData*. See the individual entries for these properties for information on valid values. ✤ The values of individual properties can be set or retrieved using set and get, respectively. ✤ Until a patch object is applied to a figure (typically by being included in a plot or plot3 command), nothing is displayed.
See also: get, set, plot, plot3, fill, fill3, line

PlotBoxAspectRatio [num$_x$, num$_y$, num$_z$]

Specifies the relative scaling of the plot box, which encloses the axes.
See also: get, set, DataAspectRatio, XLimMode, YLimMode, ZLimMode axes

PlotBoxAspectRatioMode *auto/manual*

Specifies whether the scaling of the plot box is done automatically or manually.
See also: get, set, PlotBoxAspectRatio, axes

Pointer *crosshair/arrow/watch/topl/topr/botl/botr/circle/cross/fleur/left/right/top/bottom/fullcrosshair/ibeam/custom*

Specifies the which pointer icon to use in a figure.
See also: get, set, PointerLocation, PointerShapeCData, PointerShapeHotSpot, PointerWindow, figure

PointerLocation [num$_x$, num$_y$]
Specifies the current location of the pointer on the screen.
Additional information: The measurement expresses distance from the lower-left corner of the screen. ✸ PointerLocation is a property of the *root object*, which has a handle of 0. ✸ The values num$_x$ and num$_y$ are read only, and can be accessed with get.
See also: get, Pointer, PointerShapeCData, PointerShapeHotSpot, PointerWindow, Units, root

PointerShapeCData M
Specifies a user-defined 16 × 16 pixel pointer icon.
Additional information: The Pointer property must be set to *custom* to activate PointerShapeCData. ✸ In M, values of 1 represent black pixels, values of 2 represent white pixels, and values of NaN represent transparent pixels.
See also: get, set, Pointer, PointerLocation, PointerShapeHotSpot, PointerWindow, figure

PointerShapeHotSpot [num$_r$, num$_c$]
Specifies the row and column indices for the "hot spot" in a user-defined pointer icon.
Additional information: The Pointer property must be set to *custom* to activate PointerShapeHotSpot.
See also: get, set, Pointer, PointerLocation, PointerShapeCData, PointerWindow, figure

PointerWindow hndl
Specifies the handle of the MATLAB window in which the pointer currently is located.
Additional information: If the pointer is not in a MATLAB window, the handle is 0. ✸ PointerWindow is a property of the *root object*, which has a handle of 0.
See also: get, PointerLocation, root

Position [num$_l$, num$_b$, num$_w$, num$_h$]
Specifies the position of an object with left and bottom coordinates of num$_l$ and num$_b$, and width and height of num$_w$ and num$_h$.
Additional information: The specifications are in the units specified by the Units property of the object. ✸ For light objects, Position takes a three element vector, representing a spot in three-space. ✸ For text objects, Position takes a two or three element vector, representing a spot in two- or three-space.
See also: set, get, axes, figure, light, text, Units

print -*d*str$_{dev}$ filename
Prints the contents of the current figure window in the format for device str$_{dev}$ to the file named filename.
Argument options: print to print the figure window directly to the default printer. ✸ print -*d*str$_{dev}$ -option$_1$, -option$_2$, ..., -option$_n$, filename to print according to options option$_1$ through option$_n$. -*append* appends the file to filename, rather than overwriting it. -*epsi* adds facilities for previewing an EPSF file. -*ocmyk* outputs PostScript files in "cyan/magenta/yellow/black" four color separation instead of the standard RGB three color separation. -*f*hndl prints the figure window specified by handle *hndl*. -*f*name

prints the SIMULINK model window specified by *name*. Other optional parameters that apply only to specific platforms are available. See the on-line help for details.
Additional information: For a complete list of printer types and file format supported, see the on-line help for print. ✤ Normally, MATLAB switches black backgrounds to white and white lines to black when printing a figure window. To control this behaviour, set the InvertHardCopy property to *off*. ✤ Any user interface controls or menus within a figure window are *not* printed with print. ✤ Choosing certain device types will cause an appropriate extension to be automatically added to filename (e.g., .hgl for -*dhpgl*). ✤ To change the default printing values for your system, see the entry for printopt.
See also: printopt, orient, figure

[str$_{cmd}$, str$_{dev}$] = **printopt**
Returns the current default print command and device type for your system.
Additional information: The values returned by this command can be altered by editing the M-file, `printopt.m`.
See also: print

colormap = **prism(posint)**
Creates a color map formed by cycling the six colors: red, orange, yellow, green, blue, and violet.
Output: A posint ×3 color map matrix is returned.
Argument options: colormap = prism to return a color map with a default number of rows equal to the number of rows of the current colormap. ✤ prism to redraw any lines in the current figure in varying colors of the prism. This is particularly useful for contour plots.
Additional information: There is no smooth blending between the six colors.
See also: colormap, hsv, white, contrast

Profile *on/off*
Specifies whether the files in ProfileFile are profiled.
See also: get, set, ProfileFile, ProfileInterval, root

ProfileFile *filename*
Specifies an M-file to profile.
Additional information: Profile must be turned *on* for this property to take effect.
See also: get, set, Profile, ProfileInterval, root

ProfileInterval V
Specifies how many times each line of code was encountered in a profiled function.
Additional information: Profile must be turned *on* for this property to take effect.
✤ The vector V has one entry for each line of code in the profiled function.
See also: get, set, ProfileFile, ProfileInterval, root

Projection *orthographic/perspective*
Specifies which type of projection an axes object uses.
See also: get, set, axes

refresh(hndl)
Redraws the figure window with handle hndl.
Output: Not applicable.
Argument options: refresh to redraw the current figure window.
See also: reset

Renderer *painters/zbuffer*
Specifies which rendering method is used for screen display and printing.
See also: get, set, RendererMode, figure

RendererMode *auto/manual*
Specifies whether a rendering method is chosen automatically or manually.
See also: get, set, Renderer, figure

reset(hndl)
Resets all properties, except *Position*, of the graphics object with handle hndl.
Output: Not applicable.
Additional information: Use gca or gcf to access the handles to the current axes and figure, respectively.
See also: gcf, gca, cla, clf, hold, refresh

Resize *on/off*
Specifies whether users can resize a figure window by using the mouse.
See also: get, set, ResizeFcn, figure

ResizeFcn str
Specifies a string representing a MATLAB expression, value, or function that gets evaluated when a figure is resized.
Additional information: When the button is pressed, str gets passed to the function eval.
See also: set, get, eval, figure

colormap$_{hsv}$ = rgb2hsv(colormap$_{rgb}$)
Converts a color map defined by red-green-blue values to one defined by hue-saturation-value values.
Output: A color map is returned.
Additional information: The three columns of the input represent red, green, and blue values, respectively. ♦ The three columns of the output matrix represent hue, saturation, and value, respectively. All values in both types of color maps are between 0 and 1, inclusive. ♦ For more information on the meaning of different values, see the help page for hsv2rgb.
See also: colormap, brighten, hsv2rgb

root object
The root object contains all the properties for the top-level MATLAB session.
Additional information: The root object is the parents of all other graphic objects such as figure, patch, surface, etc. ♦ The available properties include *Automatic-*

FileUpdates, BusyAction, ButtonDownFcn, CallbackObject, Children, Clipping, CreateFcn, CurrentFigure, DeleteFcn, Diary, DiaryFile, Echo, ErrorMessage, Format, FormatSpacing, HandleVisibility, Interruptible, Parent, PointerLocation, PointerWindow, Profile, ProfileCount, ProfileFile, ProfileInterval, ScreenDepth, ScreenSize, Selected, SelectionHighlight, ShowHiddenHandles, Tag, Type, Units, UserData, and *Visible*. See the individual entries for these properties for information on valid and default values. ✦ The handle for the root object is always 0. ✦ The values of individual properties can be set or retrieved using set and get, respectively.
See also: set, get, axes, figure, line, text, image, patch, surface, figure

rotate(hndl, num$_{ang}$, [num$_{az}$, num$_{el}$])

Rotates the graphics object with handle hndl by angle num$_{ang}$ about the axis described by azimuth num$_{az}$ and elevation num$_{el}$.
Output: Not applicable.
Argument options: rotate(hndl, num$_{ang}$, [num$_x$, num$_y$, num$_z$]) to define the axis by the x, y, and z directions. ✦ rotate(hndl, num$_{ang}$, [num$_{az}$, num$_{el}$], [num$_{o,x}$, num$_{o,y}$, num$_{o,z}$]) to specify an origin of rotation other than [0, 0, 0].

Rotation num

Specifies orientation of a text object is anchored at num degrees.
Additional information: On some platforms you can specify any rotation. On others num must be one of 0, ±90, ±180, ±270.
See also: set, get, text

ScreenDepth n

Specifies number of bits available per pixel on your monitor.
Additional information: If you are using an X Windows terminal, this value is automatically set to take full advantage of your monitor. Otherwise, check out the default value that is set. ✦ The number of simultaneous colors available on your terminal is then 2^n. ✦ ScreenDepth overrides the value of the BlackAndWhite property. ✦ ScreenDepth is a property of the *root object*, which has a handle of 0.
See also: set, get, ScreenSize, root

ScreenSize [0, 0, num$_w$, num$_h$]

Specifies the screen size with numeric values num$_w$ and num$_h$ representing the width and height, respectively.
Additional information: ScreenSize is a property of the *root object*, which has a handle of 0. ✦ The values num$_w$ and num$_h$ are read only, and can be accessed with get.
See also: get, Units, ScreenDepth, roots

Selected on/off

Specifies whether an object is selected.
See also: get, set, SelectionHighlight, axes, figure, light, line, patch, root, surface, text

SelectionHighlight on/off

Specifies whether an object is highlighted when selected.
See also: get, set, Selected, axes, figure, light, line, patch, root, surface, text

SelectionType *normal/extended/alt/open*
Specifies how a selection is achieved with the mouse.
Additional information: The detail of the above selection types may vary slightly for different platforms and different mouse types.
See also: get, figure

set(hndl, str$_{prop}$, str$_{value}$)
Set the property specified by string str$_{prop}$ of the object defined by handle hndl to the value str$_{value}$.
Output: Not applicable.
Argument options: set(hndl), to list all the settable properties for the object defined by hndl. ✤ set(hndl, str$_{prop}$), to list valid values for the property str$_{prop}$ of object hndl. ✤ set(hndl, str$_{prop}$, *'default'*), to set the property to the first-encountered default value for that property. ✤ set(hndl, str$_{prop}$, *'factory'*), to set the property to it factor installed value. ✤ set(hndl, *'Default*str$_{prop}$*'*, str$_{value}$) to set the default value for the property str$_{prop}$ in the object with handle hndl. ✤ set(hndl, *'Default*str$_{prop}$*'*, *'remove'*) to remove a default value and revert to the factory setting. ✤ set(hndl, str$_{prop}$, *'remove'*), to set the property to its factory installed value.
Additional information: set is particularly useful when adjusting objects with many components or properties, such as figures. ✤ set is not recursive, it only changes subobjects to a depth of *one* level. ✤ If you want to use the strings *remove*, *default*, or *factory*, as actual values for a property, preface them with a backslash character, \. ✤ Use the command get to query object properties.
See also: get, gca, gcf

ShareColors *on/off*
States whether existing color table slots should be reused for graphics objects.
Additional information: On systems with eight or less bits per pixel of color capability, this attribute should be set on.
See also: set, get

ShowHiddenHandles *on/off*
States whether handle hiding is enabled or disabled.
See also: set, get, HandleVisibility

SpecularColorReflectance num
Specifies how the color of specularly reflected light is defined.
Additional information: num must be between 0 and 1.
See also: set, get, SpecularExponent, SpecularStrength, patch, surface

SpecularExponent num
Specifies the harshness of specular reflectance.
Additional information: num must be greater than 1.
See also: set, get, SpecularColorReflectance, SpecularStrength, patch, surface

SpecularStrength num
Specifies the intensity of specular reflectance.

Additional information: num must be between 0 and 1.
See also: set, get, SpecularColorReflectance, SpecularExponent, patch, surface

spinmap(num)
Rotates the current color map in a cyclical pattern for approximately num seconds.
Output: Not applicable.
Argument options: spinmap to rotate the color map for 5 seconds. ✳ spinmap(*inf*) to rotate the color map indefinitely, until *CTRL-C* is pressed. ✳ spinmap(num, num$_{inc}$) to rotate the color map for num seconds using a rotation increment of num$_{inc}$. The default value is 2. Lower values mean slower rotation; higher values mean faster rotation. Negative values mean rotation in the opposite direction.
See also: colormap

String str
Specifies that the text of a text object is string str.
See also: set, get, text

Style *infinite/local*
Specifies whether lights are parallel (*infinite*) or divergent (*local*).
See also: set, get, light

hndl = subplot(m, n, int$_{subp}$)
Creates a new axis in the current figure, located in a m × n grid in the int$_{subp}$ position.
Output: An axes handle is assigned to hndl.
Argument options: subplot(hndl) to specify that the axes with handle hndl is now current. This is identical to axes(hndl).
Additional information: The m and n specifiers can be different for different axes within the same figure. In each invocation of subplot, a new grid is created to determine axes position. ✳ Beware, if a new axes definition overlays an existing one, however slightly, the existing one is automatically removed from the figure. ✳ There are several examples of the use of subplot throughout this book.
See also: axes, figure, plot

hndl = surface(name$_1$, expr$_1$, ..., name$_n$, expr$_n$)
Creates a surface object with the n properties specified by name$_1$ through name$_n$ to the values expr$_1$ through expr$_n$, respectively.
Output: A handle to a surface object is assigned to hndl.
Argument options: surface(M$_x$, M$_y$, M$_z$, M$_c$) to display the three-dimensional surface defined by the *x*, *y*, and *z* value matrices M$_x$, M$_y$, and M$_z$, and the color matrix M$_c$. For more information on coloring of surfaces, see the entries for CData and surf. ✳ surface(M$_x$, M$_y$, M$_z$) to display the three-dimensional surface whose colors are defined by M$_z$; that is, the surface is colored by height. ✳ surface(V$_x$, V$_y$, M$_z$), where [length(V$_y$), length(V$_x$)] = size(M$_z$), to use two vectors to create the *x* and *y* values. ✳ surface(M$_z$, M$_c$) or surface(M$_z$) to use the vectors V$_x$ = 1:size(M$_z$, 2) and V$_y$ = 1:size(M$_z$, 1) for the *x* and *y* values. ✳ hndl = surface(M$_x$, M$_y$, M$_z$, M$_c$) to assign a surface to the handle hndl.

Additional information: Typically, if you only want to display the surface, surf is used. ✦ Basically, surface objects are rectangular grids of connected patch objects. patch and surface share many similar properties. ✦ The available properties include *AmbientStrength*, *BackFaceLighting*, *BusyAction*, *ButtonDownFcn*, *CData*, *CDataMapping*, *Children*, *Clipping*, *CreateFcn*, *DeleteFcn*, *DiffuseStrength*, *EdgeColor*, *EdgeLighting*, *EraseMode*, *FaceColor*, *FaceLighting*, *HandleVisibility*, *Interruptible*, *LineStyle*, *LineWidth*, *Marker*, *MarkerEdgeColor*, *MarkerFaceColor*, *MarkerSize*, *MeshStyle*, *NormalMode*, *Parent*, *Selected*, *SelectionHighlight*, *SpecularColorReflectance*, *SpecularExponent*, *SpecularStrength*, *Tag*, *Type*, *UserData*, *VertexNormals*, *Visible*, *XData*, *YData*, and *ZData*. See the individual entries for these properties for information on valid values. ✦ The values of individual properties can be set or retrieved using set and get, respectively.
See also: plot3, surf, patch, line, size

Tag *str*
Specifies a user label for an object.
See also: get, set, axes, figure, light, line, patch, root, surface, text

hndl = text(name$_1$, expr$_1$, ..., name$_n$, expr$_n$)
Creates a text object with the n properties specified by name$_1$ through name$_n$ to the values expr$_1$ through expr$_n$, respectively.
Output: A handle to a text object is assigned to hndl.
Argument options: text(num$_x$, num$_y$, str) to add the text str to the current two-dimensional figure at the *x* and *y* coordinates num$_x$ and num$_y$. ✦ text(num$_x$, num$_y$, num$_z$, str) to add the text str to the current three-dimensional figure at the *x*, *y*, and *z* coordinates num$_x$, num$_y$, and num$_z$. ✦ text(V$_x$, V$_y$str) or text(V$_x$, V$_y$, V$_z$, str) to add the text str at the points corresponding to matching elements of the vectors. ✦ text(V$_x$, V$_y$, V$_z$, M$_{str}$) to add the text at the corresponding row of text matrix M$_{str}$ at the points corresponding to matching elements of the vectors. ✦ hndl = text(num$_x$, num$_y$, str) or hndl = text(num$_x$, num$_y$, num$_z$, str) to assign a text object to the handle hndl.
Additional information: The available properties include *BusyAction*, *ButtonDownFcn*, *Children*, *Clipping*, *Color*, *Editing*, *EraseMode*, *Extent*, *FontAngle*, *FontName*, *FontSize*, *FontUnits*, *FontWeight*, *HandleVisibility*, *HorizontalAlignment*, *Interpreter*, *Interruptible*, *Parent*, *Position*, *Rotation*, *Selected*, *SelectionHighlight*, *String*, *Tag*, *Type*, *Units*, *UserData*, *VerticalAlignment*, and *Visible*. See the individual entries for these properties for information on valid values. ✦ The values of individual properties can be set or retrieved using set and get, respectively. ✦ Until a text object is applied to a figure (typically by being included in a plot or plot3 command), nothing is displayed.
See also: plot, plot3, line, gtext, num2str

TickDir *in/out*
Specifies whether tickmarks are to be drawn inward or outward from the axes.
Additional information: The default value for two-dimensional plots is *in*. ✦ The default value for three-dimensional plots is *out*.
See also: set, get, axes, TickDirMode, TickLength

TickDirMode *auto/manual*
Specifies whether direction of tickmarks is chosen automatically or manually.

See also: set, get, axes, TickDir

TickLength [num$_{2d}$, num$_{3d}$]
Specifies the ratio of the axes tickmark lengths to their respective axis lengths for both two- and three-dimensional objects within an axes object.
Additional information: The default value of TickLength is [0.01, 0.025].
See also: set, get, axes, TickDir, XTick, YTick, ZTick

title(str)
Allows the user to specify a title of str for the current figure.
Output: Not applicable.
Additional information: The title is placed at the top of the plot.
See also: xlabel, ylabel, zlabel, text, gtext, plot, num2str, int2str

Title hndl
Allows the user to specify a figure title with the text pointed to by text handle hndl.
Additional information: Use the title command whenever possible. ✣ See the entry for text for more details on controlling text objects.
See also: xlabel, ylabel, zlabel, text, title, axes

Type str
Specifies the type of an object.
Output: Not applicable.
Additional information: You cannot implicitly set the type of an object.
See also: get, axes, figure, light, line, image, patch, root, surface, text

Units *points/pixels/inches/centimeters/normalized*
Specifies the units of measurement used in an object.
Additional information: This property greatly affects the Extent and Position properties.
See also: set, get, text, axes, figure, root, Extent, Position

UserData M
Specifies user data in matrix M which is never acted upon by any of MATLAB's built-in commands.
Additional information: The primary use of UserData is to create functions that perform specialized manipulations on objects.
See also: set, get, axes, figure, light, line, image, patch, root, surface, text

VertexNormals M
Specifies the vertex normals for a patch or surface object.
See also: set, get, patch, surface

VerticalAlignment *top/cap/middle/baseline/bottom*
Specifies the vertical justification of a text object so that the position point is at the top of the string.
See also: set, get, text, Position, HorizontalAlignment

Vertices M
specifies the $x-$, $y-$, and $z-$coordinates for each vertex.
Additional information: This property is used in conjunction with the Faces property.
See also: set, get, Faces, patch

view(num$_{az}$, num$_{el}$)
Sets the azimuth and horizontal elevation viewing angles of the current three-dimensional figure to num$_{az}$ and num$_{el}$ degrees, respectively.
Output: Not applicable.
Argument options: view([num$_{az}$, num$_{el}$]) to produce the same result. ✤ view([num$_x$, num$_y$, num$_z$]) to specify that the viewing angle is from the cartesian coordinate (num$_x$, num$_y$, num$_z$). ✤ view(2) to view the plot from directly above, giving it a two-dimensional feel. ✤ view(3) to set the viewing angle to the default of [-37.5, 30.]. ✤ view(M$_T$), where M$_T$ is a 4 × 4 transformation matrix, to set the viewing angle relative to M$_T$. ✤ [num$_{az}$, num$_{el}$] = view to assign the current viewing angle to num$_{az}$ and num$_{el}$. ✤ M$_T$ = view to assign the current viewing angle transformation matrix to M$_T$.
Additional information: The azimuth measures counterclockwise rotation about the z-axis and the horizontal elevation measures distance above (positive values) or below (negative values) the object.
See also: plot3

View [num$_{az}$, num$_{el}$]
Specifies the azimuth and horizontal elevation viewing angles of an axes to num$_{az}$ and num$_{el}$ degrees, respectively.
Output: Not applicable.
Additional information: Typically this property is set with the view command. ✤ The azimuth measures counterclockwise rotation about the z-axis and the horizontal elevation measures distance above (positive values) or below (negative values) the object.
See also: set, get, view, axes

Visible on/off
Specifies whether an object is ever displayed on the screen.
See also: set, get, Clipping, axes, figure, image, light, line, patch, root, surface image, text

colormap = white(posint)
Creates an all-white color map.
Output: A posint ×3 color map matrix is returned.
Argument options: colormap = white to return a color map with a default number of rows equal to the number of rows of the current colormap.
Additional information: This command is meant for users with monochrome screens.
See also: colormap, hsv, prism, whitebg, contrast

whitebg
Toggles the current figure's background color from white to black (and vice versa).
Output: Not applicable.

Argument options: whitebg(V_{hndl}) to change each figure listed in the vector of handles V_{hndl}. ✽ whitebg(0) to toggle the background color used for new figures. ✽ whitebg(str) to change the background to the color represented by string str. A 1×3 RGB matrix can also be used to specify the color.
Additional information: When the background color is changed, the colors of other elements in the figure are adjusted to make the plot look good. ✽ For complete control over colors, use the handles to the figure.
See also: Color, colordef, white, colormap, graymon, figure

WindowButtonDownFnc str

Specifies a string representing a MATLAB expression, value, or function that gets evaluated when the mouse button is pressed down in a figure window.
Additional information: When the button is pressed, str gets passed to the function eval. ✽ To set such a string for more general types of graphics objects, use the ButtonDownFnc property.
See also: set, get, figure, eval, WindowButtonMotionFnc, WindowButtonUpFnc, ButtonDownFnc

WindowButtonMotionFnc str

Specifies a string representing a MATLAB expression, value, or function that gets evaluated when the mouse is in motion within a figure window.
Additional information: When the mouse is in motion, str gets passed to the function eval. ✽ Typically, str gets defined by WindowButtonDownFnc function and undefined by a subsequent WindowButtonUpFnc.
See also: set, get, figure, eval, WindowButtonDownFnc, WindowButtonUpFnc

WindowButtonUpFnc str

Specifies a string representing a MATLAB expression, value, or function that gets evaluated when the mouse button is released in a figure window.
Additional information: When the mouse button is released, str gets passed to the function eval. ✽ Typically, str is used to undefine a function defined by WindowButtonDownFnc.
See also: set, get, figure, eval, WindowButtonDownFnc, WindowButtonMotionFnc

WindowStyle normal/modal

Specifies wether a figure window has normal or modal behaviour.
Additional information: Modal figures are always displayed above normal figures.
See also: set, get, figure

XAxisLocation top/bottom

Specifies whether the x-axis tickmarks and labels are displayed at the top or bottom of the axes.
See also: set, get, axes, YAxisLocation

XColor [num_r, num_g, num_b]

Specifies that the x-axis and all the marks and lines associated with it have a color defined by RGB (red, green, blue) values num_r, num_g, and num_b, respectively.

Argument options: XColor str to specify one of the preset colors. See the entry for color for a listing of the valid strings.
Additional information: Included in the coloring are the x-axis, its tickmarks, labels, and grid lines.
See also: set, get, axes, YColor, ZColor

XData M_x

Specifies the x-axis data used to create the current object.
Additional information: Typically, the x-axis data is set within plot or plot3. ✦ In some circumstances, M_x is a vector or one-dimensional matrix.
See also: plot, plot3, set, get, line, image, surface, patch, YData, ZData

XDir *normal/reverse*

Specifies the direction of increasing/decreasing values on the x-axis.
Additional information: With the *normal* option, a right-hand rule is used to determine direction.
See also: set, get, axes, YDir, ZDir

XGrid *on/off*

Specifies whether grid lines be drawn in the x direction, at each tickmark.
See also: set, get, axes, GridLineStyle, YGrid, ZGrid

xlabel(str)

Allows the user to specify a label of str for the x-axis of the current two- or three-dimensional figure.
Output: Not applicable.
Additional information: In three-dimensional plots, the label is always placed so as not to be obscured by the plot itself.
See also: ylabel, zlabel, text, title, plot, plot3, num2str, int2str

XLabel hndl

Specifies the text used to label the x-axes as the text pointed to by handle hndl.
Additional information: Typically, this value is set with the command xlabel. ✦ If a text handle is used, the location parameters of the text are ignored.
See also: set, get, axes, xlabel, text, YLabel, ZLabel

XLim [num_{min}, num_{max}]

Specifies the minimum and maximum x-axis values as num_{min} and num_{max}.
Argument options: XLim([-*inf*, num_{max}] or XLim([num_{min}, *inf*] to set the maximum or minimum value, respectively, and let the other extreme be automatically scaled.
Additional information: The value of this property affects scaling of the x dimension and the location of tickmarks on the x-axis.
See also: set, get, axes, XLimMode, YLim, ZLim

XLimMode *auto/manual*

Specifies whether the x-limit values should be computed automatically or manually.

Additional information: If specific values are set in XLim, the XLimMode property is automatically set to *manual*. ♣ If the *auto* option is chosen, the limits are calculated to round figures.
See also: set, get, axes, XData, XLim, YLimMode, ZLimMode

XScale *linear/log*
Specifies whether the *x*-axis is scaled linearly or logarithmically.
See also: set, get, axes, YScale, ZScale

XTick V
Specifies that the *x*-axis tickmarks be placed at the *x* values found in vector V.
See also: set, get, axes, XTickMode, XTickLabel, XTickLabelMode, YTick, ZTick

XTickLabel M$_{str}$
Specifies that the *x*-axis tickmark labels from the string matrix M$_{str}$ be used.
Additional information: The strings in M$_{str}$ replace the normal numeric strings that are automatically created. ♣ If M$_{str}$ doesn't contain enough labels, then some of the automatically generated numeric strings are kept. ♣ Labels can also be entered as one large string using the — separator, or as standard numeric ranges.
See also: set, get, axes, XTick, XTickMode, XTickLabelMode, YTickLabel, ZTickLabel

XTickLabelMode *auto/manual*
Specifies whether the *x*-tickmark labels should be set automatically or manually.
Additional information: If specific values are set in XTickLabel, the XTickLabelMode property is automatically set to *manual*.
See also: set, get, axes, XTickLabel, YTickLabelMode, ZTickLabelMode

XTickLabelMode *manual*
Specifies that the values in XTickLabels should be used to specify the *x*-axis tickmark labels.
Argument options: XTickLabelMode *auto* to use the default numeric strings that equal the positions of the *x*-axis tickmarks. This is the default.
Output: Not applicable.
Additional information: If specific values are set in XTickLabels, the XTickLabelMode property is automatically set to *manual*. ♣ To set the *x*-axis tickmark label mode, use set or axes. ♣ To retrieve the *x*-axis tickmark label mode, use get.
See also: set, get, axes, XTick, XTickMode, XTickLabels, YTickLabelMode, ZTickLabelMode

XTickMode *manual*
Specifies that the values in XTick should be used to specify the *x*-axis tickmark locations.
Argument options: XTickMode *auto* to pick a reasonable set of tickmark location from the range of the corresponding XData. This is the default.
Output: Not applicable.

YAxisLocation *right/left*
Specifies whether the *x*-axis tickmarks and labels are displayed at the left or right of the axes.
See also: set, get, axes, XAxisLocation

YColor [num$_r$, num$_g$, num$_b$]
Specifies that the *y*-axis and all the marks and lines associated with it have a color defined by RGB (red, green, blue) values num$_r$, num$_g$, and num$_b$, respectively.
Argument options: YColor str to specify one of the preset colors. See the entry for color for a listing of the valid strings.
See also: set, get, axes, XColor, ZColor

YData M$_y$
Specifies the *y*-axis data used to create the current object.
Additional information: Typically, the *y*-axis data is set within plot or plot3. ✤ In some circumstances, M$_y$ is a vector or one-dimensional matrix.
See also: plot, plot3, set, get, line, image, surface, patch, XData, ZData

YDir *normal/reverse*
Specifies the standard direction of increasing/decreasing values on the *y*-axis.
Additional information: With the *normal* option, a right-hand rule is used to determine direction.
See also: set, get, axes, XDir, ZDir

YGrid *on/off*
Specifies whether grid lines are drawn in the *y* direction, at each tickmark.
See also: set, get, axes, GridLineStyle, XGrid, ZGrid

ylabel(str)
Allows the user to specify a label of str for the *y*-axis of the current two- or three-dimensional figure.
Output: Not applicable.
Additional information: In three-dimensional plots, the label is always placed so as not to be obscured by the plot itself.
See also: xlabel, zlabel, text, title, plot, plot3, num2str, int2str

YLabel hndl
Specifies the text used to label the *y*-axes as the text pointed to by handle hndl.
Additional information: Typically, this value is set with the command ylabel. ✤ If a text handle is used, the location parameters of the text are ignored.
See also: set, get, axes, ylabel, text, XLabel, ZLabel

YLim [num$_{min}$, num$_{max}$]
Specifies the minimum and maximum y-axis values as num$_{min}$ and num$_{max}$.
Argument options: YLim([-*inf*, num$_{max}$] or YLim([num$_{min}$, *inf*] to set the maximum or minimum value, respectively, and let the other extreme be automatically scaled.
Additional information: The value of this property affects scaling of the y dimension and the location of tickmarks on the y-axis.
See also: set, get, axes, YLimMode, XLim, ZLim

YLimMode *auto/manual*
Specifies whether the values in YLim should be set automatically or manually.
Additional information: If specific values are set in YLim, the YLimMode property is automatically set to *manual*. ♣ If the *auto* option is chosen, the limits are calculated to round figures.
See also: set, get, axes, YData, YLim, XLimMode, ZLimMode

YScale *linear/log*
Specifies whether the y-axis is scaled linearly or logarithmically.
See also: set, get, axes, XScale, ZScale

YTick V
Specifies that the y-axis tickmarks be placed at the y values found in vector V.
See also: set, get, axes, YTickMode, YTickLabel, YTickLabelMode, XTick, ZTick

YTickLabel M$_{str}$
Specifies that the y-axis tickmark labels from the string matrix M$_{str}$ be used.
Additional information: The strings in M$_{str}$ replace the normal numeric strings that are automatically created. ♣ If M$_{str}$ doesn't contain enough labels, then some of the automatically generated numeric strings are kept. ♣ Labels can also be entered as one large string using the — separator, or as standard numeric ranges.
See also: set, get, axes, YTick, YTickMode, YTickLabelMode, XTickLabel, ZTickLabel

YTickLabelMode *auto/manual*
Specifies whether the values in YTickLabels should be set automatically or manually.
Additional information: If specific values are set in YTickLabels, the YTickLabelMode property is automatically set to *manual*.
See also: set, get, axes, YTick, YTickMode, YTickLabel, XTickLabelMode, ZTickLabelMode

YTickMode *auto/manual*
Specifies whether the values in YTick should be set automatically or manually.
Additional information: If specific values are set in YTick, the YTickMode property is automatically set to *manual*.
See also: set, get, axes, YData, YTick, YTickLabel, YTickLabelMode, XTickMode, ZTickMode

ZColor [num$_r$, num$_g$, num$_b$]
Specifies that the z-axis and all the marks and lines associated with it have a color defined by RGB (red, green, blue) values num$_r$, num$_g$, and num$_b$, respectively.

Argument options: ZColor str to specify one of the preset colors. See the entry for color for a listing of the valid strings.
See also: set, get, axes, XColor, YColor

ZData M_z

Specifies the z-axis data used to create the current three-dimensional object.
Additional information: Typically, the z-axis data is set within plot3. ✤ In some circumstances, M_z is a vector or one-dimensional matrix.
See also: plot3, set, get, line, surface, patch, XData, YData

ZDir *normal/reverse*

Specifies the standard direction of increasing/decreasing values on the z-axis.
Additional information: With the *normal* option, a right-hand rule is used to determine direction.
See also: set, get, axes, XDir, YDir

ZGrid *on/off*

Specifies whether grid lines are drawn in the z direction, at each tickmark.
Additional information:
See also: set, get, axes, GridLineStyle, XGrid, YGrid

zlabel(str)

Allows the user to specify a label of str for the y-axis of the current three-dimensional figure.
Output: Not applicable.
Additional information: The label is always placed so as not to be obscured by the plot itself.
See also: xlabel, ylabel, text, title, plot3, num2str, int2str

ZLabel hndl

Specifies the text used to label the z-axes as the text pointed to by handle hndl.
Additional information: Typically, this value is set with the command zlabel. ✤ If a text handle is used, the location parameters of the text are ignored.
See also: set, get, axes, zlabel, text, XLabel, YLabel

ZLim [num_{min}, num_{max}]

Specifies the minimum and maximum z-axis values as num_{min} and num_{max}.
Argument options: ZLim([-*inf*, num_{max}] or ZLim([num_{min}, *inf*] to set the maximum or minimum value, respectively, and let the other extreme be automatically scaled.
Additional information: The value of this property affects scaling of the z dimension and the location of tickmarks on the z-axis.
See also: set, get, axes, ZLimMode, XLim, YLim

ZLimMode *auto/manual*

Specifies whether the values in ZLim should be set automatically or manually.

Additional information: If specific values are set in ZLim, the ZLimMode property is automatically set to *manual*. ✤ If the *auto* option is chosen, the limits are calculated to round figures.
See also: set, get, axes, ZData, ZLim, XLimMode, YLimMode

zoom *on*

Enables zooming in the current figure.
Output: Not applicable.
Argument options: zoom *off* to disable zooming. ✤ zoom *out* to return to the original axis. ✤ zoom *xon* to zoom only in the *x*-direction. ✤ zoom *yon* to zoom only in the *y*-direction. ✤ zoom *reset* to clear the zoom out point.
Additional information: Once zooming is enabled, clicking on the left mouse button in the current figure lessens the range of the axes by a factor of two about that point. ✤ Clicking with the left mouse button and dragging allows you to define the new axis ranges more precisely. ✤ Clicking the right mouse button zooms out by a factor of two. You cannot, however, exceed the original axis ranges.

ZScale *linear/log*

Specifies whether the *z*-axis is scaled linearly or logarithmically.
See also: set, get, axes, XScale, YScale

ZTick V

Specifies that the *z*-axis tickmarks be placed at the *z* values found in vector V.
See also: set, get, axes, ZTickMode, ZTickLabel, ZTickLabelMode, XTick, YTick

ZTickLabel M_{str}

Specifies that the *z*-axis tickmark labels from the string matrix M_{str} be used.
Additional information: The strings in M_{str} replace the normal numeric strings that are automatically created. ✤ If M_{str} doesn't contain enough labels, then some of the automatically generated numeric strings are kept. ✤ Labels can also be entered as one large string using the — separator, or as standard numeric ranges.
See also: set, get, axes, ZTick, ZTickMode, ZTickLabelMode, XTickLabel, YTickLabel

ZTickLabelMode *auto/manual*

Specifies whether the values in ZTickLabel should be set automatically or manually.
Additional information: If specific values are set in ZTickLabels, the ZTickLabelMode property is automatically set to *manual*.
See also: set, get, axes, ZTick, ZTickMode, ZTickLabel, XTickLabelMode, YTickLabelMode

ZTickMode *auto/manual*

Specifies whether the values in ZTick should be set automatically or manually.
Additional information: If specific values are set in ZTick, the ZTickMode property is automatically set to *manual*.
See also: set, get, axes, ZData, ZTick, ZTickLabels, ZTickLabelMode, XTickMode, YTickMode

Graphical User-Interface Functions

This chapter looks at adding a graphical user interface (GUI) to a MATLAB program.

Up until now, we have used input and disp (or fprintf) to interactively read and write text and numbers in the MATLAB command window. The functions in this chapter allow users to provide input and get output in more "graphical" ways through the use of MATLAB figure windows containing sliders, buttons, menus, etc., and include:

- uicontrol
    - uicontrol('Style', 'Frame', ...)
    - uicontrol('Style', 'Text', ...)
    - uicontrol('Style', 'Edit', ...)
    - uicontrol('Style', 'Slider', ...)
    - uicontrol('Style', 'Popupmenu', ...)
    - uicontrol('Style', 'Listbox', ...)
    - uicontrol('Style', 'Radiobutton', ...)
    - uicontrol('Style', 'Checkbox', ...)
    - uicontrol('Style', 'Pushbutton', ...)
- uimenu
- errordlg, helpdlg, questdlg, warndlg, listdlg, msgbox, choices
- uigetfile, uiputfile, printdlg, uisetfont, uisetcolor
- Miscellaneous:
    - gtext, ginput
    - rbbox, waitforbuttonpress

There are two main methods of using these commands: (1) from a MATLAB script, and (2) from a MATLAB function. This chapter shows examples of both ways.

MATLAB has an interactive tool called guide that allows you to design a graphical user interface while it writes the corresponding MATLAB code for you. We postpone the discussion of guide until later in this chapter, so that you will first be able to understand what the code does that guide generates for you.

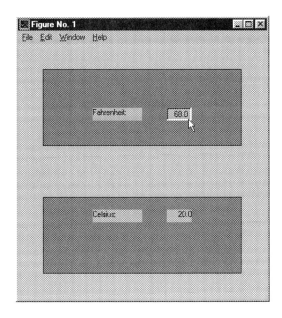

FIGURE 82. Fahrenheit-to-Celsius conversion

Inputting and Outputting Strings and Numbers Using uicontrols

Consider the following MATLAB program to read in a temperature given in degrees Fahrenheit and convert it to degrees Celsius.

```
f = input('Enter temperature (degrees F): ');
c = (f - 32) * 5 / 9;
fprintf(1, 'Temperature (degrees C) is: %g\n', c);
```

This simplistic three-line program has much in common with more realistic MATLAB programs, as they both:

- read input value(s),

- perform calculation(s), and

- write output value(s).

Let's create a graphical version of this small program, which will also illustrate techniques used in larger programs. The input screen is set up as shown in Figure 82. This example is expanded on throughout the whole chapter to explore the various GUI functions.

First Draw the Frames

The following code draws two rectangular *frames* in the current figure window, and then changes the interior color of each frame to light gray: [73]

```
set(gcf, 'DefaultUicontrolUnits', 'Normalized')

% ----------------------------------------------------------

frame1_ = uicontrol(gcf,                                    ...
         'Style',      'Frame',                             ...
         'Position',   [0.1  0.1    0.8  0.3]);

frame2_ = uicontrol(gcf,                                    ...
         'Style',      'Frame',                             ...
         'Position',   [0.1  0.6    0.8  0.3]);

% ----------------------------------------------------------

set(frame1_, 'BackgroundColor', [0.50  0.50  0.50]);

set(frame2_, 'BackgroundColor', [0.50  0.50  0.50]);
```

The *position boxes* are located with their lower left corners at (0.1, 0.1) and (0.1, 0.6), respectively, and are both 0.8 units wide and 0.3 units high.

The coordinates are specified in "normalized" units. This means that the lower-left corner of the figure window is at (0, 0), and the top-right corner is at (1, 1), and that all other measurements are based upon these values.

Inputting and Outputting Text and Numbers Using the `Edit` and `Text` Controls

The following uses a Text uicontrol to write the string "Fahrenheit" in the top frame:

```
text_f_ = uicontrol(gcf,                                    ...
         'Style',      'Text',                              ...
         'String',     'Fahrenheit: ',                      ...
                                                            ...
         'Position',   [0.3  0.7    0.2  0.05],             ...
         'HorizontalAlignment',   'Left');
```

The string "Fahrenheit" is *left*-aligned in the position box.

The following code uses an Edit uicontrol to write the string "68.0" beside "Fahrenheit". The string is *right*-aligned in its position box:[74]

```
edit_f_ = uicontrol(gcf,                                    ...
         'Style',      'Edit',                              ...
         'String',     '68.0',                              ...
                                                            ...
         'Position',   [0.6  0.7    0.1  0.05],             ...
         'HorizontalAlignment',   'right',                  ...
                                                            ...
         'Callback',   'fc_calc' );
```

---

[73] If no figure window is open, MATLAB automatically opens a new one first.
[74] On some systems, this alignment is not done due to limitations in the underlying windowing system.

This text is "editable"; that is, the user can change it. When the user changes the value and presses [Enter], MATLAB invokes the command given in the Callback string: fc_calc.

Finally, the following creates two more Text uicontrols to display the current calculation results, "Celsius" and "20.0", in the *bottom* frame:

```
text_c1_ = uicontrol(gcf,                               ...
            'Style',     'Text',                        ...
            'String',    'Celsius: ',                   ...
                                                        ...
            'Position',  [0.3  0.3   0.2  0.05],        ...
            'HorizontalAlignment',  'Left');

text_c2_ = uicontrol(gcf,                               ...
            'Style',     'Text',                        ...
            'String',    '20.0',                        ...
                                                        ...
            'Position',  [0.6  0.3   0.1  0.05],        ...
            'HorizontalAlignment',  'Right');
```

Callback Strings

The screen's appearance has been completed and now matches Figure 82. Now we have to create fc_calc.m containing the commands to be executed every time the editable text box is changed:

```
f = get(edit_f_, 'String');
f = str2num(f);

c = (f - 32) * 5 / 9;

c = num2str(c);
set(text_c2_, 'String', c);
```

The above script uses get to determine the current value of the string in our Edit uicontrol (which is pointed to by the handle edit_f_). Then the script:

1. converts the string to a number,[75]

2. converts that number from Fahrenheit to Celsius,

3. converts the resulting number to a string, and

4. stores that string in the uicontrol pointed to by the handle text_c2_.

---

[75] If something other than a number is passed to str2num, then the result is a null vector. Arithmetic on a null vector also returns a null vector. num2str converts a null vector to a null string.

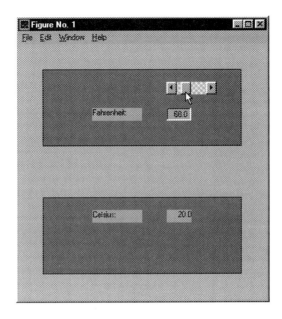

FIGURE 83. Fahrenheit-to-Celsius conversion, using a "slider"

Inputting Numbers Using a `slider`

Sliders can also be used to graphically input a numeric value.

Figure 83 shows what the screen looks like once we add a slider. (Note: MATLAB only supports *horizontal* sliders, not *vertical* sliders.)

The command to add this slider is:

```
slider_f_ = uicontrol(gcf,                       ...
            'Style',      'Slider',              ...
            'Min',         32.0,                 ...
            'Max',        212.0,                 ...
            'Value',       68.0,                 ...
                                                 ...
            'Position',   [0.6  0.8   0.2  0.05], ...
                                                 ...
            'Callback',   'fc_slider_f; fc_calc');
```

As you can see, the Callback string can actually be a list of MATLAB commands, separated by semicolons (or commas). The above Callback string invokes the script `fc_slider_f.m` to update the temperature displayed in the Edit uicontrol:

```
f = get(slider_f_, 'Value');

f = num2str(f);

set(edit_f_, 'String', f);
```

It then invokes `fc_calc.m` to re-calculate the output temperature.

You should also change the Callback string of the Edit control to move the slider when the edit value is changed by the user.

```
set(edit_f_, ...
    'Callback',    'fc_edit_f; fc_calc');
%
```

where `fc_edit_f.m` contains:

```
f = get(edit_f_, 'String');

f = str2num(f);

set(slider_f_, 'Value', f);
```

Now both uicontrols will always be synchronized with each other.

Inputting Choices (Integers) Using `Popupmenu`, `Listbox`, `Radiobutton`, and `Checkbox`

As well as asking for textual or numeric input, programs typically offer choices from lists of options. For example, let's change the following *non-GUI* version of this program so that it asks if you want the result in Celsius, Kelvin, or Rankine.

```
f = input('Enter temperature (degrees F): ');

r = f + 459.7;
c = (f - 32) * 5 / 9;
k = c + 273.15;

choice = input(['Enter 1 for Rankine,',    ...
                ' 2 for Celsius,'          ...
                ' 3 for Kelvin: '          ]);

if choice == 1
    fprintf(1, 'Temperature (degrees R) is: %g\n', r);
elseif choice == 2
    fprintf(1, 'Temperature (degrees C) is: %g\n', c);
elseif choice == 3
    fprintf(1, 'Temperature (degrees K) is: %g\n', k);
end
```

This section illustrates four alternate graphical methods of specifying the desired temperature unit to our GUI version of this program:

1. a "popup" menu
2. a "list box"
3. "radio" buttons
4. "check" boxes

The fourth method allows more than one choice to be selected at a time. A later section considers fifth and sixth GUI techniques which can also be used to specify choices; the sixth technique also allows more than one choice at a time.[76]

---

[76] Refer to the sections on Pushbuttons *and* Menus.

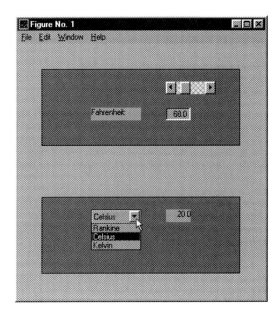

FIGURE 84. Fahrenheit-to-various conversion, using a "popup" menu

Using Popupmenu to Input a Choice

Now delete the *Celsius* text control and replace it with a popup menu:

```
delete(text_c1_);

popup_c_ = uicontrol(gcf,                           ...
             'Style',     'Popupmenu',              ...
             'String',    'Rankine|Celsius|Kelvin', ...
             'Value',              2              , ...
                                                    ...
             'Position',  [0.3  0.3    0.2  0.05],  ...
                                                    ...
             'Callback',  'fc_popup_c; fc_calc2');
```

Figure 84 shows what the screen looks.

When you click on the popup menu, the three choices appear. After you pick one, the other choices disappear and the selected choice remains displayed on the screen.

Notice that only one entry is displayed (the second) as specified by the Value property. When you click on the menu, all the choices are revealed, and you select one by clicking on it. Then the Callback string first invokes the script fc_popup_c.m, which sets the variable choice to the *index* (1, 2, or 3) of the currently selected item:

```
choice = get(popup_c_, 'Value');
```

The Callback string then invokes fc_calc2.m, which looks at the variable choice to determine which temperature unit to use:

```
f = get(edit_f_, 'String');
f = str2num(f);

r = f + 459.7;
```

```
c = (f - 32) * 5 / 9;
k = c + 273.15;

if     choice == 1,   t = r;
elseif choice == 2,   t = c;
elseif choice == 3,   t = k;
end

t = num2str(t);
set(text_c2_, 'String', t);
```

Note: Change the Edit and Slider uicontrols to use `fc_calc2.m` as well:

```
set(edit_f_, ...
    'Callback',  'fc_edit_f; fc_calc2');
%                            ^^^^^^^^
set(slider_f_, ...
    'Callback',  'fc_slider_f; fc_calc2');
%                              ^^^^^^^^
```

Notes

You can replace the 'Popupmenu' uicontrol with the 'Listbox' uicontrol. The only difference is that the 'Listbox' uicontrol can display more than one element of the list at a time, depending on how large the height is in its 'Position' property.

Using "radio" Buttons to Input a Choice

Delete the Rankine/Celsius/Kelvin popup menu uicontrol and replace it with *three* radio button uicontrols: one for each choice of temperature unit:

```
delete(popup_c_);

strings = ['Rankine'; 'Celsius'; 'Kelvin '];
show    = [    0   ;     1    ;     0    ];
ys      = [    3   ;     2    ;     1    ] * 0.075 + 0.075;

for i=1:3
    radio_c_(i) = uicontrol(gcf,                       ...
            'Style',     'Radiobutton',                ...
            'String',    strings(i),                   ...
            'Value',     show(i),                      ...
                                                       ...
            'Position',  [0.3  ys(i)   0.2  0.05],     ...
                                                       ...
            'Callback',  'fc_radio_c; fc_calc2');
end
```

Figure 85 shows what the screen looks like. The "Celsius" button is initially selected (its Value is set to 1). Notice that MATLAB fills the inside of this button, but leaves the other two buttons hollow.

When you click on one of the three buttons, MATLAB fills inside the button, and then calls the specified Callback string. In this case, we are using the *same* Callback string for all three radio buttons.

The Callback string first invokes `fc_radio_c.m`. It determines which button was chosen (1, 2, or 3) and stores this in the variable choice.

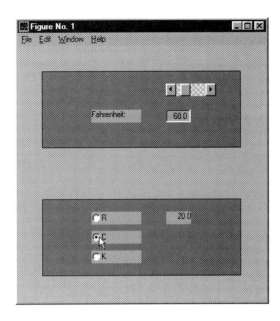

FIGURE 85. Fahrenheit-to-various conversion, using three radiobuttons

```
for i=1:3
    if gcbo == radio_c_(i)
        choice = i;
        set(radio_c_(i), 'Value', 1);
    else
        set(radio_c_(i), 'Value', 0);
    end;
end;
```

This script goes through a for loop, comparing the handle of the callback object (as returned by the gcbo function) with the handle of each button. The button that matches is turned on, while the others are turned off. [77]

The Callback string then invokes fc_calc2.m to perform the chosen conversion and display the result. Notice that fc_calc2.m does not need to be modified. [78]

Using "check" Boxes to Input One or More Choices

Now change the three Rankine/Celsius/Kelvin radio button uicontrols to three check box uicontrols - one for each choice of temperature unit:

```
delete(radio_c_);       %% All 3 handles

strings = ['Rankine'; 'Celsius'; 'Kelvin '];
show    = [    0    ;     1    ;    0     ];
```

---

[77] MATLAB automatically turns the selected button on if it isn't already on, but it does not turn off the other buttons: it is up to the Callback function to do that.

[78] fc_calc2.m is written in such a way that it is independent of the uicontrol used to make a choice.

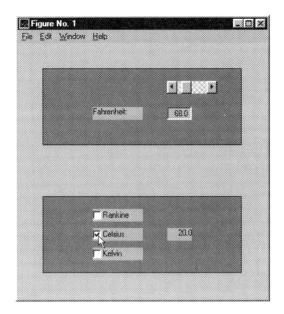

FIGURE 86. Fahrenheit-to-various conversion, using three "check" boxes

```
ys      = [   3    ;    2    ;    1    ] * 0.075 + 0.075;

for i=1:3
        check_c_(i) = uicontrol(gcf,                     ...
                'Style',     'Checkbox',                 ...
                'String',    strings(i,:),               ...
                'Value',     show(i),                    ...
                                                         ...
                'Position',  [0.3  ys(i)    0.2  0.05],  ...
                                                         ...
                'Callback',  'fc_check_c; fc_calc3');
end
```

Next to each check box will be a Text uicontrol. Initially, only *Celsius* will be visible:

```
delete(text_c2_);

for i=1:3
    text_c_(i) = uicontrol(gcf,                          ...
            'Style',     'Text',                         ...
            'String',    '20.0',                         ...
                                                         ...
            'Position',  [0.6  ys(i)    0.1  0.05],      ...
            'HorizontalAlignment',  'Right');
end;

set(text_c_(1), 'Visible', 'Off')
set(text_c_(2), 'Visible', 'On')
set(text_c_(3), 'Visible', 'Off')
```

Figure 86 shows what the screen looks like. The Celsius box is initially selected (its Value attribute is set to 1), and MATLAB fills the inside of the check box.

When you click on one of the three checkboxes, MATLAB *toggles* the check box between between being checked or not, and then executes the specified Callback string. In this example, the *same* Callback string is used for all three check boxes.

The Callback string first invokes fc_check_c.m, which determines which check box was selected (1, 2, or 3), and stores 1 or 0 in the element show(i). The other boxes are left unchanged:

```
for i=1:3
    if gcbo == check_c_(i)

        show(i) = get(check_c_(i), 'Value');

    end;
end;
```

The Callback string then invokes fc_calc3.m, which looks at the show variable to determine which temperature unit (or units) to calculate and display:

```
f = get(edit_f_, 'String');
f = str2num(f);

t(1) = f + 459.7;           % R
t(2) = (f - 32) * 5 / 9;    % C
t(3) = t(2) + 273.15;       % K

for i=1:3
    if show(i)
        ti = num2str( t(i) );
        set(text_c_(i), 'Visible', 'On', ...
                        'String', ti );
    else
        set(text_c_(i), 'Visible', 'Off');
    end
end
```

Note: Change the Edit and Slider uicontrols to use fc_calc3.m, too:

```
set(edit_f_, ...
    'Callback', 'fc_edit_f; fc_calc3');
%

set(slider_f_, ...
    'Callback', 'fc_slider_f; fc_calc3');
%
```

Initiating an Action Using "push" Buttons

The last uicontrol style to consider is Pushbutton, which is often used to initiate an action once all the data (text, numbers, and choices) have been entered.

Let's add push buttons to the program for the following three typical actions:

- "Start" calculations.

- "Reset" parameters (text, numbers, and choices) to default values.

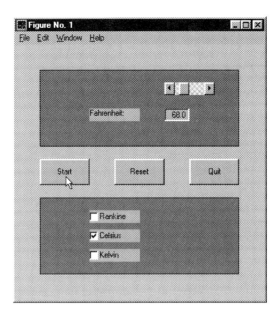

FIGURE 87. Fahrenheit-to-various conversion, using three pushbuttons

- "Quit" the program.

Figure 87 shows what the screen looks like.

Until now, this program actually started calculations whenever a choice was made, but now change it so that it does not do any calculations until the "Start" button is pressed.

The definition of the "Start" button is simply:

```
push_start_ = uicontrol(gcf,                    ...
          'Style',     'Pushbutton',            ...
          'String',    'Start',                 ...
          'Value',     0,                       ...
                                                ...
          'Position',  [0.1  0.45    0.2  0.1], ...
                                                ...
          'Callback',  'fc_calc3');
```

When the user presses the button, MATLAB invokes the Callback string, which in this case invokes the script fc_calc3.m to calculate and display the results.

Note: You should also change the Callback strings for the Edit, Slider, and Checkbox uicontrols to no longer invoke fc_calc3.m:

```
set(edit_f_, ...
    'Callback',  'fc_edit_f');

set(slider_f_, ...
    'Callback',  'fc_slider_f');

for i=1:3
    set(check_c_(i),                    ...
        'Callback',  'fc_check_c');
end
```

You should also change the text uicontrols for the output, in particular the Celsius output, so that no output is visible initially:

```
set(text_c_(2), 'Visible', 'Off');
```

The definition of the "Reset" button is:

```
push_reset_ = uicontrol(gcf,                              ...
                'Style',    'Pushbutton',                 ...
                'String',   'Reset',                      ...
                'Value',    0,                            ...
                                                          ...
                'Position', [0.4  0.45    0.2  0.1],      ...
                                                          ...
                'Callback', 'fc_reset');
```

where "fc_reset.m" contains:

```
set(edit_f_,    'String',  '68.0')
set(slider_f_,  'Value',   68.0);

show = [ 0 1 0 ];
for i=1:3
    set( check_c_(i), 'Value',   show(i) );
    set( text_c_(i),  'Visible', 'Off'   );
end
```

And finally, the definition of the "Quit" button is:

```
push_quit_ = uicontrol(gcf,                               ...
                'Style',    'Pushbutton',                 ...
                'String',   'Quit',                       ...
                'Value',    0,                            ...
                                                          ...
                'Position', [0.7  0.45    0.2  0.1],      ...
                                                          ...
                'Callback',           'fc_quit');
```

Note: When you created the Quit push button uicontrol, you specified that its Callback string was Interruptible. This is necessary whenever a Callback string invokes commands that require mouse button presses (such as questdlg).[79]

The file fc_quit.m contains:

```
my_figure = gcbf;

answer = questdlg('Do you really want to quit?',  ...
                'Quit?',                          ...
                'Yes', 'No', 'Cancel',            ...
                'Yes'                             );

if strcmp(answer, 'Yes')
    delete(my_figure);
end
```

The function questdlg opens a dialog box titled "Quit?" containing the question "Do you really want to quit?" along with three pushbuttons: "Yes," "No," and "Cancel." The default button is specified to be "Yes" (last argument in function invocation).

---

[79] Normally the Interruptible attribute is used to specify that the user can click on other uicontrols and interrupt this one's Callback string in the middle of its execution.

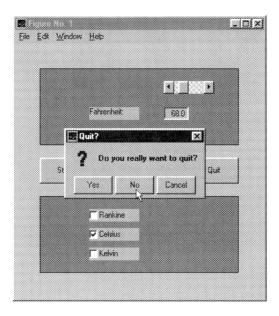

FIGURE 88. Question "dialog box" for quitting conversion program, created using questdlg

This means that the user can press Enter to choose "Yes", rather than having to explicitly click on the "Yes" button.

Figure 88 shows what the dialog box now looks like.

MATLAB also has other dialog box functions to handle special circumstances. In addition to questdlg, these include:

- helpdlg

- errordlg

- warndlg

- listdlg

- msgbox

- choices

These functions save you the trouble of creating a new figure window and placing Text and Pushbutton uicontrols on it to handle each of these common situations. helpdlg is demonstrated in a later section.

Another similar routine is choices; which puts up a window containing a vertical menu of choices.

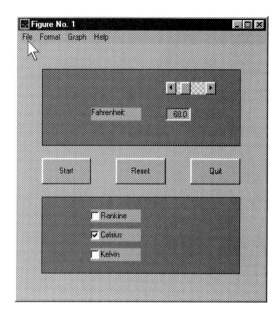

FIGURE 89. Menu added along top of figure window

Initiating an Action Using Menus — `uimenu`

As an alternative to "Pushbutton"s, you can add a menu to your program for less frequently used actions,[80] such as a menu along the top of our figure window with the following headings:

```
File   Format   Graph   Help
```

Figure 89 shows what the screen looks like when the menu is displayed.

The commands to produce this menu are as follows:

```
% --------------------------------------------------
% Remove the STANDARD figure window menu:
% --------------------------------------------------

set(gcf, 'MenuBar', 'none')

% --------------------------------------------------
% Create our OWN menu in place of it:
% --------------------------------------------------

File_   = uimenu(gcf, 'Label', 'File');
Format_ = uimenu(gcf, 'Label', 'Format');
Graph_  = uimenu(gcf, 'Label', 'Graph');

Help_   = uimenu(gcf, 'Label',       'Help', ...
                      'Callback',    'fc_help');
```

---

[80] In the next section you will see how menu entries can also be used for less frequently used choices.

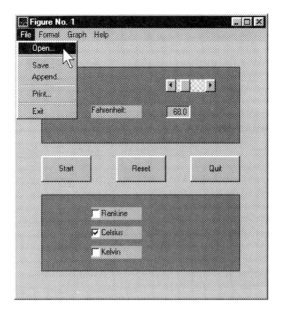

FIGURE 90. Entries in `File` submenu

Notice that the only menu item that immediately does anything when clicked on is `Help`. The other three menu items each have a submenu.

Under `File` add a submenu to perform the following actions:

Open a file containing data.
Save results to a file.
Append results to a file.
Print the figure.
Exit the program.

Figure 90 shows what the screen looks like when the `File` item is selected. The commands to produce this submenu are as follows:

```
File_Open_    = uimenu(File_, 'Label',      'Open...',    ...
                               'Callback',   'fc_open');

File_Save_    = uimenu(File_, 'Label',      'Save...',    ...
                               'Separator', 'On',         ...
                               'Callback',   'perm = ''wt''; fc_save');

File_Append_  = uimenu(File_, 'Label',      'Append...',  ...
                               'Callback',   'perm = ''at''; fc_save');

File_Print_   = uimenu(File_, 'Label',      'Print...',   ...
                               'Separator', 'On',         ...
                               'Callback',   'fc_print');

File_Exit_    = uimenu(File_, 'Label',         'Exit',    ...
                               'Separator',    'On',      ...
                               'Callback',     'fc_quit');
```

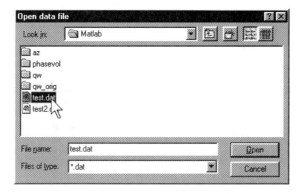

FIGURE 91. Dialog box displayed by uigetfile

Notice that the Separator attribute is turned on for the Save, Print, and Exit submenu items. This causes a separator line to appear *above* each of them in the File submenu.

The Callback for the Open submenu item invokes fc_open.m which contains:

```
[Fn, Pn] = uigetfile('*.dat', 'Open data file');

if Fn ~= 0
              fid = fopen([Pn, Fn], 'rt');
     f   = fgetl(fid);
           fclose(fid);

     set(edit_f_, 'String', f);
     fc_edit_f;
end
```

Notes

- The uigetfile function puts up a dialog box similar to that shown in Figure 91, allowing the user to select a file. The function returns a string containing the name of the file selected and the path to it (or the number 0 if no file is selected).

- The fopen, fgetl, and fclose functions in MATLAB are discussed in detail in the *File Input/Output* chapter.

The Callbacks for the Save and Append submenu items are similar to Open. First, they define the *permission* (wt or at), and then invoke the commands in fc_save.m:

```
[Fn, Pn] = uiputfile('*.out', 'Save output to file');

if Fn ~= 0
     f = get(edit_f_, 'String');

                fid = fopen([Pn,Fn], perm);
          fprintf(fid, '%s\n', f);
           fclose(fid);
end
```

The Callback for the Print submenu item is:

```
print -dps fc.ps
```

This will print the figure window along with the uicontrols. If we had wished *not* to print the uicontrols, the -noui option could have been given to print.

```
print -dps -noui fc.ps
```

The Callback for the Exit submenu item is simply the same fc_quit.m that we used for the Quit push button.

Notice that in creating the Exit submenu item, its Callback string is specified as Interruptible. Recall that this is necessary because (like the Quit push button) the Callback string for Exit invokes command(s) that require mouse button presses (i.e., questdlg).

Specifying Choices Using Menus

Under Format, let's have menu entries to select how many digits the final result(s) should have:

- three digits,

- four digits, or

- five digits,

with the default being four digits.

Figure 92 shows how the screen looks when the *Format* item is selected.

The commands to produce this submenu are as follows:

```
digits  = [ 3         ;   4         ;   5         ];
fmt     =                 2;

%- - - - - - - - - - - - - - - - - - - - - - - - - - - - - - - - - -

labels = ['3 digits';  '4 digits';  '5 digits'];
checks = [   'off'   ;   'on '    ;   'off'    ];

callbacks = [ 'fmt = 1;   fc_format';     ...
              'fmt = 2;   fc_format';     ...
              'fmt = 3;   fc_format' ];

for i=1:3
    Format_Digits_(i) = uimenu(Format_,             ...
                                                    ...
                         'Label',    labels(i,:),   ...
                         'Checked',  checks(i,:),   ...
                                                    ...
                         'Callback', callbacks(i,:));
end
```

Notice that the Checked attribute is turned on for the second label (4 digits).

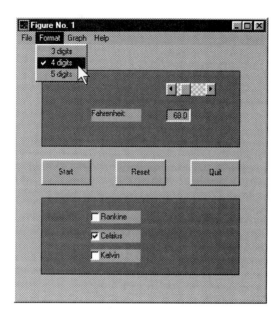

FIGURE 92. Entries in "Format" submenu

When the user clicks on one of the menu choices, MATLAB does *not* automatically toggle the submenu item to be Checked; this has to be done in the Callback string.[81]

The Callback string first records which choice was made in the variable fmt, and then invokes fc_format.m:

```
for i=1:3
    if fmt == i
         set(Format_Digits_(i), 'Checked', 'On');
    else
         set(Format_Digits_(i), 'Checked', 'Off');
    end;
end;
```

This also requires changes to fc_calc3.m when it writes out numbers. Put the new version in fc_calc4.m:

```
f = get(edit_f_, 'String');
f = str2num(f);

t(1) = f + 459.7;           % R
t(2) = (f - 32) * 5 / 9;    % C
t(3) = t(2) + 273.15;       % K

for i=1:3
    if show(i)
         ti = num2str( t(i), digits(fmt) );
                              ^^^^^^^^^^^
%
         set(text_c_(i), 'Visible', 'On', ...
```

---

[81] However, this does mean that you are free to treat a group of submenu items as being like radio buttons or like check boxes. In this example, we have chosen to treat them like radio buttons, as we only allow one choice for the number of digits in the output.

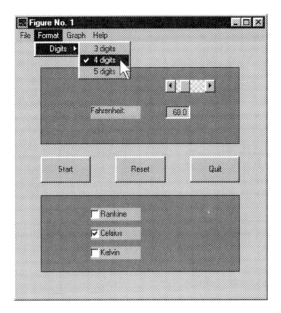

FIGURE 93. Entries in "Digits" subsubmenu

```
                            'String',   ti );
    else
        set(text_c_(i), 'Visible', 'Off');
    end
end
```

Note: You also have to change the Start push button uicontrol to now use fc_calc4.m:

```
set(push_start_,                      ...
        'Callback',   'fc_calc4');
```

Menus, Submenus, Sub-submenus, ...

In the previous two sections submenus were created under the top-level menu. You can also have submenus under the submenus.

Under Format, have a single menu item called Digits, and under it have a menu with the three choices for the number of digits.

Figure 93 shows what the screen looks like when the Digits item is selected. Notice that an arrow automatically appears beside Digits to indicate that there is a sub-submenu beside it.

The commands to produce this sub-submenu are:

```
delete(Format_Digits_(1));
delete(Format_Digits_(2));
delete(Format_Digits_(3));

%-------------------------------------------------
```

```
        callbacks = [ 'fmt = 1;    fc_digits';   ...
                      'fmt = 2;    fc_digits';   ...
                      'fmt = 3;    fc_digits'  ];
%                     ^^^^^^^^^
        Format_Digits_ = uimenu(Format_, 'Label', 'Digits');
%       ^^^^^^^^^^^^^^^^^^^^^^^^^^^^^^^^^^^^^^^^^^^^^^^^^^^
%               |
%               +---------------------------------+
%                                                 |
        for i=1:3                     %          \|/
                                      %  vvvvvvvvvvvvvv
                Format_Digits_Num_(i) = uimenu(Format_Digits_,   ...
                                                                 ...
                                        'Label',   labels(i,:),  ...
                                        'Checked', checks(i,:),  ...
                                                                 ...
                                        'Callback', callbacks(i,:));
        end
```

Notice that the "Callback" strings have been set to invoke `fc_digits.m`, which refers to the handles for the sub-submenu items:

```
        for i=1:3
            if fmt == i
                set(Format_Digits_Num_(i), 'Checked', 'On');
            else
                set(Format_Digits_Num_(i), 'Checked', 'Off');
            end;
        end;
```

Creating a Sub-window

Callback commands can themselves create separate figure windows. This is especially useful for creating a window, and then graphing your data within it.

For example, under Graph, add a menu item called Context Plot:

```
        uimenu(Graph_, 'Label',    'Context Plot',  ...
                       'Callback', 'fc_plot')
```

The Callback string invokes `fc_plot.m`. The most interesting part of its code is the first six lines, where it is determined if a new figure window has already been created. If it has, then you simply re-use that new figure window.

The handle of the new figure window is stored in the variable context_plot. The exist function in MATLAB is used to check whether that variable has been defined yet:

```
        if exist('context_plot', 'var')
            figure(context_plot)
        else
            figure
            context_plot = gcf;
        end

        fahr = get(edit_f_, 'String');
        fahr = str2num(fahr);

        f = [fahr-50, fahr, fahr+50];
```

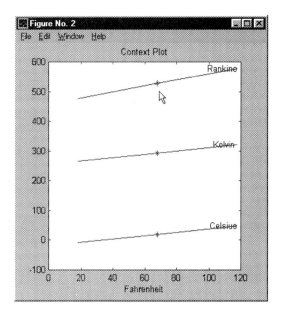

FIGURE 94. Context plot

```
t(1,1:3) = f + 459.7;        % R
t(2,1:3) = (f - 32) * 5 / 9; % C
t(3,1:3) = t(2,1:3) + 273.15; % K

names(1,:) = 'Rankine';
names(2,:) = 'Celsius';
names(3,:) = 'Kelvin ';

for i=1:3
    if show(i)
        plot(f, t(i,1:3), '-',   f(2), t(i,2), '*')
        text(f(3), t(i,3), names(i,:), ...
            'HorizontalAlignment', 'right');
        hold on
    end
end
hold off

xlabel('Fahrenheit')
title('Context Plot')
```

The resulting plot is shown in Figure 94 with a 100 degrees Fahrenheit window around the current temperature (shown as an asterisk).

Encapsulating GUI Calls in one Function

Up until now all GUI commands have been in scripts. However, sometimes it is more convenient to put them all together in a single function. In particular, this technique avoids naming conflicts with other variables in your main program.

This technique is often used when assigning three callback strings to graphical objects, one for each of the following user actions:

- clicking on an object,
- dragging the object with the mouse, and
- releasing the mouse button.

For example, let's create our own version of MATLAB's rotate3d function called az (which allows interactive rotation of 3-D plots), in order to illustrate the GUI programming involved.[82]

az.m is initially activated by the user by calling [83]

```
az on
```

to define, for the figure window, three callback strings for the three mouse situations. Each callback string consists of an invocation of the az function with different arguments:

az down is invoked when mouse is down in the window.

az motion is invoked when mouse is down and moving in the window.

az up is invoked when mouse is raised in the window.

Finally, the above meanings of mouse actions are cancelled when you issue:

```
az off
```

Here is the source to az.m. Notice the use of the global variable AZ_POINT. It is used to keep track of the previous mouse position between successive invocations of az:

```
function az(button)

global AZ_POINT

%-----------------------------------------------------------

if nargin == 0
    button = 'on';
end

if strcmp(button, 'on')
    set(gcf,'WindowButtonDownFcn',    'az down');
    set(gcf,'WindowButtonMotionFcn',  'az motion');
    set(gcf,'WindowButtonUpFcn',      'az up');

    AZ_POINT = [];       % Button up status
                         % (No starting point)

%-----------------------------------------------------------
```

---

[82] Visit our website at http://www.pracapp.com/matlab/ for further examples of how to *drag and drop* objects on a MATLAB screen.

[83] Recall that az on is the same to MATLAB as az('on').

```
    elseif strcmp(button, 'off')

        set(gcf,'WindowButtonDownFcn',    '');
        set(gcf,'WindowButtonMotionFcn',  '');
        set(gcf,'WindowButtonUpFcn',      '');

        clear global AZ_POINT

%-----------------------------------------------------------
    elseif strcmp(button, 'down')

        AZ_POINT = get(gcf, 'CurrentPoint');

%-----------------------------------------------------------
    elseif strcmp(button, 'motion')

        if AZ_POINT ~= []

            new_point = get(gcf, 'CurrentPoint');

            x = [AZ_POINT(1),  new_point(1)];
            y = [AZ_POINT(2),  new_point(2)];

            [azimuth, zenith] = view;

            position   = get(gcf, 'Position');
            AZ_WINDOW = position(3:4);                  % width, height

            azimuth = azimuth - 360. * (x(2) - x(1)) / AZ_WINDOW(1);
            zenith  = zenith  - 360. * (y(2) - y(1)) / AZ_WINDOW(2);

            view([ rem(azimuth, 360), ...
                   rem(zenith,  360) ])

            AZ_POINT = get(gcf, 'CurrentPoint');

        end

%-----------------------------------------------------------
    elseif strcmp(button, 'up')

        AZ_POINT = [];

end
```

Using GUIDE

MATLAB's GUIDE function initiates the **GUI D**esign **E**nvironment, which allows you to construct user interfaces graphically on the screen.

Let's re-create our above example using GUIDE. First we invoke GUIDE:

```
>> guide
```

The following windows appear:

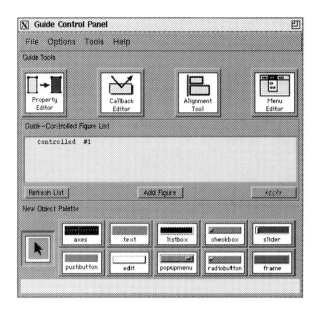

FIGURE 95. Guide Control Panel

1. Figure No. 1

2. Guide Control Panel

The Guide Control Panel is shown in Figure 95.

First let's construct the frames by clicking on the **frame** button, and then click the mouse in the figure window at the (0.1, 0.1) location - as indicated by the handy grid-lines - and then drag the mouse to the (0.9, 0.4) location and let go.

Similarly for the second frame: click on the **frame** button, and then click the mouse in the figure window at the (0.1, 0.6) location, and then drag the mouse to the (0.9, 0.9) location and let go.

You can move the frames by clicking on them and dragging. You can also resize the frames using the pull tabs in the corners or along the sides.

Now let's change the frame background colors to light gray. To do this we first click on the **Property Editor** in the **Guide Control Panel**. The **Property Editor** is shown in Figure 96.

We recommend you immediately check off the two boxes **Show Object Browser** and **Show Property List,** as shown in the figure. This will display one scrolling window with your graphics objects, and another scrolling window with their properties.

Make sure that one of our two frames (1 or 2) is selected in the object list. Then click on **BackgroundColor** in the property list, and replace the value beside it with [0.75, 0.75, 0.75]. Click on the other frame and change its color as well.

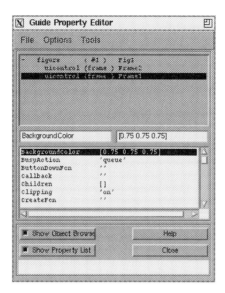

FIGURE 96. Guide Property Editor

Scroll down to the **Position** property and notice that the position is given in units of points. Now scroll down and click on the **Units** property for each frame. Click on the listbox and change the setting to **normalized** units. [84]

Scroll back up to the **Position** property, and notice that the position is now given in normalized coordinates. Now change these coordinates to round numbers (like **0.1**) so that the MATLAB code that GUIDE produces for you will be easier to read.

Now let's take a look at the code MATLAB generates for our figure so far. Click on the **File** menu of the Figure window, and select **"Save Figure As..."** and type in a filename such as "tempgui". When you click OK, MATLAB will create the file "tempgui.m" containing the following MATLAB code:

```
function tempgui()

% This is the machine-generated representation of ...

load tempgui

a = figure('Color',[0.8 0.8 0.8], ...
  'Colormap',mat0, ...
  'PointerShapeCData',mat1, ...
  'Position',[82 37 384 384], ...
  'Tag','Fig1', ...
  'DefaultuicontrolUnits','normalized');
b = uicontrol('Parent',a, ...
  'Units','normalized', ...
  'BackgroundColor',[0.5 0.5 0.5], ...
  'Position',[0.1 0.1 0.8 0.3], ...
```

---

[84] If you wish to change the units of *all* your uicontrols at once, you can issue the following from the MATLAB command window: `kids = findobj(gcf, 'Type', 'uicontrol')` and `set(kids, 'Units', 'Normalized')`.

```
'Style','frame', ...
'Tag','Frame1');
b = uicontrol('Parent',a, ...
'Units','normalized', ...
'BackgroundColor',[0.5 0.5 0.5], ...
'Position',[0.1 0.6 0.8 0.3], ...
'Style','frame', ...
'Tag','Frame2');
```

MATLAB also creates a file called "tempgui.mat" containing any data needed by "tempgui.m", such as the figure's color map.

Carrying on ...

Let's re-load the user interface described in "tempgui.m". Close the figure window, and also quit the **Guide Control Panel** and all tools.

Then invoke the function (tempgui) describing the user interface:

```
>> tempgui
```

and finally re-invoke guide:

```
>> guide
```

Now we can continue adding GUI objects to the figure.

Notes

We recommend only using GUIDE for creating the basic arrangement and appearance of your objects on the screen. Use your main MATLAB program to specify the Callback properties of the objects, so that all of your executable MATLAB code is together in one place where you can easily see it or modify it.

Let's define the remaining GUI controls in our example:

- Use GUIDE to create a text uicontrol with string "Fahrenheit" positioned at [0.3 0.7 0.2 0.05] in normalized units, left-aligned.

- Create an edit uicontrol with string "68.0" positioned at [0.6 0.7 0.1 0.05] in normalized units, right-aligned.

- Create a slider uicontrol with Min, Max and Value properties of 32, 212 and 68, respectively, positioned at [0.6 0.8 0.2 0.05] in normalized units.

- Create three (3) checkbox uicontrols with **string** properties of 'Rankine', 'Celsius', and 'Kelvin', respectively, **value** properties of 0, 1 and 0, respectively, and positioned at [0.3 0.300 0.2 0.05], [0.3 0.225 0.2 0.05], and [0.3 0.150 0.2 0.05], respectively, in normalized units.

- Create three (3) text uicontrols all with strings of '20', all right-aligned, and positioned at [0.6 0.300 0.1 0.05], [0.6 0.225 0.1 0.05], and [0.6 0.150 0.1 0.05], respectively, in normalized units. Set their **visible** property to 'off', 'on', and 'off', respectively.

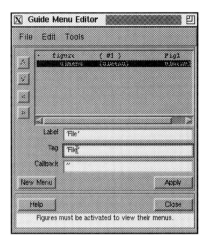

FIGURE 97. Guide Menu Editor

- Create three (3) pushbutton uicontrols with strings of Start, Reset, and Quit, respectively, all with a value of 0, and positioned at [0.1 0.45 0.2 0.1], [0.4 0.45 0.2 0.1], and [0.7 0.45 0.2 0.1], respectively, in normalized units.

We will define the **Callback**s for these uicontrols all together in our main program (which is in a later section).

Defining Menus

Click on GUIDE's **Menu Editor** to bring up the screen shown in Figure 97.

Click on **figure** (#1) in the editor, and then click on **New Menu**.

Fill in the **Label** field with 'File', and fill in the **Tag** field with 'File', then click on **Apply**.

To add a menu item below **File**, click on **uimenu** (File), and the click on **New Menu**.

Fill in the **Label** field with 'Open...', and fill in the **Tag** field with 'File_Open', then click on **Apply**.

In similar fashion you can define the remaining menu items under **File**, namely: **Save...**, **Append...**, **Print...**, and **Exit**.

In the **Guide Property Editor**, turn the **Separator** property to 'On' for the Save, Print and Exit menu items so that separator lines will appear *above* each of these items in the File menu.

Defining Callbacks

While creating or editing your objects in GUIDE, we recommend that you assign a value to the **tag** property of each object, which you can use later to refer to the object.

For example, assign the following tags to the frames in our example:

Bottom Frame   'frame1'
Top Frame   'frame2'

From our main MATLAB program, we can now invoke the tempgui function (which GUIDE produced) to create the interface, and then use findobj to return the handles of our objects based on the value of their **tag** property:

```
tempgui

edit_f_     = findobj(gcf, 'Tag', 'edit_f');
slider_f_   = findobj(gcf, 'Tag', 'slider_f');
check_c_    = findobj(gcf, 'Tag', 'check_c');
push_start_ = findobj(gcf, 'Tag', 'push_start');
push_reset_ = findobj(gcf, 'Tag', 'push_reset');
push_quit_  = findobj(gcf, 'Tag', 'push_quit');
File_Open_  = findobj(gcf, 'Tag', 'File_Open');
File_Save_  = findobj(gcf, 'Tag', 'File_Save');
File_Append_= findobj(gcf, 'Tag', 'File_Append');
File_Print_ = findobj(gcf, 'Tag', 'File_Print');
File_Exit_  = findobj(gcf, 'Tag', 'File_Exit');

set(edit_f_,      'CallBack', 'fc_edit_f')
set(slider_f_,    'CallBack', 'fc_slider_f')
set(check_c_,     'CallBack', 'fc_check_c')
set(push_start_,  'CallBack', 'fc_calc3')
set(push_reset_,  'CallBack', 'fc_reset')
set(push_quit_,   'CallBack', 'fc_quit')
set(File_Open_,   'Callback', 'fc_open')
set(File_Save_,   'Callback', 'perm = ''wt''; fc_save')
set(File_Append_, 'Callback', 'perm = ''at''; fc_save')
set(File_Print_,  'Callback', 'fc_print')
set(File_Exit_,   'Callback', 'fc_quit')
```

If one wished to simplify this code, one could create a **utility** function in tag.m:

```
function h = tag(t)

h = findobj(gcf, 'Tag', t);
```

which returns the handle of the object whose tag is passed to it. This simplifies the above main program to:

```
tempgui

set( tag('edit_f'),       'CallBack', 'fc_edit_f'   )
set( tag('slider_f'),     'CallBack', 'fc_slider_f' )
set( tag('check_c'),      'CallBack', 'fc_check_c'  )
set( tag('push_start'),   'CallBack', 'fc_calc3'    )
set( tag('push_reset'),   'CallBack', 'fc_reset'    )
set( tag('push_quit'),    'CallBack', 'fc_quit'     )

set( tag('File_Open'),    'Callback', 'fc_open'     )
set( tag('File_Save'),    'Callback', ...
                          'perm = ''wt''; fc_save'  )
set( tag('File_Append'),  'Callback', ...
                          'perm = ''at''; fc_save'  )
set( tag('File_Print'),   'Callback', 'fc_print'    )
set( tag('File_Exit'),    'Callback', 'fc_quit'     )
```

GUIDE greatly simplifies the process of creating user interfaces, but it does *not* replace the need to have both a reading and a writing knowledge of the code it subsequently produces.

Miscellaneous Topics

Miscellaneous GUI Functions - gtext, ginput, rbbox

MATLAB provides other useful functions which are commonly used to help annotate a graph:

gtext Up until now you have used the text function to add text to a graph, and using it you had to specify the (x, y) coordinates:

```
text(x, y, 'string')
```

However, using gtext you specify only the string to be plotted, and then click at the location on the graph where you want the text positioned:

```
gtext('string')
```

This greatly simplifies the task of graph annotation, as in the case of the graph produced by:

```
plot([1 2 3], [1 4 9], '-');    hold on
plot([1 2 3], [1 4 9], 'o');    hold off

h = gtext('Interesting point');
set(h, 'FontAngle', 'oblique')
```

ginput: This command allows you to click on points in a figure window, and have their coordinates saved in x and y vectors: [85]

```
[x, y] = ginput(npoints)
```

The points are readily plotted, if you wish. For example:

```
plot(x, y, '-')
```

Here are two examples of using ginput in actual practice:

pretext: The first example is a function called pretext which adds text *and* a pointer interactively to a plot. It is invoked much like gtext:

```
pretext('string')
```

The file pretext.m contains the following MATLAB code:

```
function [hl,ht] = pretext(string)

disp('Click where want pointer, then where want text.')
[x, y] = ginput(2);

%-----------------------------------------------------------

if x(1) < x(2), horiz = 'left'; end
if x(1) > x(2), horiz = 'right'; end

if y(1) < y(2), vert = 'cap'; end
if y(1) > y(2), vert = 'cap'; end
```

---

[85] Other ways of using this function are given in the *Command Listing* section.

```
%------------------------------------------------------------
h1 = line(x, y);

ht = text(x(2), y(2), [' ', string, ' '], ...
          'HorizontalAlignment', horiz,     ...
          'VerticalAlignment',   vert);
```

rect: The second example is a function called rect to interactively add a rectangular area to a plot. It is invoked with no arguments:

```
rect
```

The file rect.m contains the following MATLAB code:

```
function h = rect

disp('Click at opposite corners of rectangle.')

[x,y] = ginput(2);

xx = [x(1) x(2) x(2) x(1) x(1)];
yy = [y(1) y(1) y(2) y(2) y(1)];

h = line(xx, yy);
```

rbbox rbbox puts a so-called "rubberband box" on the screen. One corner is anchored, while the opposite corner tracks your mouse movements.

Consequently, the box changes size and shape as the mouse moves. The mouse button must be pressed down throughout this operation. When you are finished, release the mouse button and the box disappears from the screen.

The following example is similar to the previous one except here you use rbbox to define the rectangle to be drawn. This gives the user much more graphical feedback.

This function is invoked with no arguments:

```
rectbox
```

The file rectbox.m contains the following MATLAB code:

```
function h = rectbox

disp('Click and drag in graph area')
waitforbuttonpress;

a = get(gca, 'CurrentPoint');
xa = a(1,1);   ya = a(1,2);

f = get(gcf, 'CurrentPoint');
xf = f(1);     yf = f(2);
rbbox([xf, yf, 0, 0], [0, 0])

b = get(gca, 'CurrentPoint');
xb = b(1,1);   yb = b(1,2);

h = line([xa, xb, xb, xa, xa], ...
         [ya, ya, yb, yb, ya]);
```

Notes

- Note the use of waitforbuttonpress. As its name suggests, it pauses MATLAB until the mouse button is pressed down, and then execution resumes. It is used so that the mouse is in the required down position when rbbox is subsequently invoked.
- The coordinates in the call to line are in *axis units*, as we are accustomed to using. However, notice that rbbox uses *figure units*.

Command Listing

Accelerator char

Specifies the *accelerator key* character char for an interface menu object.
Output: Not applicable.
Additional information: Each platform has its own particular keystroke combination for accelerator keys. See the on-line help for those relevant to your system. ✤ In order for an accelerator key to work, the window focus must be within the relevant figure. ✤ To set the accelerator key of an interface menu object, use set or uimenu. ✤ To retrieve the accelerator key of an interface menu object, use get.
See also: set, get, uimenu

align

Initiates a uicontrol and axes alignment tool.
Output: Not applicable.
Additional information: See the on-line help for more details on align.
See also: guide, cbedit, ctlpanel, menuedit, propedit

BackGroundColor [num_r, num_g, num_b]

Specifies that an interface control object or menu object has background color defined by RGB (red, green, blue) values num_r, num_g, and num_b, respectively.
Argument options: BackGroundColor str to specify one of the preset colors. str can be any of the long or short strings *yellow*, *y*, *magenta*, *m*, *cyan*, *c*, *red*, *r*, *green*, *g*, *blue*, *b*, *white*, *w*, *black*, and *k*.
Output: Not applicable.
Additional information: The values of num_r, num_g, and num_b must all be in the range [0, 1]. ✤ To set the background color of an interface object, use set, uicontrol, or uimenu. ✤ To retrieve the background color of an interface object, use get.
See also: set, get, uicontrol, *ColorSpec*, colormap

CallBack str

Specifies a string representing a MATLAB expression, value, or function that is evaluated when an interface object is activated.
Output: Not applicable.
Additional information: When the object is activated, str is passed to the function eval. For more information on what activates the various types of interface objects,

see the entry for Style or uimenu. ✸ To set the callback string of an interface object, use set, uicontrol, or uimenu. ✸ To retrieve the callback string of an interface object, use get.
See also: Style, set, get, uicontrol, uimenu, eval, WindowButtonDownFnc

cbedit
Initiates a callback string editing tool.
Output: Not applicable.
Additional information: See the on-line help for more details on cbedit.
See also: guide, align, ctlpanel, menuedit, propedit

Checked *off*
Specifies that an interface menu item is *not* checked.
Argument options: Checked *on* to specify that an item does have a check next to it.
Output: Not applicable.
Additional information: There is no way to specify that a menu item is not able to be checked. ✸ To set the check status, use set, or uimenu. ✸ To retrieve the check status, use get.
See also: set, get, uimenu

choices(str_n, str, M_{lab}, M_{call})
Creates a "list of choices" dialog box with name str_n, text str, button labels specified by string matrix M_{lab}, and callback strings specified by string matrix M_{call}.
Output: Not applicable.
See also: dialog, questdlg, inputdlg, errordlg, warndlg, helpdlg, listdlg, msgbox

ctlpanel
Initiates the control panel for the guide tool.
Output: Not applicable.
Additional information: See the on-line help for more details on ctlpanel.
See also: guide, align, cbedit, menuedit, propedit

hndl = dialog(str_1, M_1, ..., str_n, M_n)
Creates a dialog box with options defined by the given parameter/value pairs.
Output: A figure handle is assigned to hndl.
Additional information: The dialog function accepts all parameters available to the figure function.
See also: figure, choices, questdlg, inputdlg, errordlg, warndlg, helpdlg, listdlg, msgbox

dragrect(M_{rect})
Creates dragable rectangles at positions defined by the $n \times 4$ matrix M_{rect}.
Output: Not applicable.
Argument options: dragrect(M_{rect}, num) to specify the step size of rectangle movements, num. ✸ M_{out} = dragrect(M_{rect}) to return the location of the rectangles after dragging has been completed.
Additional information: Rectangles can be located anywhere on the screen.
See also: rbbox

Enable *off*
Specifies that an interface menu item is dimmed, and therefore cannot be selected.
Argument options: Enable *on* to specify an item is able to be chosen. This is the default.
Output: Not applicable.
Additional information: To set the enabledness of an object, use set, or uimenu. ✤ To retrieve the enabledness of an object, use get.
See also: set, get, *Visible*, uimenu

hndl = **errordlg**(str$_{err}$, str$_{title}$)
Creates an error dialog box titled str$_{title}$, containing the message str$_{err}$.
Output: The handle for the dialog is assigned to hndl.
Argument options: errordlg(str$_{err}$, str$_{title}$, *'on'*) to specify that if an identically titled error dialog already exists, that it simply be brought to the front.
Additional information: The user must click on the OK button before continuing with the current session.
See also: dialog, helpdlg, warndlg, questdlg, inputdlg, listdlg, msgbox, choices

ForeGroundColor [num$_r$, num$_g$, num$_b$]
Specifies that the text in an interface control or menu control object has color defined by RGB (red, green, blue) values num$_r$, num$_g$, and num$_b$, respectively.
Argument options: ForeGroundColor str to specify one of the preset colors. str can be any of the long or short strings *yellow*, *y*, *magenta*, *m*, *cyan*, *c*, *red*, *r*, *green*, *g*, *blue*, *b*, *white*, *w*, *black*, and *k*.
Output: Not applicable.
Additional information: The values of num$_r$, num$_g$, and num$_b$ must all be in the range [0, 1]. ✤ To set the text color of an interface object, use set, uicontrol, or uimenu. ✤ To retrieve the text color of an interface object, use get.
See also: set, get, uicontrol, *ColorSpec*, colormap

guide
Initiates a GUI design tool.
Output: Not applicable.
Additional information: See the on-line help or the above tutorial for more details on guide.
See also: align, cbedit, ctlpanel, menuedit, propedit

hndl = **helpdlg**(str$_{help}$, str$_{title}$)
Creates a help dialog box titled str$_{title}$, containing the message str$_{help}$.
Output: The handle for the dialog is assigned to hndl.
Additional information: If a help dialog with the same title already exists, it is simply brought to the front. ✤ The user must click on the OK button before continuing with the current session.
See also: dialog, errordlg, warndlg, questdlg, inputdlg, listdlg, msgbox, choices

HorizontalAlignment *center*
Specifies the horizontal justification of the labeling text as centered with respect to the interface object.

Argument options: HorizontalAlignment *right* to set the labeling text to right justification. This is the default. ✦ HorizontalAlignment *left* to set the labeling text to left justification.
Output: Not applicable.
Additional information: To set the justification of the labeling text, use set or uicontrol. ✦ To retrieve the justification of the labeling text, use get.
See also: set, get, String, Position

$M_{c,out}$ = inputdlg($M_{c,in}$)
Creates an input dialog box with input fields specified by cell array $M_{c,in}$.
Output: A cell array of the values input by the user is assigned to $M_{c,out}$.
Additional information: See the on-line help file for details on optional parameters.
See also: dialog, questdlg, errordlg, warndlg, helpdlg, listdlg, msgbox, choices

Label str
Specifies the string to be printed on the interface menu object.
Output: Not applicable.
Argument options: Label 'str_1 — str_2 — ... — str_n' to specify n lines of text. The quotes must go around the entire string.
Additional information: To set the string of an interface menu object, use set or uimenu. ✦ To retrieve the string of an interface menu object, use get.
See also: set, get, uimenu

[V, int] = listdlg(str_1, M_1, ..., str_n, M_n)
Creates a list selection dialog box whose parameters are defined by parameter/value pairs str_1/M_1 through str_n/M_n.
Output: A vector of indices of the selections is assigned to V. If the "OK" button is pushed, a value of 1 is assigned to int. Otherwise, a value of 0 is assigned to int.
Additional information: See the on-line help file for the complete list of valid parameter/value pairs.
See also: dialog, questdlg, inputdlg, errordlg, warndlg, msgbox, choices

Max num
Specifies the maximum value allowed for the Value property of the interface object.
Output: Not applicable.
Additional information: The Max property is closely tied to the Value property. ✦ The Max property is only relevant for interface objects of type *radiobutton*, *checkbox*, *edit*, *slider*, and *popupmenu*. For more information, see the entry for Style. ✦ The default value of Max is 1. ✦ To set the maximum value of the interface object, use set or uicontrol. ✦ To retrieve the maximum value of the interface object, use get.
See also: set, get, Min, Value, Style

menuedit
Initiates a menu editing tool.
Output: Not applicable.
Additional information: See the on-line help for more details on menuedit.
See also: guide, align, cbedit, ctlpanel, propedit

Min num

Specifies the minimum value allowed for the Value property of the interface object.
Output: Not applicable.
Additional information: The Min property is closely tied to the Value property. ✦ The Min property is only relevant for interface objects of type *radiobutton*, *checkbox*, *edit*, *slider*, and *popupmenu*. For more information, see the entry for Style. ✦ The default value of Min is 0. ✦ To set the minimum value of the interface object, use set or uicontrol. ✦ To retrieve the minimum value of the interface object, use get.
See also: set, get, Max, Value, Style

msgbox(str)

Creates a message box containing the message str.
Output: Not applicable.
Argument options: msgbox(str, str_{title}) to apply a title to the message box. ✦ msgbox(str, str_{title}, str_{icon}), where str_{icon} is *'none'*, *'error'*, *'help'*, *'warn'*, or *'custom'*, to alter the icon used in the message box. ✦ hndl = msgbox(str) to return the handle of the message box.
Additional information: If the *'custom'* option is used, then icon data must be supplied. See the help file for details.
See also: dialog, questdlg, inputdlg, errordlg, warndlg, helpdlg, listdlg, choices

Position [num_l, num_b, num_w, num_h]

Specifies the size and position of the interface control within its figure, by giving values for the distance from the left side and bottom edge of the figure, the object's width, and the object's height, respectively.
Output: Not applicable.
Additional information: The specifications are in the units specified by the Units property of the object. ✦ To set the position of an interface object, use set, or uicontrol. ✦ To retrieve the position of an object, use get.
See also: set, get, uicontrol, *figure*, Units

Position num

Specifies the position of the interface menu object relative to the other such objects on its level.
Output: Not applicable.
Additional information: The lower valued objects are placed to the left of the menu bar and the top of individual menus or submenus. ✦ To set the position of an interface menu object, use set, or uimenu. ✦ To retrieve the position of an interface menu object, use get.
See also: set, get, uimenu

propedit

Initiates a property editing tool.
Output: Not applicable.
Additional information: See the on-line help for more details on propedit.
See also: guide, align, cbedit, ctlpanel, menuedit

str = questdlg(str$_q$, str$_1$, str$_2$, str$_3$)

Creates a question dialog box containing the question str$_q$, with up to three response buttons labelled str$_1$, str$_2$, and str$_3$.
Output: The chosen response is assigned to str.
Additional information: The user must click on one of the response buttons before continuing with the current session.
See also: dialog, warndlg, errordlg, listdlg, msgbox, choices

rbbox([num$_l$, num$_b$, num$_w$, num$_h$], [num$_x$, num$_y$])

Creates a rubberband box in the current figure, whose initial location is defined by left, bottom, width and height values in the first parameter, and initial tracking point [num$_x$, num$_y$].
Output: Not applicable.
Additional information: The mouse button must be pressed down when rbbox is called. We recommend that you call waitforbuttonpress to ensure a mouse down condition. ✶ Typically, the values num$_w$, num$_h$, num$_x$, and num$_y$ are all 0. ✶ To determine the final dimensions of the box, use the CurrentPoint property of the current figure.
See also: dragrect, waitforbuttonpress, CurrentPoint

Separator *on*

Specifies that horizontal line be drawn above the text for an interface menu object.
Argument options: Separator *off* to specify that no separator line is drawn. This is the default.
Output: Not applicable.
Additional information: To set the separator line, use set, or uimenu. ✶ To retrieve the separator line, use get.
See also: set, get, uimenu

String *str*

Specifies the string to be printed on the interface object.
Output: Not applicable.
Argument options: String 'str$_1$ — str$_2$ — ... — str$_n$' to specify n lines of text. The quotes must go around the entire string.
Additional information: For interface objects of type *radiobutton*, *pushbutton*, *checkbox*, *slider*, and *popupmenu*, str specifies the label or title on the object. ✶ For interface objects of type *edit*, str represents the string in the edit field, and is updated with the text entered by the user. ✶ To set the string of an interface object, use set or uicontrol. ✶ To retrieve the string of an interface object, use get.
See also: set, get, HorizontalAllignment, uicontrol, Style

Style *slider*

Specifies that the interface object is a *slider*.
Argument options: Style *checkbox* to create a check box control. ✶ Style *pushbutton* to create a push button control. ✶ Style *radiobutton* to create a radio button control. ✶ Style *popupmenu* to create a popup menu control. ✶ Style *edit* to create an editable text field.
Output: Not applicable.

Additional information: Style must be explicitly set for each interface object; there is no default value. ✤ A slider allows you to move a small box between a maximum and minimum value. In most cases, up and down arrows are provided at the ends of the slider to allow you to move in increments of 1/100 of the total range. ✤ A check box allows you toggle a switch between on (1) and off (0). ✤ A push button allows you to activate a process each time you push it. ✤ A radio button allows you toggle a switch between on (1) and off (0). You must code yourself the mechanism that keeps one and only one button activated at any time. ✤ A popup menu displays a list of choices. The currently chosen option is the label for the unactivated menu. ✤ An editable text field is just that: a place to enter text. If the related values Max - Min ≤ 1 then single line mode is used, and a carriage return completes the input. In all other cases, clicking off of the text field or entering Control-Return completes the input. ✤ After any object is activated, its CallBack property is entered. ✤ To set the interface object style, use set or uicontrol. ✤ To retrieve the interface object style, use get.
See also: set, get, uicontrol, CallBack, Value, String

hndl = **uicontrol**(name$_1$, expr$_1$, ..., name$_n$, expr$_n$)

Creates a user interface control object with n properties specified by name$_1$ through name$_n$ set to the values expr$_1$ through expr$_n$, respectively.
Output: A handle to an interface control object is assigned to hndl.
Argument options: hndl$_{sub}$ = uicontrol(hndl, name$_1$, expr$_1$, ..., name$_n$, expr$_n$) to create a subitem to the interface control object represented by handle hndl.
Additional information: The most important property of an interface control object is its *Style*. There are six styles available, including push buttons, check boxes, sliders, etc. For more information, see the entry for Style. ✤ To create a standard menu, use uimenu. ✤ Interface control objects are the children of figure objects and have no children of their own. ✤ Some of the available properties include *BackGroundColor, CallBack, Children, ForeGroundColor, HorizontalAllignment, Max, Min, Parent, Position, String, Style, Type, Units, UserData, Value,* and *Visible*. See the individual entries for these properties for information on valid values. ✤ The values of individual properties can be set or retrieved using set and get, respectively. ✤ Until an interface control object is applied to a figure, nothing is displayed.
See also: uimenu, uigetfile, uiputfile, figure, *set, get*

[str$_{file}$, str$_{path}$] = **uigetfile**(str$_{filter}$, str$_{title}$, num$_x$, num$_y$)

Creates a dialog box beginning at pixel location [num$_x$, num$_y$] with title str$_{title}$, which allows the user to choose an existing file in the group defined by filter str$_{filter}$.
Output: Two strings representing the file name and path of the chosen file are assigned to str$_{file}$ and str$_{path}$.
Argument options: uigetfile to allow the system to use default values for all input parameters.
Additional information: A typical value for str$_{filter}$ is '.m'. On most systems, this filtering string can be changed within the dialog created. ✤ Some platforms do not allow you to set the initial location of the dialog. ✤ If a valid file is not selected,

then both output strings are set to 0. ✤ There are no special properties that can be set with this dialog box, as there are with uicontrol and uimenu.
See also: uiputfile, uicontrol, uimenu

hndl = **uimenu**(name$_1$, expr$_1$, ..., name$_n$, expr$_n$)

Creates a user interface menu object with n properties specified by name$_1$ through name$_n$ set to the values expr$_1$ through expr$_n$, respectively.
Output: A handle to an interface menu object is assigned to hndl.
Argument options: hndl$_{sub}$ = uimenu(hndl, name$_1$, expr$_1$, ..., name$_n$, expr$_n$) to create a subitem to the interface menu represented by handle hndl.
Additional information: When called at the top level, uimenu creates a one element menu placed along the top of the current figure. ✤ In order to have choices for a top-level menu, submenu items must be created. ✤ Remember that each call to uimenu can create only one menu or submenu item. ✤ The order in which the menu or submenu items are displayed is control by the Position property. ✤ To create more complex objects, use uicontrol. ✤ Interface menu objects are the children of figure objects or other uimenu objects. ✤ Some of the available properties include *Accelerator, BackGroundColor, CallBack, Checked, Children, Enable, ForeGroundColor, Label, Parent, Position, Separator, Type, UserData,* and *Visible*. See the individual entries for these properties for information on valid values. ✤ The values of individual properties can be set or retrieved using set and get, respectively. ✤ Until an interface menu object is applied to a figure, nothing is displayed.
See also: uicontrol, uigetfile, uiputfile, figure, *set, get*

[str$_{file}$, str$_{path}$] = **uiputfile**(str$_{filter}$, str$_{title}$, num$_x$, num$_y$)

Creates a dialog box beginning at pixel location [num$_x$, num$_y$] with title str$_{title}$, which allows the user to choose an existing file or create a new file in the group defined by filter str$_{filter}$.
Output: Two strings representing the file name and path of the file are assigned to str$_{file}$ and str$_{path}$.
Argument options: uiputfile to allow the system to use default values for all input parameters.
Additional information: This command is typically used when you want to write to a file. ✤ A typical value for str$_{filter}$ is '.m'. On most systems, this filtering string can be changed within the dialog created. ✤ Some platforms do not allow you to set the initial location of the dialog. ✤ If a valid file is not selected, then both output strings are set to 0. ✤ There are no special properties that can be set with this dialog box, as there are with uicontrol and uimenu.
See also: uigetfile, uicontrol, uimenu

uiresume(hndl)

Resume execution of halted uicontrol hndl.
Output: Not applicable.
Additional information: The uiwait function halts execution of the uicontrol.
See also: uiwait, waitfor

uiwait(hndl)
Halts/blocks execution of uicontrol hndl.
Output: Not applicable.
Argument options: uiwait to halt the current figure.
Additional information: The uiresume function resumes execution of the uicontrol.
See also: uiresume, waitfor

Value num
Specifies the current (or starting) value of the interface object.
Output: Not applicable.
Additional information: The Value property is closely tied to the Min and Max properties. ✦ The Value property is only relevant for interface objects of type *radiobutton*, *checkbox*, *slider*, and *popupmenu*. For more information, see the entry for Style. ✦ The default value of Value depends on the style of interface object created. ✦ To set the value of the interface object, use set or uicontrol. ✦ To retrieve the value of the œinterface object, use get.
See also: set, get, Min, Max, Style

waitfor(hndl)
Returns control once the graphic object hndl is deleted.
Output: Not applicable.
Argument options: waitfor(hndl, str_p), where str_p is a property name, to return when the value of that property for hndl changes. ✦ waitfor(hndl, str_p, M) to return when the value of the str_p property for hndl changes to M.
Additional information: Nested calls to waitfor are supported.
See also: uiresume, uiwait

hndl = warndlg(str_{warn}, str_{title})
Creates a warning dialog box titled str_{title}, containing the message str_{warn}.
Output: The handle for the dialog is assigned to hndl.
Additional information: The user must click on the OK button before continuing with the current session.
See also: dialog, questdlg, inputdlg, errordlg, listdlg, msgbox, choices

Programming in MATLAB

MATLAB possesses the *traditional* programming constructs found in other languages. These include for-loops, while-loops, if-elseif-else blocks, and case statements.

MATLAB *also* possesses many *matrix-oriented* functions and operators not found in traditional languages. They provide functionality similar to the above, but in a more concise, mathematical notation. Another advantage to using matrix-oriented functions and operators is that they generally execute more quickly. However, since they cannot be used in all instances, both the traditional constructs and the matrix-oriented constructs are covered here, and some guidance is given on when to use each.

This chapter is divided into four sections:

1. Comprehensive Example

2. Looping (for, while, ...)

3. Testing (if, find, case, ...)

4. User-defined Functions

The Comprehensive Example program in the first section is referred to in the later sections, with its variations illustrating alternative programming constructs.

MATLAB also supports *object-oriented* programming constructs. An object-oriented version of this example is presented in the *Object-Oriented Programming* chapter for comparison.

Comprehensive Example - Planetary Motion

Consider the plots shown in Figures 98 and 99. The entire elliptical paths of the planets are shown as well as how far they moved in 1995 after 0.5 Earth years.

The mainline program (called a *script* in MATLAB) to produce the plot is shown below:

```
%------------------------------------------------
% Declare global variables (from data and starts):
%------------------------------------------------

      globals

%------------------------------------------------
```

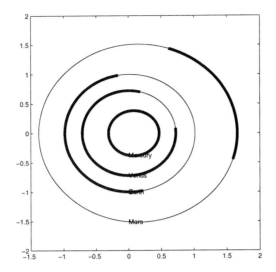

FIGURE 98. Planetary motion of 0.5 Earth years

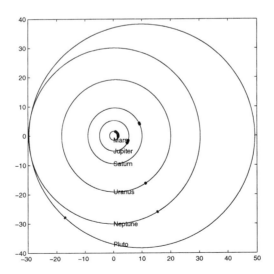

FIGURE 99. Planetary motion of 0.5 Earth years

```
% Read in planetary information:
%------------------------------------------------

    data

%------------------------------------------------
% Read in known Start time and angular positions:
%------------------------------------------------

    starts          %%%  Start_t = 1995 + 3/365
                    %%%  Start_th(Mercury) = ...

%------------------------------------------------
% Prompt for Final time and solve for positions:
%------------------------------------------------

    finals          %%%  Final_t = prompts
                    %%%  Final_th(Mercury) = ...

%------------------------------------------------
% Plot Start and Final Times and Positions:
%------------------------------------------------

    plots

%------------------------------------------------
% Repeat "finals" and "plots" as desired:
%------------------------------------------------

    repeats
```

Notice that it consists of no MATLAB commands. Rather, it invokes a series of special-purpose MATLAB scripts, which in turn invoke MATLAB commands to perform each task.

The "call tree" showing which scripts are called by the mainline script is shown in Figure 100: It also shows which scripts or functions that they in turn call. Notice that function names are shown in italics. Also note that the mainline script as well as several functions invoke the script in globals.m. It declares global many of the variables that are used by more than one function.

Reading Data

The data needed by this program includes:

1. Planet name (e.g., Earth) and position (e.g., 3).
2. Period for each planet to revolve around the Sun (e.g., Earth = 1 year).
3. Details on each planet's elliptical orbit, such as their mean distance from the Sun (e.g., Earth = 1 astronomical unit).

The script to read the data (data.m) in turn invokes five other scripts:

```
%------------------------------------------------

    idnums          %%%  Earth = 3
```

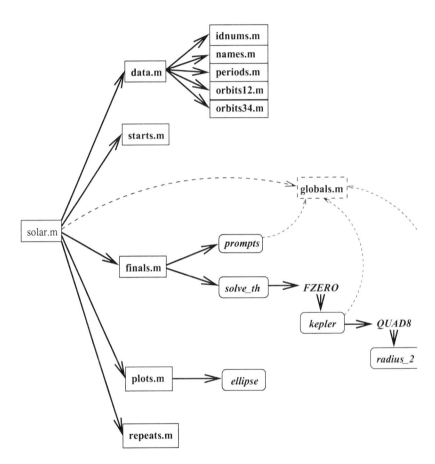

FIGURE 100. Tree diagram showing scripts and functions used by solar.m

```
names              %%%  Name(Earth,:) = 'Earth  '

%-------------------------------------------------

periods            %%%  Period(Earth) = 1.0

orbits12           %%%  Orbit(Earth,1:2) = [dist, ecc]
orbits34           %%%  Orbit(Earth,3:4) = [ang, incl]

%-------------------------------------------------
```

Here is a typical script, periods.m, which contains each planet's period:

```
% -----------------------------------------------
Period = zeros(9,1);

Period(Mercury) =   0.24;    % rel. to Earth
Period(Venus)   =   0.62;
Period(Earth)   =   1.00;

Period(Mars)    =   1.88;
Period(Jupiter) =  11.86;
Period(Saturn)  =  29.46;

Period(Uranus)  =  84.0 ;
Period(Neptune) = 164.8 ;
Period(Pluto)   = 247.7 ;

% -----------------------------------------------
% Source: _Pictorial_Astronomy_, 3rd rev. ed.,
% by Alter et al. New York, Crowell, 1969, p. 93.
% -----------------------------------------------
```

Notice that the data was entered directly using MATLAB assignment statements, i.e., as opposed to using load to load a data file containing only a matrix of numbers. The latter approach would have been easier to type, but more difficult to check without the names, so the former approach was taken. [86]

Notice that we use the planet names as indices into the Period vector. This was made possible by first creating a scalar variable for each planet with its corresponding position from the Sun (in idnums.m):

```
Mercury = 1;
Venus   = 2;
Earth   = 3;

Mars    = 4;
Jupiter = 5;
Saturn  = 6;

Uranus  = 7;
Neptune = 8;
Pluto   = 9;
```

These indices are used in the remaining data files as well:

1. names.m stores the planet names in a string array (called Name), for use in labeling textual and graphical output:

---

[86] However, two ways of using "pure" data files are illustrated in the final section. One of these ways allows you to store the names as comments alongside the data, to simplify checking.

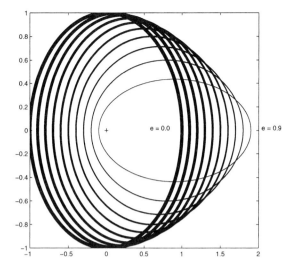

FIGURE 101. Effect of increasing eccentricity, e, on the shape of an ellipse

```
Name = '';

Name(Mercury,:)  =  'Mercury';
Name(Venus,   :) =  'Venus  ';
Name(Earth,   :) =  'Earth  ';

Name(Mars,    :) =  'Mars   ';
Name(Jupiter,:)  =  'Jupiter';
Name(Saturn, :)  =  'Saturn ';

Name(Uranus, :)  =  'Uranus ';
Name(Neptune,:)  =  'Neptune';
Name(Pluto,  :)  =  'Pluto  ';
```

2. orbits12.m stores into the first two columns of the 9 × 4 Orbit matrix:

   - the *mean distance from the Sun*, a, and
   - the *eccentricity*, e.

Together, these two quantities define the elliptical path of each planet around the Sun.

Refer to Figure 101 which shows the effect of increasing the eccentricity, e, from 0 (a circle) to higher values. For example, Halley's Comet has a very eccentric orbit for which e = 0.967.

Also refer to Figure 102 which plots an ellipse showing how the two quantities a and e relate to other traditional graphical representations for ellipses.

```
%  -------------------------------------------------
Orbit = zeros(9,4);

%                 Mean Distance from Sun
%                      (Earth = 1.00)
%                            |
%                            |        Eccentricity
```

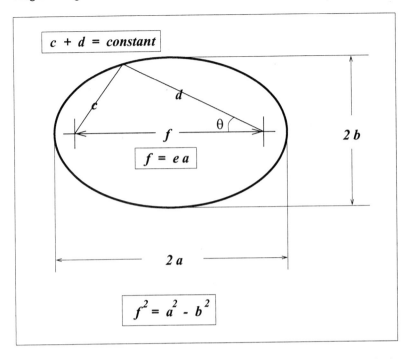

FIGURE 102. Relationship between eccentricity (e) and minor axis length (b)

```
%
%                                       |           |
                                       \|/         \|/
Orbit(Mercury,1:2)  =  [   0.39,    0.206 ];
Orbit(Venus,  1:2)  =  [   0.72,    0.007 ];
Orbit(Earth,  1:2)  =  [   1.00,    0.017 ];

Orbit(Mars,   1:2)  =  [   1.52,    0.093 ];
Orbit(Jupiter,1:2)  =  [   5.20,    0.048 ];
Orbit(Saturn, 1:2)  =  [   9.54,    0.056 ];

Orbit(Uranus, 1:2)  =  [  19.19,    0.047 ];
Orbit(Neptune,1:2)  =  [  30.07,    0.009 ];
Orbit(Pluto,  1:2)  =  [  39.46,    0.249 ];

% ----------------------------------------------
% Source: _Pictorial_Astronomy_, 3rd rev. ed.,
% by Alter et al.  New York, Crowell, 1969, p. 93.
% ----------------------------------------------
```

3. orbits34.m stores into the last two columns of the same 9 × 4 Orbit matrix. The *angle* and *inclination* are also important in plotting the elliptical paths of the planets:

- The *angle* is used to rotate the major axis of each planet's ellipse relative to that of Earth's.

- The *inclination* is not used by this program which only presents a *two-dimensional* view of the solar system.

```
%  --------------------------------------------------
%               Angle of elliptic orbit's
%                 major axis to Earth's
%                (treating as 0 for now)
%                           |
%                           |         Inclination of
%                           |         orbital plane
%                           |           to Earth's
%                           |              |
%                          \|/            \|/

Orbit(Mercury,3:4)   =   [ 0.00,         7.0 ];
Orbit(Venus,  3:4)   =   [ 0.00,         3.4 ];
Orbit(Earth,  3:4)   =   [ 0.00,         0.0 ];

Orbit(Mars,   3:4)   =   [ 0.00,         1.9 ];
Orbit(Jupiter,3:4)   =   [ 0.00,         1.3 ];
Orbit(Saturn, 3:4)   =   [ 0.00,         2.5 ];

Orbit(Uranus, 3:4)   =   [ 0.00,         0.8 ];
Orbit(Neptune,3:4)   =   [ 0.00,         1.8 ];
Orbit(Pluto,  3:4)   =   [ 0.00,        17.1 ];

%  --------------------------------------------------
% Source: _Pictorial_Astronomy_, 3rd rev. ed.,
% by Alter et al. New York, Crowell, 1969, p. 93.
%  --------------------------------------------------
```

4. starts.m stores the angular positions (Start_th) of the planets at some starting time (Start_t). It is from these starting positions that the final positions are calculated.

   It is expected that the user may want to substitute this file with new starting positions, perhaps calculated by this program itself.

```
Start_t = 1995.0 + 3/365;

%-----------------------

Start_th = zeros(9,1);

Start_th(Mercury)  = 345.7083;
Start_th(Venus)    = 138.8194;
Start_th(Earth)    = 102.1532;

Start_th(Mars)     = 119.2770;
Start_th(Jupiter)  = 242.7067;
Start_th(Saturn)   = 348.7888;

Start_th(Uranus)   = 291.6482;
Start_th(Neptune)  = 293.7496;
Start_th(Pluto)    = 231.6706;

Start_th = Start_th * pi / 180;

%-----------------------------------
% Source: Astronomical Almanac, 1995
%-----------------------------------
```

The main-line script, solar.m, then invokes the script in globals.m:

```
global Name Period Orbit
```

```
global Start_t Start_th

global Final_t
```

`globals.m` declares some of our data variables as global, plus one other variable, Final_t, discussed below.

Calculating Results

Once the data is all read in, `solar.m` then invokes `finals.m` to:

1. prompt the user for the final time of the planetary motion, Final_t, by invoking prompts, and

2. solve for the final position of the nine planets, Final_th, by invoking solve_th.

`finals.m` contains the following:

```
Final_t = prompts;

%-----------------------

Final_th = zeros(9,1);

Final_th(Mercury) = solve_th(Mercury);
Final_th(Venus)   = solve_th(Venus);
Final_th(Earth)   = solve_th(Earth);

Final_th(Mars)    = solve_th(Mars);
Final_th(Jupiter) = solve_th(Jupiter);
Final_th(Saturn)  = solve_th(Saturn);

Final_th(Uranus)  = solve_th(Uranus);
Final_th(Neptune) = solve_th(Neptune);
Final_th(Pluto)   = solve_th(Pluto);

%-----------------------

Final_th
```

Prompting for the Final Time

The prompting for Final_t is done by the function in `prompts.m`. It first uses disp to display several character strings of information, then it uses MATLAB's input function to prompt for Final_t:

```
function Final_t = prompts

globals

disp('-----------------------------------------');
disp(['    Start time is: ', num2str(Start_t, 10)]);
disp('-----------------------------------------');

disp(['Final time can be a number: ', ...
                    num2str(Start_t+5)]);
```

```
            disp(['or it can be an expression: ', ...
                                 'Start_t+5']);

            Final_t = input('Enter Final time: ');
            disp('----------------------------------------');
            disp(['     Final time is: ', num2str(Final_t, 10)]);
            disp('----------------------------------------');
```

Here is how the screen looks after the data corresponding to Figures 101 and 102 is entered:

```
    >> solar
    ----------------------------------------
         Start time is: 1995
    ----------------------------------------
    Final time can be a number: 2000
    or it can be an expression: Start_t+5

    Enter Final time: 1995.5
    ----------------------------------------
         Final time is: 1995.5
    ----------------------------------------
```

Solving for the Final Position

The final angular positions are computed by solve_th, which is invoked once for each planet, as shown above.

solve_th makes use of MATLAB's fzero function. fzero finds the value of Final_th such that kepler(Final_th) = 0. Refer back to Figure 100 to view the call tree diagram.

```
            function f_th = solve_th(id)

            global cur_id

            cur_id = id;
                                      th_guess = pi;
            f_th = fzero('kepler', th_guess);
%                          ^^^^^^
```

kepler.m is based on one of Kepler's laws of planetary motion:

> "A line joining any planet to the sun sweeps out equal areas in equal times." [87]

Or put another way:

> The ratio of the *area swept out to the total area of the ellipse* is equal to the ratio of the *time taken to the total period of the planet*.

which can be expressed as

$$\frac{A}{\pi ab} = \frac{t}{T}$$

---

[87] *Physics*, Paul A. Tipler, Worth Publishers, 1976, p. 393.

More explicitly, in terms of our program, fzero is trying to find Final_th (shown here as θ_{Final}) such that the following equation holds true:

$$Kepler(\theta_{final}) = \frac{A}{\pi ab} - \frac{t}{T} = 0$$

$$\frac{\frac{1}{2} \int_{\theta_{Start}}^{\theta_{Final}} r(\theta)^2 d\theta}{\pi ab} - \frac{t_{Final} - t_{Start}}{Period} = 0$$

where $r(\theta)$ is

$$r(\theta) = \frac{a(1-e^2)}{1 - e\cos(\theta - angle)}$$

Notice that kepler.m must calculate the area swept between Start_th and fzero's current guess at Final_th. To do this, kepler.m invokes MATLAB's quad8 function:

```
function diff = kepler(Final_th)

global cur_id
globals

% ----------------------------------------------------
% Trying to find "Final_th" such that:
%
%          A        t
%         -----  -  -   =   0
%         Pi a b   T
%
% where:
%
%    A is the area swept between Start_th and Final_th
%
%    t is the time taken between Start_t  and Final_t
% ----------------------------------------------------

tol = 1e-6;
A = (1/2) * quad8('radius_2', Start_th(cur_id), ...
                             Final_th,          ...
                             tol);

a = Orbit(cur_id,1);      %%% semi-major axis

e = Orbit(cur_id,2);      %%% eccentricity
b = a * sqrt(1-e^2);      %%% semi-minor axis

% ----------------------------------------------------

t = Final_t - Start_t;

T = Period(cur_id);

% ----------------------------------------------------

diff = A / (pi * a * b)  -  t / T;
```

The function in radius_2.m is used to return $r(\theta)^2$ when invoked by quad8 with different values of θ:

```
function r_2 = radius_2(th)
```

```
global cur_id
globals            % for "Orbit"

% ------------------------------------------------

a     = Orbit(cur_id, 1);
e     = Orbit(cur_id, 2);
angle = Orbit(cur_id, 3);

% ------------------------------------------------

r = a*(1 - e^2) ./ ...
    (1 - e*cos(th-angle));

r_2 = r.^2;

% ------------------------------------------------
```

Reporting Results

The final positions of all nine planets end up in the vector Final_th, and are displayed by entering the name of the vector as a command in `finals.m`:

```
Final_th
```

For the above example, the screen would show the final angular positions (in radians) as follows:

```
Final_th =

    5.5470
    2.0142
    1.2245
    0.5630
    0.0964
    0.0382
    0.0136
    0.0075
    0.0031
```

In the chapter *File Input/Output*, we will get more elaborate by formatting a table with titling and labelling for each planet.

Plotting Results

The next script invoked by `solar.m` is `plots.m`. It plots the starting and final positions of the planets (shown in Figures 101 and 102):

```
for ifig=1:2
    figure(ifig)

    if ifig==1, from=Mercury; to=Mars;  end
    if ifig==2, from=Mars;    to=Pluto; end

    for id=from:to

        a    = Orbit(id, 1);
```

```
            e     = Orbit(id, 2);
            angle = Orbit(id, 3);

            ellipse(    a, e, 0, 2*pi, angle)
            hold on
            h = ellipse(a, e, Start_th(id),  ...
                                 Final_th(id), ...
                                      angle);
            set(h, 'LineWidth', 4.0)

            % - - - - - - - - - - - - - - - - -

            th = 3 * pi / 2;
            r = a*(1 - e^2) ./ ...
                    (1 - e*cos(th-angle));

            [x, y] = pol2cart(th, r);
            text(x, y, Name(id,:))

        end

        hold off

        axis equal
        axis('square')

        fname = ['solar', num2str(ifig), '.eps'];
%
        print('-deps', fname)

    end
```

Notice that there are two nested for loops:

1. The outer for loop covers which of two figure windows is used: Mercury through Mars are plotted in the first window, and Mars through Pluto are plotted in the second window. [88]

2. The inner for loop plots all the elliptical paths in a given figure window. First, the entire elliptical path is plotted. Then just the segment of interest between Start_th and Final_th is plotted with a thicker line. Finally, text is used to place the planet names next to their elliptical orbits.

Two important general features of MATLAB programming are illustrated here as well:

1. Notice the two forms of invocation of the axis function:

    ```
    axis equal
    axis('square')
    ```

    This illustrates that MATLAB treats the statement:

    function arg1 arg2 ...

---

[88] If all the planets were plotted in one figure window, then the larger orbits would dwarf the smaller orbits considerably.

exactly the same as if typed in the more *traditional* way, complete with parentheses, commas and quotes:

function('arg1', 'arg2', '...')

Hence, you could as easily have coded:

```
axis('equal')
axis square
```

2. Notice the format of the print statement. A simpler, but less general, way of coding it is to use:

```
if ifig==1,  print -deps solar1.eps,  end
if ifig==2,  print -deps solar2.eps,  end
```

however, we wanted MATLAB to automatically use the value of ifig in the print command, so we made use of the fact that the above two lines are actually equivalent to:

```
if ifig==1,  print('-deps', 'solar1.eps'),  end
if ifig==2,  print('-deps', 'solar2.eps'),  end
```

and then generalized this in the script, which builds up the filename in three pieces in the fname character array. It takes care to convert the value of ifig to a string in the process:

```
            fname = ['solar', num2str(ifig), '.eps'];
    %                 ^^^^^   ------           ^^^^
            print('-deps', fname)
```

This program uses our `ellipse` function to plot the paths of the planets:

```
function [o1, o2] = ellipse(a, e, th1, th2, ...
                            angle, x1, y1, npts)
%
%ELLIPSE  2-D plot of an ellipse.
%
%         ELLIPSE(A,E) plots an ellipse whose major
%         axis is A, and whose eccentricity is E.
%
%         ELLIPSE(A,E,TH1,TH2) plots a partial
%         ellipse from TH1 to TH2 (in radians).
%         Default is 0 to 2*pi.
%
%         ELLIPSE(A,E,TH1,TH2,ANGLE) plots the
%         (partial) ellipse with its major axis
%         rotated ANGLE radians above the X-axis,
%         pivoting about left focus.  Default is 0.
%
%         ELLIPSE(A,E,...,X1,Y1) plots the pivotal
%         focus at (X1,Y1).  Default is (0.0, 0.0).
%
%         ELLIPSE(A,E,...,NPTS) plots NPTS points
%         to draw the ellipse.  Default is calculated
%         as 1 point for every degree of arc plus 1.
%
%         H = ELLIPSE(...) returns the handle from
%         the invocation of Matlab's PLOT function.
```

```
%
%            [X,Y] = ELLIPSE(...) returns the (x,y)
%            co-ordinates of the points on the ellipse.
%            NO PLOTTING IS DONE IN THIS CASE.  Commonly
%            used to capture points so as to shade in
%            elliptical pie segments using Matlab's FILL.

%-------------------------------------------------

if nargin < 3,   th1   = 0.0;    end
if nargin < 4,   th2   = 2*pi;   end

if nargin < 5,   angle = 0.0;    end
if nargin < 6,   x1    = 0.0;    end
if nargin < 7,   y1    = 0.0;    end

if nargin < 8,
    npts  = 1 + 360 * (th2-th1) / ...
                    (  2*pi );
end

if nargin < 2  |  nargin > 8
    error('Wrong number of input arguments')
end

%-------------------------------------------------

th = linspace(th1, th2, npts);

r = a*(1 - e^2) ./ ...
      (1 - e*cos(th-angle));

%-------------------------------------------------

[x, y] = pol2cart(th, r);

x = x + x1;
y = y + y1;

%-------------------------------------------------

if nargout == 0
    plot(x, y, '-');
end

if nargout == 1
    o1 = plot(x, y, '-');
end

if nargout == 2
    o1 = x;
    o2 = y;
end

if nargout > 2
    error('Wrong number of output arguments')
end
```

Further Results

The last script invoked by solar.m is repeats.m. It provides the opportunity to enter a new final time and get textual and graphical results.

The user has the option of continuing from where they left off:

```
another = input(['Do you wish to trace ', ...
                 'another orbit? [y/n] '], 's');

while strcmp(another, 'y')

      cont = input(['Do you wish to start from ', ...
                    'where you left off? [y/n] '], 's');

      if strcmp(cont, 'y')
          Start_t  = Final_t;
          Start_th = Final_th;
      end

      finals
      plots

      another = input(['Do you wish to trace ', ...
                       'another orbit? [y/n] '], 's');

end
```

In the following section on *Looping*, a variation on this while loop is shown.

Looping (for, while, ...)

As you saw in the comprehensive example, for and while loops are useful for performing operation(s) on some or all elements of an array. One of the main design goals of this example program was to be as readable as possible—hence a little efficiency was sacrificed in places. Some of these places are re-visited in the subsections that follow.

Also, the discussion of for and while loops will be completed through using the example program; and then the advantages of using *vectorized* functions such as sum will be examined.

for Loops Re-visited

To solve for the final position of the planets you could have replaced the nine assignment statements with a for loop as follows:

```
Final_th = zeros(9,1);

for id=1:9
    Final_th(id) = solve_th(id);
end
```

Note: In all examples the increment between successive values of id has been 1. You can specify another increment (or decrement), using the syntax:

for id = *from* : *increment* : *to*

For example, you can use two for loops to do the calculations, one for the odd-numbered planets, and the other for the even-numbered planets:

```
for id=1:2:9
    Final_th(id) = solve_th(id);
end
```

is the same as:

```
Final_th(1) = solve_th(1);
Final_th(3) = solve_th(3);
Final_th(5) = solve_th(5);
Final_th(7) = solve_th(7);
Final_th(9) = solve_th(9);
```

and

```
for id=2:2:9
    Final_th(id) = solve_th(id);
end
```

is the same as:

```
Final_th(2) = solve_th(2);
Final_th(4) = solve_th(4);
Final_th(6) = solve_th(6);
Final_th(8) = solve_th(8);
```

while Loops Re-visited

Consider the following variation on repeats.m (called repeats2.m). It is a little more elegant:

```
while 1
    another = input(['Do you wish to trace ', ...
                     'another orbit? [y/n] '], 's');
    if strcmp(another, 'n')
        break
    end

    cont = input(['Do you wish to start from ', ...
                  'where you left off? [y/n] '], 's');
    if strcmp(cont, 'y')
        Start_t  = Final_t;
        Start_th = Final_th;
    end

    finals
    plots
end
```

The three distinguishing features are:

1. The loop condition no longer depends on another, but instead is simply 1 (meaning *true* to MATLAB):

    ```
    while 1
    ```

    Matlab repeatedly executes the lines inside the while loop until it executes a break statement contained therein.[89]

2. The first instance of the following prompt for input was eliminated:

    ```
    another = input(['Do you wish to trace ', ...
                     'another orbit? [y/n] '], 's');
    ```

    The second instance has been moved to be the **first** statement inside the while loop.

3. Immediately following this line the value of another is tested:

    ```
    if strcmp(another, 'n')
        break
    end
    ```

    If it is "n" we *break* out of the loop. In other words, we do not execute any further lines inside the while loop.

sum, prod, etc. as Alternatives

For simple operations, such as summing all elements of an array, MATLAB has functions (e.g., sum) which are easier and more efficient to use than for and while loops.

For example, sum the masses and diameters of the planets. Here is masses.m:

```
Mass = zeros(9,1);

Mass(Mercury) =    0.04;
Mass(Venus)   =    0.8;
Mass(Earth)   =    1.0;

Mass(Mars)    =    0.1;
Mass(Jupiter) =  318;
Mass(Saturn)  =   95;

Mass(Uranus)  =   15;
Mass(Neptune) =   17;
Mass(Pluto)   =  NaN;

% ------------------------------------------
% Source: _Pictorial_Astronomy_, 3rd rev. ed.,
% by Alter et al.  New York, Crowell, 1969, p. 93.
% ------------------------------------------
```

and here is diams.m:

```
% ------------------------------------------
Diameter = zeros(9,1);

Diameter(Mercury) =   5000;     % kilometers
```

---

[89]The break statement can also be used inside of for loops.

```
Diameter(Venus)   =  12400;
Diameter(Earth)   =  12757;

Diameter(Mars)    =   6800;
Diameter(Jupiter) = 142900;
Diameter(Saturn)  = 120900;

Diameter(Uranus)  =  50000;
Diameter(Neptune) =  43000;
Diameter(Pluto)   =    NaN;

% ----------------------------------------------
% Let's normalize diameters about Earth = 1.0
% ----------------------------------------------

Diameter = Diameter / Diameter(Earth);

% ----------------------------------------------
% Source: _Pictorial_Astronomy_, 3rd rev. ed.,
% by Alter et al.  New York, Crowell, 1969, p. 93.
% ----------------------------------------------
```

Notice that in both cases the source for the data was missing values for Pluto. Therefore, NaN was specified, which stands for *Not A Number*.[90]

To find the total mass and volume of the planets you could write a for loop as follows:

```
Total_Mass     = 0;
Total_Diameter = 0;

for id=1:9
    Total_Mass     = Total_Mass     + Mass(id);
    Total_Diameter = Total_Diameter + Diameter(id);
end

Total_Mass
Total_Diameter
```

which results in:

```
Total_Mass =
   NaN

Total_Diameter =
   NaN
```

Notice that computations involving NaN also result in NaN.

A more concise and efficient way of performing the summation is as follows, making use of MATLAB's built-in sum function:

```
Total_Mass     = sum(Mass)
Total_Diameter = sum(Diameter)
```

The result is the same, NaN.

sum can also be used with matrices, where it sums the elements in each column. For example, let's create a matrix holding Mass in the first column and Diameter in the second column:

---

[90] Usually NaN arises from invalid computations, such as: $\frac{0}{0}$, $\frac{\inf}{\inf}$, or $0 \times \inf$.

```
MD = [Mass, Diameter]
```

which results in:

```
MD =
    0.0400    0.3919
    0.8000    0.9720
    1.0000    1.0000
    0.1000    0.5330
  318.0000   11.2017
   95.0000    9.4771
   15.0000    3.9194
   17.0000    3.3707
       NaN       NaN
```

Then apply sum to the **columns** of the MD matrix; the result is a row vector with the total in each column.

```
Totals = sum( MD )
```

results in:

```
Totals =
   NaN   NaN
```

Now consider just the first eight rows of MD:

```
MD = MD(1:8, :)
```

namely:

```
MD =
    0.0400    0.3919
    0.8000    0.9720
    1.0000    1.0000
    0.1000    0.5330
  318.0000   11.2017
   95.0000    9.4771
   15.0000    3.9194
   17.0000    3.3707
```

Furthermore, you can conveniently sum the columns of this submatrix:

```
Totals = sum(MD)
```

which results in:

```
Totals =
  446.9400   30.8660
```

If sum is further applied to this resultant row vector, Totals, it sums across the row vector.

```
Grand = sum(Totals)
```

which results in the following result (which, of course, is physically meaningless):

```
Grand =
  477.8060
```

To calculate the grand total in one step, we could also do

```
Grand = sum(sum(MD))
```

with the same result.

Similar Functions

Functions similar to sum are as follows:

prod(v) product of the elements of v,

cumsum(v) cumulative sum of the elements of v,

cumprod(v) cumulative product of the elements of v.

They are all built-in to MATLAB, and so execute as quickly as compiled C code.

Testing (if, find, case, ...)

As you saw in the comprehensive example, if blocks are useful in checking the answers to questions before acting accordingly.

However, in instances where if blocks are used in loops to find elements of an array which match certain criteria, MATLAB has functions, such as find, which are simpler and more efficient to use.

Let us first complete our discussion of if blocks by re-examining our example code and looking at how it can be made more efficient.

if Blocks Re-visited

Consider the following variation on the plotting section of `ellipse.m`:

```
if nargout == 0 | nargout == 1

    h = plot(x, y, '-');

    if nargout == 1
        o1 = h;
    end

elseif nargout == 2

    o1 = x;
    o2 = y;

else
    error('Wrong number of output arguments')
end
```

This is less readable than the original, but more efficient and compact. There is only a single if block, with at most three conditions for MATLAB to check. In the original there were four if blocks and MATLAB had to check each one. There is also only a single plot function invocation, whereas in the original there were two. This was achieved by placing an if within an if, and at the expense of readability.

Using case for Improved Readability

The above example could have been made even more compact and readable, using the switch/case/otherwise programming structure.

```
switch nargout

    case {0, 1}
        h = plot(x, y, '-');

        if nargout == 1
            o1 = h;
        end

    case 2
        o1 = x;
        o2 = y;

    otherwise
        error('Wrong number of output arguments')
end
```

Using find

Continuing with the Mass and Diameter example data from the previous section, let's look at various ways of determining those planets whose diameter is the same or greater than Earth's.

The obvious way is to use a for loop and pull out the wanted planets into a vector called Keep:

```
n=0;
for id=1:9
    if Diameter(id) >= Diameter(Earth)
        n = n + 1;
        Keep(n) = Diameter(id);
    end
end
```

Or a MATLAB feature to simplify this code can be used, as follows:

```
Keep = Diameter
for id=9:-1:1
    if Diameter(id) < Diameter(Earth)
        Keep(id) = []
    end
end
```

Notice that we simply stored [] (null) in the *row* we wanted to *remove* from keep.

The most efficient technique, however, is to use the find function, as follows:

```
Keep = Diameter( find( Diameter >= Diameter(Earth) ) )
```

Now break this down into three steps from the inside out:

1. The result of:

```
                      vvvvvvvvvvvvvvvvvvvvvvvvv
                      Diameter >= Diameter(Earth)
```

is an array of the same dimensions as Diameter with 1 where this relation is true, and 0 where it is false:

```
[0 0 1 0 1 1 1 1 0]
```

2. The result of:

```
      find( Diameter >= Diameter(Earth) )
           ^^^^^^^^^^^^^^^^^^^^^^^^^^^^^^
```

is a vector containing the indices of the non-zero elements of the above 0/1 vector, i.e., those that are true:

```
[3 5 6 7 8]
```

3. Finally, the results of:

```
                vvvvvvvvvvvvvvvvvvvvvvvvvvvvvvvv
      Diameter( find( Diameter >= Diameter(Earth) ) )
                     ^^^^^^^^^^^^^^^^^^^^^^^^^^^^^
```

is an array containing only the above list of elements from Diameter.

```
[12757 142900 120900 50000 43000]
```

This array is then stored in Keep.

User-defined Functions, Scripts and Global Variables

The comprehensive example had several user-defined functions, as well as scripts and global variables. We will consider `ellipse.m` in detail.

`ellipse.m` is particularly interesting because it illustrates a function with varying numbers of input and output arguments. Consequently we created a detailed help description with it, which the user can see by issuing:

```
help ellipse
```

MATLAB then displays the first uninterrupted block of comment lines that it encounters in `ellipse.m`. (MATLAB also strips off the comment characters):

```
ELLIPSE   2-D plot of an ellipse.

          ELLIPSE(A,E) plots an ellipse whose major
          axis is A, and whose eccentricity is E.

          ELLIPSE(A,E,TH1,TH2) plots a partial
          ellipse from TH1 to TH2 (in radians).
          Default is 0 to 2*pi.

          ELLIPSE(A,E,TH1,TH2,ANGLE) plots the
          (partial) ellipse with its major axis
          rotated ANGLE radians above the X-axis,
          pivoting about left focus.  Default is 0.
```

```
            ELLIPSE(A,E,...,X1,Y1) plots the pivotal
            focus at (X1,Y1).  Default is (0.0, 0.0).

            ELLIPSE(A,E,...,NPTS) plots NPTS points
            to draw the ellipse.  Default is calculated
            as 1 point for every degree of arc plus 1.

            H = ELLIPSE(...) returns the handle from
            the invocation of Matlab's PLOT function.

            [X,Y] = ELLIPSE(...) returns the (x,y)
            co-ordinates of the points on the ellipse.
            NO PLOTTING IS DONE IN THIS CASE.  Commonly
            used to capture points so as to shade in
            elliptical pie segments using Matlab's FILL.
```

Command Listing

% str
Transforms *everything* on line after % to a comment.
Output: Not applicable.
Additional information: Any object after the % is ignored, including strings, expressions, values, etc. ♣ The % need not appear at the beginning of a line; it can come anywhere in a line. Only the information to the right of the % is ignored.
See also: !, help

! str
Enters string str as a command to the current operating system.
Output: Not applicable.
Additional information: The ! need not appear at the beginning of a line; it can come anywhere in a line as long as it is the first non-blank character of its statement. Only the information to the right of the ! is sent to the system.
See also: %

addpath(str)
Prepends the directory str to the current search path.
Output: No output is displayed.
Argument options: addpath(str_1, ..., str_n) to add n directories to the current search path. ♣ addpath(str_1, ..., str_n, 1) or addpath(str_1, ..., str_n, '-begin') to append the directories to the search path. The default action is to prepend, and can be specified by 0 or '-end'.
Additional information: Make sure that the strings representing the directories are in the valid syntax for your particular system.
See also: path, rmpath, editpath, getenv, dir, pwd

num = all(V == expr)
Determines if *all* elements of vector V equal the value expr.
Output: If the condition holds, a value of 1 is assigned to num. Otherwise, a value of 0 is assigned to num.

Argument options: num = all(expr₁ == expr₂) to test whether expressions expr₁ and expr₂ are equal. ♣ V = all(M == expr) to test if the elements in each column of $m \times n$ matrix M are equal to expr. A vector with n elements (either 0 or 1) is assigned to V. ♣ V = all(M₁ == M₂) to test if the elements in each column of matrix M₁ are equal the related elements in the related column of matrix M₂. ♣ num = all(V) or V = all(M) to test if elements are nonzero. ♣ M_{new} = all(M₁ == M₂) to test if each element of M₁ is equal to its corresponding element in M₂. A matrix of equal size, containing 0's and 1's is assigned to M_{new}. ♣ num = all(ineq) to test other valid inequalities for validity. ♣ num = all(logeq) to test a valid logical equation for validity.
Additional information: all is particularly useful when used as the conditional statement in an if statement.
See also: any, *if*

ans

Holds the latest result when no assignment was done.
Output: Not applicable.
Additional information: Whenever a MATLAB command is performed or an expression is created, but the result is *not* assigned to a user-defined variable, the result is automatically assigned to ans. ♣ ans holds each value assigned to it until the next time it is assigned.
See also: for/from/by/to/do/od, for/from/by/while/do/od, for/in/do/od, next, proc/local/options/end

num = any(V == expr)

Determines if *any* elements of vector V equal the value expr.
Output: If the condition holds, a value of 1 is assigned to num. Otherwise, a value of 0 is assigned to num.
Argument options: V = any(M == expr) to test if any of the elements in each column of $m \times n$ matrix M are equal to expr. A vector with n elements (either 0 or 1) is assigned to V. ♣ num = any(V) or V = any(M) to test if any of the elements are nonzero. ♣ num = any(ineq) to test other valid inequalities for validity. ♣ num = any(logeq) to test a valid logical equation for validity.
Additional information: any is particularly useful when used as the conditional statement in an if statement.
See also: all, find, *if*

assignin('caller', var, M)

Assigns the value M to the *caller* workspace variable var.
Output: No output is displayed.
Argument options: assignin('base', var, M) to assign to a *base* workspace variable.
See also: evalin

break

A programming construct for exiting prematurely from a loop.
Output: Not applicable.
Additional information: This construct forces an exit from the innermost for or while loop in which it is contained. The statement directly after the looping structure is then

evaluated. ♣ If break is found within any other type of construct, an error message is returned.
See also: for, while

calendar(int$_{yr}$, int$_{mn}$)
Displays a 6×7 matrix containing the calendar for month int$_{mn}$ of year int$_{yr}$.
Output: Not applicable.
Argument options: calendar(str$_{date}$), where str$_{date}$ is a string as returned by the date command.
See also: date, datenum, datestr, datevec, eomday, now, weekday

clc
Clears the session's Command window.
Output: Not applicable.
Additional information: This command has no effect on what is currently in memory.
♣ To clear the current plot display window, use clf.
See also: clf, home

clear name
Clears the variable or function name from memory.
Output: Not applicable.
Argument options: clear name$_1$, name$_2$, ..., name$_n$ to clear the n variables of functions name$_1$ through name$_n$ from memory. ♣ clear *variables* or clear to clear all variables from memory. ♣ clear *global* to clear all global variables from memory. ♣ clear option, where option is one of *functions* or mex (for MEX-files), to clear all of the corresponding objects from memory. ♣ clear *all* to empty MATLAB's memory, completely resetting the session.
Additional information: There are special rules for clearing global variables. See the on-line help for more information.
See also: pack

V = clock
Returns the current year, month, day, hour, minute, and second, in that order.
Output: A vector with five integers and a decimal value is assigned to V.
See also: date, etime, cputime, tic, toc, flops

str = computer
Returns what type of computer on which MATLAB is currently being run.
Output: An all-caps string representing the computer type is assigned to str.
Argument options: [str, posint] = computer to also assign the maximum number of elements allowed in a matrix to posint. This value may vary between platforms and versions of MATLAB.
Additional information: Possible results of computer include *PCWIN*, *SUN4*, *HP700*, etc.

num = cputime
Returns the elapsed CPU time for the current session.

Output: A numerical value is assigned to num.
Additional information: This command is useful for timing other MATLAB commands.
See also: etime, clock, tic, toc, flops

str = date
Returns the date as a string of the form day-month-year.
Output: A string is assigned to str.
See also: calendar, datenum, datestr, datevec, eomday, weekday, now, clock, etime, cputime, tic, toc

int = datenum(int$_{yr}$, int$_{mn}$, int$_{dy}$)
Computes the serial date number for the day int$_{dy}$ of month int$_{mn}$ of year int$_{yr}$.
Output: An integer value is assigned to int.
Argument options: V_{int} = datenum(V_{yr}, V_{mn}, V_{dy}) to return a vector of serial dates for the dates in the given vectors. ✦ V_{num} = datenum(V_{yr}, V_{mn}, V_{dy}, V_{hr}, V_{mi}, V_{sc}), where V_{hr}, V_{mi} and V_{sc} represent hours, minutes and seconds, to return a vector of numeric serial dates. ✦ datenum(str$_{date}$), where str$_{date}$ is a string as returned by the date command.
Additional information: Single integer values may also be entered for hours, minutes, and seconds. ✦ Any values passed to datenum that are over the normal limits (e.g., 20 months) are cycled to the next high category. For example, 95 − 20 − 11 is equal to 96 − 08 − 11. ✦ See the on-line help file to find valid formats for str$_{date}$.
See also: date, calendar, datestr, datevec, eomday, now, weekday

str = datestr(num, int)
Converts serial date number num into date string format int.
Output: A string is assigned to str.
Additional information: Serial date numbers are returned by the datenum command. ✦ See the on-line help file for descriptions of the 16 available date string formats.
See also: datenum, date, calendar, *datetick*, datevec, eomday, now, weekday

V = datevec(str)
Convert the date string str into a vector containing the year, month, day, hour, minute, and second.
Output: A six-element vector is assigned to V.
Argument options: [int$_{yr}$, int$_{mn}$, int$_{dy}$, int$_{hr}$, int$_{mi}$, int$_{sc}$] = datevec(str) to return individual values in six variables.
Additional information: See the on-line help file for datestr to find valid formats for str.
See also: datestr, datenum, date, calendar, eomday, now, weekday

demo
This is identical to the expo command.
See also: tour

disp(M)
Displays only the elements of matrix M.
Output: Not applicable.

Argument options: disp(str) to display only the characters in str.
Additional information: disp is special because it does not display the name of the object being shown and the equals sign.
See also: warning, sprintf, setstr, num2str

dos(str)

Executes the DOS command represented by str from within MATLAB.
Output: Not applicable.
Additional information: If you are running on a UNIX-based system, the unix command is automatically called with the given string.
See also: unix, !

echo *on*

Turns on echoing of M-files during execution.
Output: Not applicable.
Argument options: echo *off* to turn off echoing of M-files during execution. This is the default. ❈ diary to toggle the echoing on and off. ❈ echo fnc turns on echoing for only the function fnc. ❈ echo *on all* or echo *off all* to control echoing of all function files.
Additional information: The effects of this command are straightforward when dealing with script files (i.e., files of MATLAB commands as they would be typed in a session). ❈ When using echo with function files, whether or not the functions are compiled or interpreted is affected. This can cause great differences in efficiency, and therefore echoing should be on only during debugging and only for the functions being investigated.
See also: diary, Echo

edit *filename*

Opens MATLAB's built-in editor/debugger with the file filename loaded.
Output: Not applicable.
Argument options: edit to open the editor/debugger with a new file.

editpath

Displays a GUI interface for altering MATLAB's current search path.
Output: A dialog is displayed.
Additional information: This command allows you to do everything the existing path command does, but through an easy-to-use dialog box.
See also: path, addpath, rmpath

int = eomday(int$_{yr}$, int$_{mn}$)

Returns the last day of the month int$_{mn}$ of year int$_{yr}$.
Output: An integer $(28 - 31)$ value is returned.
See also: date, datenum, datestr, datevec, calendar, now, weekday

eps

A variable representing the distance between 1.0 and the next largest floating-point number.

Output: Not applicable.
Argument options: eps = num to assign the value num to eps.
Additional information: Changing the value of eps affects the default tolerance on many MATLAB commands, including pinv and rank. ✦ On any machine with IEEE floating-point arithmetic, eps is equal to 2^{-52}.
See also: realmin, realmax, Inf, isieee

error(str)

Forces an explicit return from within an M-file because of an error and displays the error message str.
Output: Not applicable.
Additional information: The remaining statements in the function's statement block, and the function that called it, and the function that called it, and so on, are ignored and control is immediately passed to the prompt.
See also: warning, lasterr, break, return, while/end, for/end

num = etime(V_1, V_2)

Returns the difference in seconds between two vectors returned by the clock command.
Output: A numerical value is assigned to num.
Additional information: This command is useful for timing other MATLAB commands.
✦ etime currently fails if the vectors stretch across month and year boundaries.
See also: cputime, clock, tic, toc, flops

evalin('caller', var)

Evaluates the *caller* workspace variable var.
Output: Whatever output is expected for var is returned.
Argument options: evalin('base', var) to evaluate in the *base* workspace. evalin('caller', var_1, var_2) or evalin('base', var_1, var_2) to evaluate var_2 if the evaluation of var_1 fails.
See also: eval, assignin

str = filesep

Returns the file separator character for your system.
Output: A string is assigned to str.
See also: fullfile, pathsep

var = find(V == expr)

Returns the indices of all the elements of vector V that equal expression expr.
Output: If there are any such elements, an expression sequence is assigned to var. Otherwise, an empty matrix is assigned to num.
Argument options: [var_1, var_2] = find(M == expr) to assign the row and column indices of elements of matrix M equaling expr to var_1 and var_2, respectively. ✦ [var_1, var_2, V] = find(M == expr) to also assign a column vector of the passing elements, or results of the inequality or logical statement, to V. ✦ var = find(V) to test if any of the elements of V are nonzero. ✦ var = find(ineq) to test other valid inequalities for validity. ✦ var = find(logeq) to test a valid logical equation for validity.
Additional information: find is a particularly useful tool for examining sparse vectors and matrices.

See also: isempty, sparse, nonzeros, any

num = **flops**
Returns the accumulated number of floating-point operations in the current session.
Output: A numerical value is assigned to num.
Argument options: flops(0) to reset the floating-point operation counter to 0.
Additional information: This command is useful for determining efficiency of MATLAB commands. ♣ While it is not possible to count every floating-point operation, most of them are counted.
See also: cputime, etime, clock, tic, toc

for/end
A programming construct for repetition of statements.
Output: Not applicable.
Additional information: This construct allows you to repeat a block of statements for each element in the object following for. Typically, the object after for is created with the colon operator, with a statement such as $num_1:num_2$. This structure must be terminated with end. ♣ To break out of a for/end structure prematurely, use break or return.
See also: while/end, if/elseif/else/end, switch/case/otherwise/end, break, return

filename$_{out}$ = **fullfile**(dir$_1$, ..., dir$_n$, filename)
Creates a full filename from n directories and a filename.
Output: A filename is assigned to filename$_{out}$.
Additional information: This command takes into consideration the needs of your system when building the full filename.
See also: filesep, pathsep

function [out$_1$, ..., out$_m$] = fnc(in$_1$, ..., in$_n$)
A programming construct used when defining a new function.
Output: Not applicable.
Additional information: function tells MATLAB that the remaining lines in this .m file fnc.m define a function called fnc that takes input parameters in$_1$ through in$_n$ and returns output values out$_1$ through out$_m$. ♣ The function name fnc must be the same as the root of the .m file name. ♣ All variables used within the body of the function are local variables. When a function is read into the MATLAB session it is complied, unless echo is enabled for the function. ♣ To clear a function from the session's memory, use clear. This does not affect the .m file.
See also: echo, clear, type, nargin, nargout

str$_{out}$ = **getenv**(str)
Retrieves the environment variable associated with the special string str.
Output: A string is assigned to str$_{out}$.
Additional information: A typical value for str is *'MATLABPATH'*. ♣ Most environment variables are created and set by the user.
See also: path, addpath, rmpath, get

global name$_1$, ..., name$_n$

Defines the variables named name$_1$ through name$_n$ to be global variables.
Output: Not applicable.
Additional information: All variables within functions are local, unless specified as global. ♦ In order for multiple functions to share access to a variable, *each* function must declare that variable as global. Any function not declaring the variable cannot access it. ♦ It is the standard convention in MATLAB that global variable names be in all-caps (e.g., MYVARIABLE, CLASSAVERAGE).
See also: clear, isglobal, who, whos

help str

Brings up the on-line help file for the MATLAB topic represented by string str.
Output: Not applicable.
Argument options: help to display a list of the main help topics. ♦ help dir to display the contents file for the directory dir.
Additional information: You can extend MATLAB's help facilities to include help and content files for your own functions and directories. ♦ In a function M-file, the first block of contiguous comments (lines beginning with %) form the automatic help file for that function. ♦ For more information on accessing help files, see the on-line file for help help.
See also: lookfor, dir, more, what, where, %

helpdesk

Brings up the hypertext documentation for MATLAB.
Output: Not applicable.
Additional information: The helpdesk command starts up your default Web browser (e.g., Netscape, Explorer). Information similar to that available with the help command is then accessible through hypertext links.
See also: help, lookfor

home

Repositions the cursor to the upper left corner of the Command window.
Output: Not applicable.
Additional information: This command allows you to display different results over top of one another.
See also: clc, clf

if/elseif/else/end

A programming construct for selection of different paths.
Output: Not applicable.
Additional information: This construct allows you to branch off the contents of objects placed after if and/or elseif. If these objects contain no zero elements, the block of statements directly following is executed and the construct is exited. If the objects contain any zero elements, the block of statements following else is executed. ♦ This structure must be terminated with end. ♦ Keep in mind that this structure is valid without an else and/or elseif.
See also: <, >, ==, $\tilde{=}$, switch/case/otherwise/end, while/end, for/end

fcn = **inline**(str)

Creates an inline function out of the expression string str.
Output: An inline function is assigned to fnc.
Argument options: fcn = inline(str, str_1, ..., str_n) to directly specify the variables to be pulled from str.
Additional information: The characters i and j are not used as variables. ✦ If no valid variable names are found, inline tries to use x.

M_c = **inmem**

Returns a cell array of all the M-files that are currently in memory.
Output: A cell array is assigned to M_c.
See also: cell

str = **inputname**(int)

Returns the variable name of the int^{th} parameter to the current function.
Output: A variable name is assigned to str.
Additional information: Remember that the *name* of the variable, not its value, is returned. ✦ If the examined parameter is passed directly as a value, then the empty string is returned.
See also: nargin, nargout, nargchk

int = **isempty**(M)

Determines if matrix M is an empty matrix.
Output: If M is empty a value of 1 is assigned to int. Otherwise, a value of 0 is assigned to int.
Additional information: Empty matrices have at least one zero dimension.
See also: find, size

int = **isglobal**(var)

Determines if variable var is global variable.
Output: If var is defined as a global variable in the current scope, a value of 1 is assigned to int. Otherwise, a value of 0 is assigned to int.
Additional information: Global variables are defined with global.
See also: global, who, clear

int = **isieee**

Determines if your computer supports IEEE arithmetic.
Output: If your computer does support IEEE arithmetic, a value of 1 is assigned to int. Otherwise, a value of 0 is assigned to int.
See also: computer

int = **islogical**(M)

Determines if matrix M is logical array.
Output: If var is a logical array, a value of 1 is assigned to int. Otherwise, a value of 0 is assigned to int.
See also: logical

keyboard/return

Temporarily halts execution of a M-file and takes input from the keyboard.
Output: Not applicable.
Additional information: When keyboard is encountered within an M-file, reading from the file is halted and subsequent input is taken directly from the keyboard until the command return, followed by the Return key, is entered. At that time, input is again taken from the M-file, starting at the command directly following keyboard. ♣ When in keyboard mode, any valid MATLAB command can be entered. ♣ The prompt is prepended with a K when in keyboard mode.
See also: input, debug, dbstop

str = lasterr

Returns the last error message generated by MATLAB in the current session.
Output: A string is assigned to str.
Argument options: lasterr("), where " represents the empty string, to reset lasterr.
Additional information: This command is typically used with the two-parameter version of eval. See the entry for eval for more information.
See also: eval, error

int = length(V)

Determines the number of elements in vector V.
Output: An integer value is assigned to int.
Argument options: int = length(M) to return the maximal dimension of matrix M.
See also: size, max

M_l = logical(M)

Converts numeric matrix M to a logical array.
Output: A logical array is assigned to M_l.
See also: islogical, all, any

lookfor str

Searches for all help files containing the string str in their first line.
Output: No value is assigned, but the first line of each appropriate help file is displayed.
Additional information: All M-files in the current search path are tested. ♣ For a more localized or specialized search, use the commands what and which.
See also: help, dir, ls, pwd, what, where

matlabrc.m

The automatically read in initialization file for MATLAB.
Output: Not applicable.
Additional information: If it exists, the file matlabrc.m is automatically executed when a MATLAB session starts up. ♣ This file is normally reserved for the use of the system manager. Individual users should create a startup.m file (which is typically called by matlabrc.m).
See also: quit, path, exist

str = **matlabroot**

Returns the name of the directory in which your version of MATLAB was installed.
Output: A string is assigned to str.
Additional information: pwd returns the name as specified by your system.
See also: ls, cd, dir, delete, what, lookfor, type, !

posint = **menu**(str$_{title}$, str$_1$, str$_2$, ..., str$_n$)

Displays a menu titled str$_{title}$ with numbered options str$_1$ through str$_n$ and allows the user to select one.
Output: The positive integer corresponding to the user's selection is assigned to posint.
Additional information: This command is very useful when stepping the user through an intricate problem. It represents about the greatest amount of "hand-holding" you can perform. ✦ The quality of the display of the menu depends on the capabilities of your terminal.
See also: input, demo, uicontrol

str = **mexext**

Returns the MEX filename extension for your platform.
Output: A string is assigned to str.
See also: mfilename

str = **mfilename**

Returns the currently executing (or most recently executed) M-file name.
Output: A string is assigned to str.
Additional information: When called from the command line, mfilename always returns an empty matrix.
See also: mexext

more on

Enables paging of output in the command window.
Output: Not applicable.
Argument options: more *off* to disable paging of output. This is the default.
✦ more(posint) to set the number of lines that define a "page" to posint. The default value is 23.
Additional information: When more is enabled, the space key displays the next page, the Return key displays the next line, and the q key terminates display of the output. ✦ Paging of output works similarly for results of computations and displays of information and/or help files.
See also: diary

M = **nargchk**(int$_{low}$, int$_{high}$, int)

Determines if int is between int$_{low}$ and int$_{high}$, inclusive.
Output: If it is between. then an empty matrix is assigned to M. Otherwise, an error message is assigned to M.
Additional information: This command is meant to be used to check the value of nargin within a function.
See also: nargin, nargout, inputname

n = nargin

Represents, within a function, the number of arguments (parameters) passed to that function.
Output: A non-negative integer is assigned to n.
Additional information: More parameters than are stipulated in the definition of a function can be passed to that function. nargin is useful in determining how many arguments were passed.
See also: nargout, varargin, function, inputname

n = nargout

Represents, within a function, the number of arguments (outputs) passed from that function.
Output: A non-negative integer is assigned to n.
Additional information: The number of arguments passed from a function depends upon how the user calls the function (i.e., what the function is assigned to).
See also: nargin, varargout, function, inputname

num = now

Returns the date and time as a serial date number.
Output: A numeric value is assigned to num.
Additional information: The date and time can be separated by using floor(now) and rem(now, 1).
See also: datenum, date, calendar, datestr, datevec, eomday, weekday, clock, etime, cputime, tic, toc

pack

Packs together workspace variables to minimize memory usage.
Output: Not applicable.
Argument options: pack filename to save the variables in temporary file filename. The default file name is pack.tmp.
Additional information: New variables in MATLAB must be stored in contiguous memory blocks. pack helps free up larger blocks when memory fragmentation is present.
✦ For more information on how pack works, see the on-line help file.
See also: clear

path

Displays MATLAB's current search path.
Output: The directory names in the current search path are displayed.
Argument options: str = path assign the current search path to string str. ✦ path(str) to replace the current search path with the path represented by str. ✦ path(str_1, str_2) to set the search path to the concatenation of the strings str_1 and str_2, in that order.
Additional information: When name is entered, MATLAB first checks if it is a current variable, then if it is a built-in function, and then if name.mex or name.m are in the current directory. If none of these is true, then the files name.mex and name.m are checked for in the current search path, directory by directory, in the order they appear. ✦ Make sure that the strings representing the directories are in the valid syntax for your particular system.

See also: editpath, addpath, rmpath, getenv, what, dir, ls, matlabroot, pwd

str = pathsep
Returns the path separator character for your system.
Output: A string is assigned to str.
See also: fullfile, filesep

pause
Causes a halt of execution until the user presses any key.
Output: Not applicable.
Argument options: pause(num) to stop execution for num seconds. ✸ pause *off* to specify that subsequent pauses have no effect. ✸ pause *on* to specify that subsequent pauses have effect. This is the default state.
Additional information: One important use of pause is to give the user time to view multiple sets of data (graphics) that are displayed one on top of the other.
See also: clc

pcode
Creates a preparsed, pseudo-code file for function fnc and places it in the current directory.
Output: No output is displayed.
Argument options: pcode fnc_1, ..., fnc_n to create pseudo-code files for n functions. ✸ pcode *.m to create pseudo-code files for every function in the current directory. ✸ pcode fnc_1, ..., fnc_n, '-*inplace*' to place the pseudo-code in the same directory as the .m files.
Additional information: All pseudo-code files have a .p extension. ✸ The .m files can be located anywhere on the current path.
See also: path, dir, pwd

quit
Terminates the MATLAB session.
Output: Not applicable.
Additional information: Because the workspace is not saved, you should use quit with caution.
See also: save, startup

return
A programming construct for returning to the invoking function.
Output: Not applicable.
Additional information: This construct forces an exit from a function construct or any subconstruct within a function construct. ✸ If the function being exited is at the top level, control is returned to the keyboard.
See also: break, error, function/end

rmpath
Removes the directory str from the current search path.
Output: No output is displayed.

Argument options: rmpath(str$_1$, ..., str$_n$) to remove n directories from the current search path.
Additional information: Make sure that the strings representing the directories are in the valid syntax for your particular system.
See also: path, addpath, editpath

[m, n] = size(M)

Determines the dimensions of matrix M.
Output: Two non-negative integers representing the number of rows and columns are assigned to m and n, respectively.
Argument options: V = size(M, int) to assign a two-element vector with the dimensions to V. ✸ m = size(M, 1) to assign the number of rows to m. ✸ n = size(M, 2) to assign the number of columns to n.
See also: length, whos

switch/case/otherwise/end

A programming construct for selection of different paths.
Output: Not applicable.
Additional information: This construct allows you to branch on the value of the variable placed after switch. Each case statement is followed by a single value or a cell of multiple values. Whichever value or cell of values contains a match, the following block of code is executed. If None of the cases are matched, then the code after otherwise is executed. Only one block of code is executed for any one call to switch/case/otherwise/end. ✸ This structure must be terminated with end. ✸ Keep in mind that this structure is valid without an otherwise.
See also: if/elseif/else/end, while/end, for/end

tic

Commences timing.
Output: Not applicable.
Additional information: To display or return the elapsed time, use toc.
See also: cputime, etime, clock, toc

toc

Displays the elapsed time (in seconds) since the tic command was last used.
Output: Not applicable.
Argument options: num = toc to assign the elapsed time to num.
Additional information: toc can be called several times for each tic, returning a different value each time.
See also: cputime, etime, clock, tic

unix(str)

Executes the UNIX command represented by str from within MATLAB.
Output: Not applicable.
Argument options: [int, str$_{out}$] = unix(str) to assign the status and the output resulting from execution of str to int and str$_{out}$, respectively.

Additional information: If you are running on a DOS-based system, the dos command is automatically called with the given string. work.
See also: dos, !

varargin
Represents, within a function, a variable number of input arguments to be passed to that function.
Output: Not applicable.
Additional information: varargin must be the last input "parameter" declared in the function definition.
See also: varargout, nargin, function

varargout
Represents, within a function, a variable number of output arguments to be passed from that function.
Output: Not applicable.
Additional information: varargout must be the last output "argument" declared in the function definition.
See also: varargin, nargout, function

ver
Lists the version information for MATLAB and its associated toolboxes on your system.
Output: Not applicable.
Argument options: ver(str), where str is the name of one of MATLAB's toolboxes, to display the version information for that application.
Additional information: To return the version number as a string, use the version command. ✦ If you do not have a certain toolbox on your system, you cannot access its version information.
See also: version, whatsnew

str = version
Returns the version of MATLAB installed on your system.
Output: A string is assigned to str.
Additional information: To display more information or determine the version of certain toolboxes, use the ver command.
See also: ver, whatsnew

warning(str)
Displays the textual warning message str.
Output: Not applicable.
Argument options: warning 'off' to turn off all further warning messages. ✦ warning 'on' to turn on all further warning messages. ✦ warning 'backtrace' to display the filename and line number that generates the warning. ✦ warning 'debug' to trigger the debugger when a warning is encountered. ✦ [str_s, str_f] = warning to return the current warning state and warning frequency, respectively.
See also: disp, error, dbstop

web str$_{url}$

Opens a web browser on your system, initialized to the URL str$_{url}$.
Output: Not applicable.
Argument options: int = web(str$_{url}$) to return the browser status in int. See the on-line help file for possible status variables.
See also: doc, helpdesk

[int, str] = weekday(num)

computes which day of the week corresponds to serial date num.
Output: An integer $(1-7)$ value is assigned to int and a three-character string (Mon—Sun) is assigned to str.
Argument options: [int, str] = weekday(str) to apply the command to a date string. See the on-line help for datestr to see acceptable string formats.
See also: date, datenum, datestr, datevec, calendar, now, eomday

whatsnew *matlab*

Lists the contents of the general README file for MATLAB.
Output: Not applicable.
Argument options: whatsnew str, where str is the name of one of MATLAB's toolboxes, to display the README information for that application.
Additional information: The README files contain new information that is not contained in the latest version of the hardcopy documentation. ♣ If you do not have a certain toolbox on your system, you cannot access its README file.
See also: version, what, lookfor, !

which fnc

Determines which directory the MATLAB function file for fnc is in.
Output: If the function file can be found, the directory name is displayed. Otherwise, a *not found* message is displayed.
Argument options: which('fnc') to get the same result. ♣ str = which('fnc') to assign the file location to the string str.
Additional information: which searches the current search path for the file fnc.m, fnc.mat, or fnc.mex. Also searched for are SIMULINK graphical functions. ♣ which only searches for exact matches with fnc. A more general search of the synopses of functions can be performed with lookfor.
See also: dir, help, lookfor, what, exist, matlabroot, findpath

while/end

A programming construct for repetition of statements.
Output: Not applicable.
Additional information: This construct allows you to repeat a block of statements until the object following while has all nonzero elements. Typically, the object after while is created with the boolean operators ==, <, etc. operating on two expressions. ♣ This structure must be terminated with end. ♣ To break out of a while/end structure before the boolean expression is violated, use break or return.
See also: for/end, if/elseif/else/end, switch/case/otherwise/end, break, return

who
Lists all variables currently in memory.
Output: A list of variables is displayed.
Argument options: who *global* to list only the global variables.
Additional information: The values assigned to the variable names are *not* listed.
♣ For a more complete description of the variables, use whos.
See also: whos, exist

whos
Lists all variables currently in memory, their sizes, and whether they are complex valued or not.
Output: A list of variables and their related information is displayed.
Argument options: whos *global* to list only the global variables and their information.
Additional information: For a less complete listing of the variables, use who.
See also: who, exist

File Input/Output

MATLAB can read and write data files in any format from simple ASCII tables to complex binary formats. The following sections outline several of the applicable functions, often referring to the comprehensive example used in the *Programming in MATLAB* chapter.

We first look at ASCII data and then deal with binary data.

Reading and Writing ASCII Data

This section considers the primary functions used in reading and writing data in ASCII files. They fall into two categories:

1. High-level: load, save
2. Low-level: fscanf, fprintf, fopen, fclose, ...

High-level Functions: load and save

load and save are easy to use, but they require the data to be in matrix format.

load

The example program solar.m used assignment statements to define its data. However, in typical practice the data will come from a file. For example, here is the data file `orbits1.dat`:

```
        0.39        0.206
        0.72        0.007
        1.00        0.017

        1.52        0.093
        5.20        0.048
        9.54        0.056

       19.19        0.047
       30.07        0.009
       39.46        0.249
```

Data in matrix form like this (even with blank lines), can be read by the load function:

```
load orbits1.dat
```

which loads the data into the 9 × 2 matrix called orbits1. That is, it is named after the filename with the extension removed.

If the filename does not have an extension, then you must specify the -ascii option as in:

```
load orbits1 -ascii
```

otherwise MATLAB assumes;

1. that the file *has* an extension of ".mat", and
2. that the file is in MATLAB's binary "MAT-file" format (discussed later).

save

As it stands, the example program displays the output to the screen. However, in actual practice, you often want to write certain output to a file, e.g., for use by another program.

Matrices can be written out in matrix format using the save function. For example, to write the orbits1 matrix and the results vector, Final_th, to the file my.out, specify:

```
save my.out orbits1 Final_th -ascii
```

Notes

1. Even if the filename does have an extension, you still must specify the -ascii option, otherwise the output will be in binary form.

2. The above command creates an 18-line file. The first nine lines hold the 9 × 2 matrix, orbits1, and the last nine lines hold the 9 × 1 matrix, Final_th. Unfortunately, the load command will not be able to read the file because the data is not rectangular. The following load attempt:

   ```
   load my.out
   ```

   results in:

   ```
   ??? Error using ==> load
   Number of columns on line 10 of ASCII file my
   must be the same as previous lines.
   ```

   A better strategy is to save the two variables side-by-side as a single 9 × 3 matrix:

   ```
   save my.out [orbits1 Final_th] -ascii
   ```

   then load the matrix and extract the two sub-matrices:

   ```
   load my.out

   orbits1  = my(:, 1:2)
   Final_th = my(:, 3)
   ```

Low-level Functions: fscanf and fprintf

Often the data file is not in matrix form and/or it has text characters in it. For example, we can add some text to label the rows and columns of our data set, and call it `orbits2.dat`:

```
             Mean      Eccent-
             Dist.:    ricity:

Mercury      0.39      0.206
Venus        0.72      0.007
Earth        1.00      0.017

Mars         1.52      0.093
Jupiter      5.20      0.048
Saturn       9.54      0.056

Uranus      19.19      0.047
Neptune     30.07      0.009
Pluto       39.46      0.249
```

Thus load cannot be used, and the general-purpose fscanf function must be turned to. Similarly, fprintf can be used to write out this sort of file.

fscanf

Before using fscanf to read `orbits2.dat`, however, let's show how it would be used to read the simpler `orbits1.dat`. Recall from the previous section that `orbits1.dat` contains a 9 × 2 matrix (with blank lines). Here is how you can read it with fscanf:

```
fid = fopen('orbits1.dat', 'rt');

Orbit = fscanf(fid, '%f', [2 Inf]);
Orbit = Orbit';

fclose(fid);
```

Notes

1. You must first open the file using fopen. The filename is specified first, and then the "permission." Here r means you can only *read* from the file. The r is followed by a t to indicate that it is a "text" file (i.e., ASCII) as opposed to binary.

   (See the *Command Listing* section for a description of other permissions.)

2. fopen returns a unique integer as the *file identifier* for each file opened. This in turn is passed to other Input/Output functions to specify which file to act upon.

3. fscanf reads data from the specified file. In this example it reads floating-point numbers ('%f') into the matrix Orbit. (See the *Command Listing* section for a description of other format codes.)

   The last argument specifies how much data is to be read, and how it will be stored in Orbit. In this case, fscanf reads an Inf × 2 block of floating-point numbers from `orbits1.dat`. In this context, "Inf" means "read as much data as it finds." Then fscanf transposes the block and stores it as a 2 × Inf matrix in Orbit.

4. Then the program transposes the matrix back to the way you want it, i.e., how it looks in the file, as a 9 × 2 matrix.

5. If fscanf does not find enough data on a line, then it skips to the next non-blank line. Likewise, if there is data left on a line after fscanf has performed a read, then it leaves it to be read by the next fscanf.

Now that the basics of fscanf have been covered, let's re-visit `orbits2.dat`:

```
              Mean      Eccent-
              Dist.:    ricity:

Mercury       0.39      0.206
Venus         0.72      0.007
Earth         1.00      0.017

Mars          1.52      0.093
Jupiter       5.20      0.048
Saturn        9.54      0.056

Uranus       19.19      0.047
Neptune      30.07      0.009
Pluto        39.46      0.249
```

The following program extracts not only the Orbit matrix, but also the Title words, and the Name for each planet:

```
    fid = fopen('orbits2.dat', 'rt');

%-------------------------------------------------

    Title = '';

    for i=1:4
        word  = fscanf(fid, '%s', 1);
        Title = str2mat(Title, word);
    end

    Title(1,:) = ''

%-------------------------------------------------

    i = 1;
    Name = '';

    while 1
        word = fscanf(fid, '%s', 1);
        Name = str2mat(Name, word);

        column_vector = fscanf(fid, '%f %f', 2);
        Orbit(i,:)    = column_vector';

        if feof(fid), break, end
        i = i + 1;
    end

    Name(1,:) = '';
    Name(i,:) = '';
```

```
        n = i - 1;
%- - - - - - - - - - - - - - - - - - - - - - - - - - - - - - - - - - - - - - - - - -
        fclose(fid);
```

Notes

1. The first fscanf reads a single character string ("%s"). We then use str2mat to append the title rows together, one on top of the other, to form the Title matrix:

    ```
    Title =

    Mean
    Eccent-
    Dist.:
    ricity:
    ```

    Notice that (unlike numbers) character strings are read into rows rather than columns.

2. Often data files will contain a title line, and you won't know beforehand how many words to expect. In that instance, you can use the fgetl function which reads the whole line into a single string variable. For example, you can read our two header lines using:

    ```
    Title = '';

    for i=1:2
        line  = fgetl(fid);
        Title = str2mat(Title, line);
    end

    Title(1,:) = '';
    ```

    Another potentially useful feature of fgetl is that it does not skip over null or blank lines. For example, if at this point you issue:

    ```
    line = fgetl(fid)
    ```

    the result would be a null string or a string of blanks, depending on what was stored in the file.

3. The next section of this program reads the actual data by invoking fscanf two times on each line: once for the planet name, and once for two pieces of orbit data.

    In the second fscanf, the format is specified as '%f %f'. However, it would suffice to use '%f', because MATLAB automatically repeats format strings when it runs out of codes.

4. After the program has read the last line of data, then any further invocations of fscanf simply return a null string ('') or null matrix ([]). So, to break out of the loop, employ the feof function. It returns a 1 (true) if the program has already tried to read beyond the last line of the file.

fprintf

In this section, fprintf is used to write out files identical to the plain `orbits1.dat` file and the annotated `orbits2.dat`.

First, re-create the plain file:

```
fid = fopen('solar1.out', 'wt');

for i=1:9
    fprintf(fid, '    %5.2f    %5.3f\n', ...
                Orbit(i,1), Orbit(i,2));
    if rem(i,3) == 0 & i ~= 9
        fprintf(fid, '\n');
    end
end

fclose(fid);
```

Most of the commands are the same as before (i.e., fopen and fclose). The new command, of course, is fprintf. Here it is used to write out the first two elements of every row of Orbit. You can express the command more succinctly as:

```
fprintf(fid, '    %5.2f    %5.3f\n', ...
            Orbit(i,1:2));
```

The second argument to fprintf specifies the format of the output. Here five blanks are left before the first number, and five blanks before the second number.

It is specified that the numbers are floating-point "f," and that they should occupy a total of "5" spaces. The first number should have "2" digits to the right of the decimal point and the second number should have "3."

The "\n" specifies that a line is to end so that any subsequent output goes on a new line, rather than being appended to the current line.

Other format codes are given in the *Command Listing* section.

If you don't insist on the blank lines in the output, the for loop can be omitted and the matrix written with one invocation of fprintf:

```
fid = fopen('solar11.out', 'wt');

fprintf(fid, '    %5.2f    %5.3f\n', Orbit');

fclose(fid);
```

Notice that we *had* to specify Orbit'—the transpose of the Orbit matrix—in the call to fprintf.

Now that the basics of fprintf have been covered, let's re-visit `orbits2.dat` and re-create it using fprintf:

	Mean Dist.:	Eccent- ricity:
Mercury	0.39	0.206
Venus	0.72	0.007
Earth	1.00	0.017

```
Mars        1.52    0.093
Jupiter     5.20    0.048
Saturn      9.54    0.056

Uranus     19.19    0.047
Neptune    30.07    0.009
Pluto      39.46    0.249
```

The following program prints out not only the Orbit matrix, but also the Title words, and the Name for each planet:

```
fid = fopen('solar2.out', 'wt');

fprintf(fid, '%s\n', '              Mean      Eccent-');
fprintf(fid, '%s\n', '              Dist.:    ricity:');
fprintf(fid, '\n');

for i=1:9
    fprintf(fid, '%7s     %5.2f     %5.3f\n', ...
        Name(i,:), Orbit(i,1), Orbit(i,2));

    if rem(i,3) == 0  &  i ~= 9
        fprintf(fid, '\n');
    end
end

fclose(fid);
```

Reading and Writing ASCII Data Interactively

You can also use many of the above functions for interactive input/output, most notably fprintf and fscanf.

Using prompt and disp

But first re-visit the example program and the prompts.m function. There input and disp were used to read from the keyboard and write to the screen, respectively:

```
function Final_t = prompts

globals

disp('-------------------------------------');
disp(['     Start time is: ', num2str(Start_t, 10)]);
disp('-------------------------------------');

disp(['Final time can be a number: ', ...
                   num2str(Start_t+5)]);
disp(['or it can be an expression: ', ...
                   'Start_t+5']);

Final_t = input('Enter Final time: ');
disp('-------------------------------------');
disp(['     Final time is: ', num2str(Final_t, 10)]);
disp('-------------------------------------');
```

disp displays the contents of a variable (scalar, matrix, string) without displaying its name. It also displays the result of expressions without displaying ans = before it.

input prompts for a MATLAB expression which is then evaluated and returned by input. The expression may contain variables; but in most cases, simply a number is entered.

If you want to enter character data, then you must enclose it in quotes; otherwise, it will be evaluated as a MATLAB expression. Alternatively, you can specify the 's' option to the input function. For example:

```
planet = input('Enter planet name', 's')
```

This tells MATLAB that your input will be a string without your having to put quotes around it.

Using fprintf and fscanf

fprintf and fscanf can also be used for interactive input/output.

With fprintf (and fscanf), you can make use of the fact that MATLAB assigns "file identifiers" to the screen and keyboard as follows: [91]

```
0    keyboard input
1    screen output
```

Thus you can write to the screen as follows:

```
fprintf(1, 'Start time is: %10.4f\n', Start_t);
```

fprintf actually allows you to omit the "file identifier" argument altogether, in which case it assumes a value of 1:

```
fprintf('Start time is: %10.4f\n', Start_t);
```

Similarly with fscanf you can specify the "file identifier" as 0, and MATLAB reads from the keyboard:

```
Final_t = fscanf(0, '%f');
```

See the chapter *Graphical User Interfaces* for more information on functions that provide **graphical** interactive input and output of text and numbers.

---

[91] MATLAB follows the Unix convention of defining three input/output streams called: standard input (stdin for short), standard output (stdout), and standard error (stderr) for error messages. Normally, these are assigned to the keyboard (stdin), and the MATLAB command screen (stdout, stderr), but Unix allows you to redirect these to other devices.

MATLAB also follows the Unix numbering scheme, and so stdin is 0, stdout is 1, and stderr is 2. This explains why file identifiers assigned by fopen start at 3.

Reading and Writing Binary Data (MAT-files, etc.)

Binary data files are much more efficient than ASCII data files in terms of storage space, and often more importantly in terms of execution time when reading and writing files. The drawback is that binary files cannot be conveniently read or edited by humans, and so usually only very large data sets are stored this way.

You have two choices in writing binary files:

1. Binary files can be written in MATLAB's own MAT-file format, which is easily manipulated from within MATLAB using load and save. The MAT-file format is also portable between operating environments.

2. Binary files can also be written (and read) in your own custom format using fwrite (and fread). This allows you to exchange data in binary formats possibly required by other commercial software or your own personal software.

High-level Functions: save and load

save

Write the orbits1 matrix and the results vector, Final_th, to the file my.mat. The following three commands are equivalent:

```
save my      orbits1 Final_th
save my.mat orbits1 Final_th
save my.mat orbits1 Final_th -mat
```

As you can see, the default extension, if you don't specify one, is .MAT. Further, the default storage mode for save is as a MAT-file (binary).

load

You can reload the data using load. The following three commands are equivalent:

```
load my
load my.mat
load my.mat -mat
```

The original variable names are stored with the data in the MAT-file; MATLAB uses them automatically. Furthermore, you don't have to worry about whether the variables have been saved as one big matrix or not, as was necessary when saving and loading ASCII data files.

Also notice that MATLAB assumes the extension is .mat if you don't specify one. If you specify an extension other than .mat the file is assumed to be ASCII, unless you specify the -mat option.

Low-level Functions: fwrite and fread

fwrite and fread are the binary equivalents of the ASCII functions fprintf and fscanf, respectively.

fwrite

Write out the orbits1 matrix and the results vector Final_th, to the file `my.bin`:

```
fid = fopen('my.bin', 'w');

    precision = 'double';

    fwrite(fid, orbits1, precision);
    fwrite(fid, Final_th, precision);

fclose(fid);
```

With fwrite you can only specify **one** matrix (or vector or scalar) at a time, hence it was called twice in this program.

You also must specify the binary format you want to use. Here we have specified 'double' for *double precision*, the precision MATLAB uses internally for floating-point computations.

A precision you might choose for portability between systems is 'float64'. MATLAB then converts from the double precision format of your computer to its own portable double precision format ('float64') before writing out the data.

Refer to the *Command Listing* section for a list of other precision choices.

fread

Read back the orbits1 matrix and the results vector Final_th, from the file `my.bin`:

```
fid = fopen('my.bin', 'r');

    precision = 'double';

    orbits1  = fread(fid, [9, 2], precision);
    Final_th = fread(fid, [9, 1], precision);

fclose(fid);
```

Again, with fread you can only specify **one** matrix, etc., at a time; hence it was invoked twice in this program as well.

Some notes:

1. Both fread and fwrite transpose the matrix before it is either read or written. In other words, with fwrite, first column 1 is written, then column 2, and so on. This doesn't affect the final result here because fread undoes the transposition performed by fwrite; but it is important to keep in mind when reading or writing binary data from *other* software.

Command Listing

cd directory
Changes the current directory on your computer to directory.
Output: Not applicable.
Argument options: cd('directory') to get the same result. ✤ cd .. to change to the directory one level above the current one. ✤ cd to print the current directory. No value is assigned. ✤ str = cd to assign the current directory to the string str.
Additional information: Of course, directory must be in the proper syntax for your particular system. If you are unsure of the proper form, consult your system administrator.
See also: dir, ls, what, lookfor, path, pwd

delete filename
Delete the M-file filename from the filespace.
Output: Not applicable.
Argument options: delete hndl to delete the object with handle hndl.
Additional information: No confirmation is requested when delete is used, so exercise caution.
See also: close, cd, dir, type, who, what, !

diary filename
Saves all subsequent input and output (excluding graphics) to the ASCII file filename.
Output: Not applicable.
Argument options: diary off to stop saving to filename. ✤ diary on to restart saving to filename. If no diary filename has been used in the current session, the filename *diary* is used. ✤ diary to toggle the diary on and off, as above.
Additional information: When the diary facility is on, it does not affect what is displayed on your terminal. ✤ If the more facility is switched on during a diary session, the more prompts are also saved to file. ✤ Do not use the filenames *on* and *off*.
See also: Diary, save, fprintf, disp, echo, more

dir
Lists the files in the current directory.
Output: Not applicable.
Argument options: dir directory to list the files in directory.
Additional information: directory can be any valid directory path for your system.
See also: cd, ls, delete, what, pwd, lookfor, type, !

M = dlmread(filename, char)
Reads an ASCII data file whose values are delimited by the character char.
Output: A matrix is assigned to M.
Argument options: M = dlmread(filename, char, int_r, int_c) to specify the upper-left corner of the data, by row and column, within filename.
See also: dlmwrite

dlmwrite(filename, M, char)
Writes matrix M to an ASCII data file whose values are delimited by the character char.
Output: Not applicable.
Argument options: dlmwrite(filename, M, char, int_r, int_c) to specify the upper-left corner of the file position, by row and column.
See also: dlmread

int = exist(name)
Determines whether an object name already exists in the current session.
Output: If name is a built-in MATLAB function, 5 is assigned to int. If name is a compiled Simulink function, 4 is assigned to int. If name is an MEX-file, 3 is assigned to int. If name or name.m is the name of file on disk in the path specified by MATLABPATH, 2 is assigned to int. If name is variable in the current workspace, 1 is assigned to int. If name is none of the above, 0 is assigned to int.
See also: which, dir, help, lookfor

int = fclose(fileid)
Closes the file with file identifier fileid.
Output: If file fileid is open and fclose is successful, a value of 0 is assigned to int. If file fileid is not open or fclose is unsuccessful, a value of -1 is assigned to int.
Argument options: int = fclose('all') to close *all* currently open files.
Additional information: fileid is an integer identification number given to a file when it is opened with fopen.
See also: fopen, ferror, fread, fwrite, fseek, fscanf

num = feof(fileid)
Determines whether the end-of-file indicator has been set for file fileid.
Output: If the end-of-file indicator is set, 1 is assigned to num. If the end-of-file indicator is not set, 0 is assigned to num.
Additional information: Before feof can operate, the appropriate fopen command must be issued.
See also: fopen, fgetl, fgets, frewind

[str, int] = ferror(fileid)
Determines the error status of the most recent file operation.
Output: If the latest file operation was successful, an empty string is assigned to str and 0 to int. If the latest file operation was unsuccessful, an error message is assigned to str and a nonzero integer to int.
Argument options: [str, int] = ferror(fileid, 'clear') to clear the error indicator for the file.
Additional information: fileid is an integer identification number given to a file when it is opened with fopen. ♣ The value of the error number returned in int comes from the C language library *stdio*.
See also: fopen, fclose, ferror, fread, fwrite, fseek, fscanf, fprintf

str = **fgetl**(fileid)

Returns the next line of file fileid without the preceding newline or carriage return character.
Output: A string is assigned to str.
Additional information: Use this command with *text* files only—not with binary files.
✢ If the end-of-file character is the only one on the line, −1 is returned. ✢ To get the string with the newline or carriage return characters, use fgets. ✢ For a simple example of usage, see the on-line help file.
See also: fopen, feof, fgets, frewind

str = **fgets**(fileid)

Returns the next line of file fileid including the preceding newline or carriage return character.
Output: A string is assigned to str.
Additional information: Use this command with *text* files only—not with binary files.
✢ If the end-of-file character is the only one on the line, −1 is returned. ✢ To get the string without the newline or carriage return character, use fgetl.
See also: fopen, feof, fgetl, frewind

[fileid, str] = **fopen**(filename)

Opens the file with file name filename.
Output: If file filename opens successfully, a file identifier is assigned to fileid and the empty string is assigned to str. If file filename does not open successfully, a value of −1 is assigned to fileid and an error message is assigned to str.
Argument options: fileid = fopen(filename) to dispense with error messages. ✢ fileid = fopen(filename, str) to open filename in the mode specified by permission string str. 'r' opens an existing file for reading. 'w' creates a new file or deletes the contents of an existing one, and opens it for writing. 'a' creates a new file or opens an existing file for writing and appends to the end of it. Adding a + to the end of any of the above options opens the file for reading *and* writing. ✢ fileid = fopen(filename, str_1, str_2) to specify the architecture of the file with str_2. For more information on valid architecture strings, see the on-line help file. ✢ [fileid, str_p, str_a] = fopen(filename) to assign the relevant permission string and architecture string for filename to str_p and str_a, respectively. ✢ V = fopen('all') to assign to vector V the file identifiers of all currently open files.
Additional information: On systems that can specify between text and binary files, a b can be added to the permission string. ✢ fileid is an integer identification number given to a file when it is opened with fopen.
See also: fclose, ferror, fread, fwrite, fseek, fscanf

format *short*

Sets the default output format for numerical values to 5-digit scaled fixed-point.
Output: Not applicable.
Argument options: format *long* to set the default format to 15-digit scaled fixed-point.
✢ format *short e* to set the default format to 5-digit scaled floating-point *e*-notation.
✢ format *long e* to set the default format to 16-digit scaled floating-point *e*-notation.
✢ format *hex* to set the default format to hexadecimal notation. ✢ format *bank* to set the default format to dollars and cents notation. ✢ format *rat* to set the default

format to the ratio of small integers. ✤ format + to set the default format to display instead of the values themselves a + for positive values, a - for negative values, and a blank for zero values. ✤ format *compact* to set the default display format to eliminate excess line feeds. ✤ format *loose* to encourage line feeds. ✤ format to set the default format to the overall defaults, *short* and *loose*.

Additional information: The display formats *compact* and *loose* have no effect on the numerical results, but simply alter their presentation.

See also: rat, spy, num2str, sprintf, fprintf, Format

fprintf(fileid, str$_{fmt}$, M)

Prints the contents of matrix M to file fileid using a format specified by the string str$_{fmt}$.
Output: Not applicable.
Argument options: fprintf(1, str$_{fmt}$, M) or fprintf(str$_{fmt}$, M) to print the contents of M to the screen. ✤ fprintf(2, str$_{fmt}$, M) to print the contents of M to standard error. ✤ fprintf(fileid, V$_{str}$, M) to print each successive column of matrix M with a successive format string from V$_{str}$, a vector of format strings. The strings in V$_{str}$ are cycled through until all columns of M are printed. ✤ int = fprintf(fileid, str$_{fmt}$, M) to assign a count of the number of bytes written to int.
Additional information: This command is modelled on the C language command of the same name. ✤ The format string is made up of special flags, justification information, width specifications, precisions, and type details for the expressions to be printed. ✤ For more information on fprintf and for examples of its uses, see the on-line help file or a C language manual.
See also: fscanf, sprintf, fopen, fclose, sscanf

M = fread(fileid, [posint$_r$, posint$_c$])

Reads binary data from the file fileid into a matrix with posint$_r$ rows and posint$_c$ columns.
Output: A matrix is assigned to M.
Argument options: V = fread(fileid, n) to read n values into a column vector. ✤ V = fread(fileid, *inf*) to read all the values in fileid into a column vector. ✤ M = fread(fileid, [posint$_r$, posint$_c$], str$_p$) to use a precision specified by str$_p$ when reading in values. For more information on precision options available, see the on-line help file. ✤ M = fread(fileid, [posint$_r$, posint$_c$], str$_p$, int$_s$) to skip int$_s$ bytes between each read access. ✤ [M, int] = fread(fileid, [posint$_r$, posint$_c$]) to assign the number of elements transferred to count.
Additional information: Before fread can operate, the appropriate fopen command must be issued. ✤ ferror can be used to check on the status of the most recent fread command. ✤ Several fread commands can be called on a single file if it contains enough information. ✤ Remember to close each file with fclose when you are finished with it.
See also: fopen, fclose, ferror, fwrite, fseek, fscanf

frewind(fileid)

Resets the file pointer of file fileid to the beginning of the file.
Output: Not applicable.

Additional information: With some tape devices frewind will not work, and will not return an error message, either.
See also: fopen, feof, fgetl, fgets

M = fscanf(fileid, str$_{fmt}$, [posint$_r$, posint$_c$])

Reads the contents of file fileid using a format specified by the string str$_{fmt}$.
Output: A matrix with posint$_r$ rows and posint$_c$ columns is assigned to M.
Argument options: V = fscanf(fileid, str$_{fmt}$, n) to read n values into a column vector.
* V = fread(fileid, str$_{fmt}$, inf) to read all the values in fileid into a column vector. * M = fscanf(fileid, V$_{str}$, [posint$_r$, posint$_c$]) to read each successive column of matrix M with a successive format string from V$_{str}$, a vector of format strings. The strings in V$_{str}$ are cycled through until all columns of M are read. * [M, int] = fscanf(fileid, str$_{fmt}$, [posint$_r$, posint$_c$]) to assign a count of the number of elements read to int.
Additional information: This command is modelled on the C language command of the same name. * The format string is made up of special flags, justification information, width specifications, precisions, and type details for the expressions to be read. * For more information on fscanf and for examples of its uses, see the on-line help file or a C language manual.
See also: fprintf, sprintf, ferror, fopen, fclose, sscanf, load

int = fseek(fileid, posint, 'cof')

Moves the file position pointer in fileid ahead posint bytes from its current position.
Output: If the move is successful, a value of 0 is assigned to int. If the move is unsuccessful, a value of −1 is assigned to int.
Argument options: int = **fseek**(fileid, negint, 'cof') to move negint bytes towards the beginning of the file. * int = **fseek**(fileid, 0, 'cof') to make no move at all. * int = **fseek**(fileid, posint, 'bof') to move relative to the beginning of the file. * int = **fseek**(fileid, posint, 'eof') to move relative to the end of the file.
Additional information: Before fseek can operate, the appropriate fopen command must be issued. * Use ftell to get the current file position pointer. * ferror can be used to check on the status of the most recent fseek command. * You cannot move past the last byte in the file using fseek. * Remember to close each file with fclose when you are finished with it.
See also: ftell, fopen, fclose, ferror, fwrite, fread

int = ftell(fileid)

Returns the current file position pointer for file fileid.
Output: If the query is successful, an integer value representing the number of bytes from the beginning of the file is assigned to int. If the query is unsuccessful, a value of −1 is assigned to int.
Additional information: Before ftell can operate, the appropriate fopen command must be issued. * Use fseek to move the current file position pointer. * ferror can be used to check on the status of the most recent ftell command. * Remember to close each file with fclose when you are finished with it.
See also: fseek, fopen, fclose, ferror, fwrite, fread

int = **fwrite**(fileid, M)
Writes binary data from the matrix M to the file fileid in column order.
Output: The number of successfully written elements is assigned to int. M = fwrite(fileid, M, str$_p$) to use a precision specified by str$_p$ when writing values. For more information on precision options available, see the on-line help file for fread.
Argument options: M = fwrite(fileid, M, str$_p$) to use a precision specified by str$_p$ when writing out values. For more information on precision options available, see the on-line help file. ✤ M = fwrite(fileid, M, str$_p$, int$_s$) to skip int$_s$ bytes between each write.
Additional information: Before fwrite can operate, the appropriate fopen command must be issued. ✤ ferror can be used to check on the status of the most recent fwrite command. ✤ Remember to close each file with fclose when you are finished with it.
See also: fopen, fclose, ferror, fread, fseek, fscanf

var = **input**(str)
Prompts the user for input from the keyboard by displaying the prompt str.
Output: The value entered is assigned to var.
Argument options: str$_{in}$ = input(str, 's') to specify that the value entered be treated as a string.
Additional information: Any type of MATLAB expression can be returned in var, but multiple expressions are not allowed in a single invocation of input. ✤ If the 's' option is not specified, the value entered is fully evaluated before the assignment is done.
✤ Hitting the Return key without entering any input returns an empty matrix.
See also: keyboard, ginput, uicontrol, menu

load filename
Reads in the entire file filename from disk.
Output: Not applicable.
Argument options: load to load in the file named *matlab.mat*.
Additional information: If filename ends in a *.mat* extension, then that specifies that the file is in internal MATLAB format. Do not name a file with such an extension, unless it is saved in MATLAB format! ✤ If filename ends in any other extension, that specifies that the file is in ASCII format. ✤ The major difference between the load command and commands like fscanf is that fscanf allows you to read parts of a file only.
See also: save, fscanf

ls
Lists the contents of the current directory.
Output: Not applicable.
Additional information: ls lists the contents as specified by your system.
See also: cd, dir, delete, what, pwd, lookfor, type, !

save filename
Saves all the variables in the current workspace in binary format in a file named filename.mat.
Output: Not applicable.

Argument options: save to save all the variables in a file named *matlab.mat*. ✤ save *stdio* to send all the information to standard output. ✤ save filename var$_1$, var$_2$, ..., var$_n$ To save the variables var$_1$ through var$_n$ only in filename.mat. ✤ save filename *-ascii* to save the variables in 8-digit ASCII format in a file named filename. ✤ save filename *-ascii -double* to save the variables in 16-digit ASCII format in a file named filename. ✤ save filename *-ascii -tabs* or save filename *-ascii -double -tabs* to separate data elements with tabs instead of spaces.

Additional information: For more information on how binary files are saved, see the on-line help file. ✤ To load in a saved file, use the save command. ✤ Any file created with save can be read on any platform MATLAB supports (though some accuracy may be lost). ✤ ASCII based files can also be read in by other applications.

See also: load, fwrite, fprintf

str = sprintf(str$_{fmt}$, M)

Converts the contents of matrix M to a string using a format specified by the string str$_{fmt}$.

Output: A string is assigned to str.

Argument options: str = sprintf(str$_{fmt}$, M$_1$, M$_2$, ..., M$_n$) to convert the contents of matrices M$_1$ through M$_n$ all to a single string.

Additional information: This command is modelled on the C language command *fprintf*. ✤ The format string is made up of special flags, justification information, width specifications, precisions, and type details for the expressions to be printed. ✤ For more information on sprintf and for examples of its uses, see the on-line help files for fprintf and sprintf or a C language manual.

See also: int2str, num2str, fprintf, sscanf

[M, int] = sscanf(str, str$_{fmt}$)

Reads data from MATLAB string str, which is in a format specified by the string str$_{fmt}$, into a matrix.

Output: A matrix of values is assigned to M and a count of elements successfully read is assigned to int.

Argument options: [M, int] = sscanf(str, str$_{fmt}$, int$_{in}$) to read at most int$_{in}$ elements.

Additional information: This command is modelled on the C language command *sscanf*. ✤ The format string is made up of special flags, justification information, width specifications, precisions, and type details for the expressions to be printed. ✤ For more information on sscanf and for examples of its uses, see the on-line help files for fscanf and sprintf or a C language manual.

See also: fscanf, sprintf

filename = tempname

Returns a unique name that can be used as a temporary filename.

Output: A string is assigned to filename.

Additional information: filename has no extension. ✤ filename is of the form tp#######, where the # characters are replaced by a random 6-digit positive integer.

See also: what, lookfor

terminal(str)
Sets the session to using the type of graphics terminal indicated by str.
Output: Not applicable.
Argument options: terminal to be presented with a menu of terminal choices from which to choose. This works differently on different platforms.
Additional information: Currently supported terminal types include: *tek401x, tek4100, tek4105, retro, sg100, sg200, vt240tek, ergo, graphon, citoh, xtermtek, wyse, kermit, hp2647, versa, versa4100, versa4105,* and *hds.* For more information on these terminal types, see the on-line help file for terminal.

type filename
Displays the contents of file filename without actually running it.
Output: Not applicable.
Additional information: If filename has no extension, an extension of .m is assumed.
✦ type searches the entire search path for filename.
See also: cd, dir, path, what, dbtype

what
Lists the files with .m, .mat, or .mex extensions in the current directory.
Output: Not applicable.
Argument options: what dirname lists the files with .m, .mat, or .mex extensions in the directory dirname.
Additional information: If the directory dirname cannot be found with only the given name, it is searched for in the current search path. Therefore, it is not always necessary to provide the full pathname of a directory. ✦ Of course, the syntax of a pathname depends on your host system.
See also: dir, ls, lookfor, which, path, pwd, matlabroot, findpath

M = wk1read(filename)
Reads Lotus 1-2-3 WK1 spreadsheet file.
Output: A matrix is assigned to M.
Argument options: M = wk1read(filename, int_r, int_c) to specify the upper-left corner of the data, by row and column, within filename.
See also: wk1write

wk1write(filename, M)
Writes matrix M to Lotus 1-2-3 WK1 spreadsheet file.
Output: Not applicable.
Argument options: wk1write(filename, M, int_r, int_c) to specify the upper-left corner of the output file, by row and column.
See also: wk1read

Debugging MATLAB Programs

There are two *main* methods of debugging MATLAB programs. One is to use the individual debugging commands built into MATLAB.

The commands covered are:

1. Breakpoints:
    - dbstop - set (three ways)
    - dbclear - remove
    - dbstatus - display

2. Execution:
    - *usual command* - start
    - dbcont - continue until breakpoint or end
    - dbstep - execute current line in function then stop again
    - dbquit - quit debug mode *and* stop execution

3. Context:
    - dbstack - display "call list"
    - dbup - go up a function (so you can display variables in its workspace)
    - dbdown - go back down

The other is to use the graphical debugger built into the MATLAB editor:

- edit

In this chapter we will show examples of both of these ways, and how they can be used together.

Additionally, there are four other debugging techniques available in MATLAB:

1. Use echo to display commands as they are executed. You can specify which functions are to be traced. (For details see the *Command Listing* section.)

2. Remove the ";" from the end of commands so that you can see intermediate results.

3. Add statements like the following to display intermediate results:

   ```
   >> x
   >> disp(['x is:', num2str(x)])
   >> fprintf(1, 'x is: %f', x);
   ```

4. Insert a keyboard command in a function file so that MATLAB will pause there during execution, and take commands from the keyboard. (For details see the *Command Listing* section.)

5. (*Best!*) Use the MATLAB Debugger. With it you can tell MATLAB to stop at a certain line (or when an error occurs), display the values of variables, and resume execution until another "breakpoint" is reached, or simply execute the next line.

   The MATLAB Debugger is the subject of this chapter.

Sample Debugging Session 1

The following is a sample debugging session. First we type out our mainline program (area_f.m), and the functions it calls (crude.m and f.m):

```
>> dbtype area_f

1     % Program to invoke "crude" function to estimate
2     % the area under "f" function between 0 and pi.
3
4     a = 0;
5     b = pi;
6
7     q = crude('f', a, b);
8
9     fprintf(1, 'Area is: %f\n', q);

>> dbtype crude

1     % Function to make crude estimate of area
2     % under a function between x=a and x=b.
3
4     function q = crude(func, a, b)
5
6     c      = (a - b) / 2;           % BUG: should be "+"
7     func_c = feval(func, c);
8
9     q = func_c * (b - a);

>> dbtype f

1     function y = f(x)
2
3     y1 = sin(x);
4     y2 = cos(x);
5
6     y = y1 + y2;
```

The bug we are trying to stomp out is on line 6 of crude, but let's pretend we don't know that yet and use MATLAB's debugger to track it down.

First we set breakpoints at the lines where we want the MATLAB debugger to suspend execution. Here we set one at line 9 in crude, and then remove it in favor of line 6:

```
>> dbstop at 9 in crude
>> dbclear at 9 in crude
>>
>> dbstop at 6 in crude
```

We chose line 6 because it was the *first* executable line in crude.m. We could therefore have issued the following shorter and equivalent dbstop command:

```
>> dbstop in crude
```

Then dbstatus tells us what breakpoints we have set for crude.

```
>> dbstatus
Breakpoint for ...crude.m is on line 6.
```

Then we invoke our mainline M-file. When execution gets to line 6 in crude we can verify where we are, using dbstack:

```
>> area_f

6   c      = (a + b) / 2;

K>> dbstack

In ...crude.m at line 6
```

(Notice that the prompt has changed from the familiar MATLAB prompt, ">>", to the MATLAB Debugger prompt, "K>>".)

On Windows and Macintosh, MATLAB *Editor/Debugger* window also appears with an arrow pointing to the line we are on.

Then we can display the values of variables, such as a and b:

```
K>> [a, b]

ans =
         0    3.1416
```

Remember that breakpoints cause execution to stop *before* the specified line is executed. Therefore, c has not been defined yet.

Now let's execute this line and stop before executing the next line using dbstep, and then display the value of c:

```
K>> dbstep

7   func_c = feval(func, c);

K>> c

c =
   -1.5708
```

At this point, if we again issue dbstep, the MATLAB Debugger executes line 7, and then stops at the next line in *this* function.

However, we want the Debugger to stop at the first executable line in the *function* f, so we issue: dbstep in.

```
K>> dbstep in
```

We can again use dbstack if we wish to confirm where we are:

```
K>> dbstack

In ...f.m at line 3
In ...crude.m at line 7
In ...area_f.m at line 7
```

Then we can step through the calculations in f, and display intermediate results:

```
K>> dbstep

K>> dbstep

K>> [y1, y2]

ans =
     -1.0000    0.0000
```

At this point we display the variables that have been defined in f:

```
K>> who

Your variables are:
  ans         x         y1        y2
```

Let's leave execution suspended in f, but go back up to crude and issue a who command to display the variable names there:

```
K>> dbup

In workspace belonging to ...crude.m.

K>> who

Your variables are:
  a           b         func
  ans         c
```

Let's go right back up to the level of the mainline ("base" workspace) and issue who again:

```
K>> dbup

In workspace belonging to ...area_f.m.

K>> who

Your variables are:
  a           ans       b
```

Now let's work our way back down from the mainline, through crude and back to f:

```
K>> dbdown
```

            In workspace belonging to ...crude.m.

    K>> dbdown

            In workspace belonging to ...f.m.

    K>> who

            Your variables are:
            ans   x   y1   y2

Now let's just remind ourselves of where we are in f, and what that line actually looks like:

    K>> dbstack

            In ...f.m at line 6
            In ...crude.m at line 7
            In ...area_f.m at line 7

    K>> dbtype f

            1   function y = f(x)
            2
            3   y1 = sin(x);
            4   y2 = cos(x);
            5
            6   y = y1 + y2;

Now invoke dbstep again so that the MATLAB Debugger executes line 6 (the last line), and then stops *before* returning to the calling function. This enables us to examine the value calculated for y:

    K>> dbstep

    K>> y

    y =
            -1.0000

Now let's step back to the calling function (crude), and display the result of the calculation on line 7:

    K>> dbstep

    K>> func_c

    func_c =
            -1.0000

Now lets execute the last line in crude, line 9, and display the result:

    K>> dbstep

    K>> q

    q =
            -3.1416

Finally, let's step back to the calling mainline.

```
K>> dbstep
```

We are now on the last line of the mainline program. To execute it, let's issue dbstep.

```
K>> dbstep
Area is: -3.141593
```

The MATLAB Debugger has finished executing the last command but has not exited the program file. To do this, issue dbstep once again.

```
K>> dbstep
>>
```

Notice that the prompt has changed back to the regular MATLAB input prompt.

Example 1: Using dbcont

Our breakpoint is still in effect:

```
>> dbstatus
Breakpoint for ...crude.m is on line 6.
```

Let's run our program again:

```
>> area_f
```

and it will stop again so that we can check the value of the arguments to crude:

```
K>> [a, b]
ans =
         0    3.1416
```

That is all we want to do, so we can let execution resume until the next breakpoint (there isn't one), or to completion:

```
K>> dbcont
Area is: -3.141593
>>
```

Alternatively, if we don't want execution to resume, but rather want it to stop at this point with no further processing, we could issue:

```
K>> dbquit
>>
```

Notice that this returns us directly to the base workspace prompt and no results are returned.

Finally, let's remove our breakpoint. As this is the only breakpoint, and it is on the first executable line in crude.m, either of the following have the same effect, in this case:

```
>> dbclear all
>> dbclear at 6 in crude
```

Sample Debugging Session 2

Let's invoke the second form of the dbstop command:

>> dbstop if error

and now let's invoke our function crude with the name of the wrong function:

>> crude('ff', 0, pi)

```
??? Can not find function 'ff'.

Error in ==> ...crude.m
On line 7   ==> func_c = feval(func, c);
```

The MATLAB Debugger stops execution *inside* crude so that we can examine variables, such as func:

K>> func

```
func =
ff
```

K>>

You cannot resume execution with either dbstep or dbcont; both return you to the base workspace prompt:

K>> dbcont

>>

Finally, let's remove our breakpoint:

>> dbclear if error

Sample Debugging Session 3

Let's invoke the third form of the dbstop command:

>> dbstop if infnan

or:

>> dbstop if naninf

This causes execution to be suspended if an Inf or NaN (infinity or Not-a-Number) is encountered during execution.

Now let's invoke our function crude with Inf as one of the limits of integration:

>> crude('f', 0, Inf)

K>>

Naturally if we continue execution, we get another break when line 6 is executed:

K>> dbcont

K>>

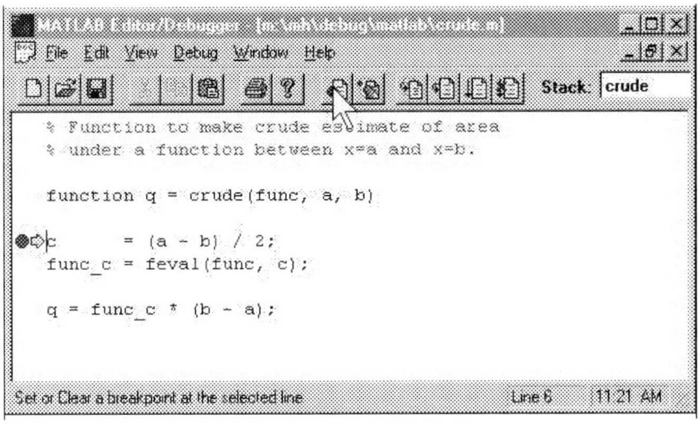

FIGURE 103. MATLAB *Editor/Debugger* window.

To prevent further breaks, let's change b and c before resuming execution:

K>> b = pi;

K>> c = pi/2;

K>> dbcont

ans =
    3.1416

Finally, let's remove our breakpoint:

>> dbclear if infnan

Graphical Debugging

As mentioned earlier, on Windows and Macintosh systems, the MATLAB *Editor/Debugger* window automatically opens when your program encounters a breakpoint. In this window, you can also set (and clear) breakpoints, by clicking on the desired line and clicking on the "stop sign" icon on the toolbar.

You can also step through your program by clicking on the "step in" and "single step" icons on the toolbar.

For example, let's set a breakpoint in crude and run our program:

>> dbstop in crude

>> area_f

The MATLAB *Editor/Debugger* window appears looking something like Figure 103.

Then click on the "step in" button repeatedly to step through the program. Notice that the MATLAB *Editor/Debugger* window switches between the lines and functions of your program as it executes.

Command Listing

dbclear *at* posint *in* filename

Removes the breakpoint from M-file filename that was set at line posint.
Output: Not applicable.
Argument options: dbclear *in* filename to remove the breakpoint located at the first executable line in filename. ✦ dbclear *all in* filename to remove all the breakpoints in filename. ✦ dbclear *all* to remove all breakpoints in all the current M-file functions. ✦ dbclear *if* option to remove a general breakpoint in case of special occurrences. option can be one of *error* for when a runtime error occurs, or *infnan* or *naninf* for when either a *Not-A-Number* or *Infinity* is generated.
Additional information: To set a breakpoint, use the dbstop command. Both are very helpful in debugging M-file functions.
See also: dbstop, dbcont, dbstep, dbstatus, dbstack, dbdown, dbup, dbtype, dbquit

dbcont

Continues execution of the current session after a breakpoint is encountered.
Output: Not applicable.
Additional information: Each time a dbstop command is encountered in an M-file, execution of the function halts. At these points one option is to recommence execution with dbcont. ✦ You cannot resume execution of an function after a runtime error breakpoint is encountered.
See also: dbstop, dbclear, dbstep, dbstatus, dbstack, dbdown, dbup, dbtype, dbquit

dbdown

Moves forward one step, towards the place where a breakpoint was encountered, in the current workspace stack.
Output: Not applicable.
Additional information: By stepping through the current workspace stack, you can examine values at various functions that lead to a dbstop command being executed. ✦ In any one stack investigation, you must always have used at least as many dbup calls as dbdown calls. ✦ When you are ready to continue from the breakpoint using dbcont or dbstep, you do not have to use dbdown to reach the breakpoint first.
See also: dbstop, dbclear, dbcont, dbstep, dbstatus, dbstack, dbup, dbtype, dbquit

dbquit

Returns control of the session to the base prompt, without completing the current function.
Output: Not applicable.
Additional information: Using dbquit does not affect any of the breakpoints that were set.
See also: dbstop, dbclear, dbcont, dbstep, dbstatus, dbstack, dbdown, dbup, dbtype

dbstack

Lists the M-file names and line numbers of the current workspace stack.
Output: Not applicable.

Additional information: The first function listed is the one most recently called. The second function listed is the function that called it. And so on.
See also: dbstop, dbclear, dbcont, dbstep, dbstatus, dbdown, dbup, dbtype, dbquit

dbstatus filename

Lists the line numbers of all the breakpoints set in M-file filename.
Output: Not applicable.
Additional information: To set a breakpoint in an M-file, use the dbstop command.
See also: dbstop, dbclear, dbcont, dbstep, dbstack, dbdown, dbup, dbtype, dbquit

dbstep

Fully executes the next line only, after encountering a breakpoint.
Output: Not applicable.
Argument options: dbstep posint to fully execute the next posint lines only. ✤ dbstep *in*, when the next executable line is a call to another function, to specify that that function is to be stepped through as well.
Additional information: To move back through the stack to determine what *lead* to the breakpoint being triggered, use dbup and dbdown. ✤ Using dbcont continues execution from the current line to which you have stepped.
See also: dbstop, dbclear, dbcont, dbstatus, dbstack, dbdown, dbup, dbtype, dbquit

dbstop *at* posint *in* filename

Sets a breakpoint in M-file filename just prior to line posint.
Output: Not applicable.
Argument options: dbstop *in* filename to set a breakpoint at the first executable line in filename. ✤ dbstop *if* option to set a general breakpoint in case of special occurrences. option can be one of *error* for when a runtime error occurs, or *infnan* or *naninf* for when either a *Not-A-Number* or *Infinity* is generated.
Additional information: After a breakpoint is encountered, execution is halted and the condition that caused the halt is displayed. ✤ To continue execution after a breakpoint is encountered or to return control to the base prompt, use the dbcont or dbquit commands, respectively. ✤ To remove a breakpoint, use the dbclear command. Both are very helpful in debugging M-file functions. ✤ Editing or clearing an M-file removes all breakpoints set at specific line numbers automatically.
See also: dbclear, dbcont, dbstep, dbstatus, dbstack, dbdown, dbup, dbtype, dbquit, warning

dbtype filename

Lists the lines of M-file filename with the corresponding line numbers beside them.
Output: Not applicable.
Argument options: dbtype filename $posint_1:posint_2$ to list only from line $posint_1$ to $posint_2$ from filename.
Additional information: This command helps you know where to place the appropriate breakpoints with dbstop.
See also: dbstop, dbclear, dbcont, dbstep, dbstack, dbdown, dbup, dbstatus, dbquit

dbup

Moves backward one step, away from the place where a breakpoint was encountered, in the current workspace stack.
Output: Not applicable.
Additional information: By stepping through the current workspace stack, you can examine values at various functions that lead to a dbstop command being executed. ✦ In any one stack investigation, you must always have used at least as many dbup calls as dbdown calls. ✦ When you are ready to continue from the breakpoint using dbcont or dbstep, you do not have to use dbdown to reach the breakpoint first.
See also: dbstop, dbclear, dbcont, dbstep, dbstatus, dbstack, dbdown, dbtype, dbquit

profile fnc

Turns on execution profiling for function fnc.
Output: Not applicable.
Argument options: profile *report* to display a summary report of the function currently being profiled. ✦ profile *posint* to display only the posint lines that use the most time. ✦ profile *num*, where num is between 0 and 1, to display only those line taking more than num of the total time. posint lines that use the most time. ✦ function currently being profiled. ✦ profile *reset* to reset the profiling information for the function currently being profiled. ✦ profile *done* to turn off profiling for the current function. ✦ var = profile to return information fields about the current status of profiling.
Additional information: It is recommended that you end all calls to functions being profiled with the ; character, to avoid voluminous printouts. ✦ Some built-in MATLAB functions cannot be profiled. ✦ Subfunctions and functions within classes, as well as private functions and methods can also be profiled. See the on-line help for more details. ✦ When creating reports, profile calls the profsumm function.
See also: profsumm, debug

profsumm

Prints a report on the function currently being profiled.
Output: A Profiling summary is printed out.
Argument options: profsumm(posint) to display only the posint lines that use the most time. ✦ profsumm(num), where num is between 0 and 1, to display only those line taking more than num of the total time. profsumm(str) to display only the lines that contain the string str.
Additional information: Most of the functionality of profsumm command is available through the profile command.
See also: profile, debug

Advanced Data Structures

In addition to matrices and vectors, MATLAB also has *advanced* data structures similar to those found in other high-level languages (but easier to use, of course). These are:

- multi-dimensional arrays
- "cell" arrays
- structure variables

The functions we will cover in this chapter are:

- cell
- struct

Multi-dimensional Arrays

Consider the following data set stored in the file `house.dat`.

```
0 0         1 1 1       2 2 2       3 3 3       4 4 4
0 0         1 2 3       1 2 3       1 2 3       1 2 3

1 1         3 4 5       4 4 4       4 4 4       4 4 4
1 2         4 5 6       4 4 5       4 4 4       4 4 4
1 3         5 6 7       4 5 6       4 4 5       4 4 4

2 1         2 3 4       3 4 4       4 4 4       4 4 4
2 2         3 4 5       4 4 4       4 4 4       4 4 4
2 3         4 5 6       4 4 5       4 4 4       4 4 4

3 1         1 2 3       2 3 4       3 4 4       4 4 4
3 2         2 3 4       3 4 4       4 4 4       4 4 4
3 3         3 4 5       4 4 4       4 4 4       4 4 4
```

It consists of a 3×4 array of 3×3 matrices. Each 3×3 matrix contains temperature values in rooms on one floor of a house. The house in question has three floors and the temperatures throughout the house are sampled four times.

The first two rows and the first two columns of the data in the file are not actual temperature readings, but rather are the indices into the 3×4 array and the individual 3×3 matrices.

Let's load in this file:

438   Data Structures

```
load house.dat
```

Now, let's extract just the temperature values:

```
house = house(3:end, 3:end);
```

And now let's create a 4-dimensional matrix called temp containing these values.

```
%                    x  y
%                    /  /  z = floor
%                   /  /  /  t = time period
%                  /  /  /  /
temp = zeros(3, 3, 3, 4);

for z=1:3
    for t=1:4
        for x=1:3
            for y=1:3
                i = (z-1)*3 + x;
                j = (t-1)*3 + y;
                temp(x,y,z,t) = house(i, j);
            end
        end
    end
end
```

Let's have a look at temp.

```
temp

temp(:,:,1,1) =
     3     4     5
     4     5     6
     5     6     7
temp(:,:,2,1) =
     2     3     4
     3     4     5
     4     5     6
temp(:,:,3,1) =
     1     2     3
     2     3     4
     3     4     5

temp(:,:,1,2) =
     4     4     4
     4     4     5
     4     5     6
temp(:,:,2,2) =
     3     4     4
     4     4     4
     4     4     5
temp(:,:,3,2) =
     2     3     4
     3     4     4
     4     4     4

temp(:,:,1,3) =
     4     4     4
     4     4     4
     4     4     5
temp(:,:,2,3) =
     4     4     4
     4     4     4
```

```
                    4         4         4
temp(:,:,3,3) =
                    3         4         4
                    4         4         4
                    4         4         4

temp(:,:,1,4) =
                    4         4         4
                    4         4         4
                    4         4         4
temp(:,:,2,4) =
                    4         4         4
                    4         4         4
                    4         4         4
temp(:,:,3,4) =
                    4         4         4
                    4         4         4
                    4         4         4
```

MATLAB's vectorized functions such as sum and prod support multi-dimensional arrays. For example, let's calculate the sum of all the temperature values in each time period.

```
total = sum(sum(sum(temp)));
```

MATLAB sums up the first three dimensions, and creates a $1 \times 1 \times 1 \times 4$ matrix in total. Let's use the reshape function to convert this to a 4×1 matrix.

```
total = reshape(total, [4,1])
```

which results in:

```
ans =
     108
     108
     108
     108
```

"Cell" Arrays

MATLAB supports two types of arrays: *numeric* arrays and *cell* arrays. Numeric arrays can only have numbers or characters as elements, whereas "cell" arrays can have numbers, characters, and other arrays as elements.

Let's reload the data from the previous example and create a 3×4 "cell" array (called temp) where each element is a 3×3 numeric array.

```
load house.dat
house = house(3:end, 3:end);

%                    x  y
%                   /  /   z = floor
%                  /  /  / t = time period
%                 /  /  / /
floor = zeros(3, 3      );
temp  = cell(      3, 4);
```

```
      for z=1:3
          for t=1:4
              for x=1:3
                  for y=1:3
                      i = (z-1)*3 + x;
                      j = (t-1)*3 + y;
                      floor(x,y)=house(i,j);
                  end
              end
              temp{z,t} = floor;
          end
      end
```

The cell function was used to initialize the "cell" array to be 3 rows by 4 columns. Notice that when indexing a "cell" array, you have to use {} brackets instead of () brackets for the indices.

Now let's display our result.

```
      temp
```

```
      temp =
          [3x3 double]    [3x3 double]    [3x3 double]    [3x3 double]
          [3x3 double]    [3x3 double]    [3x3 double]    [3x3 double]
          [3x3 double]    [3x3 double]    [3x3 double]    [3x3 double]
```

To display one of the elements of temp, we can index temp as follows:

```
      temp{1,1}
```

which results in

```
      ans =
          3    4    5
          4    5    6
          5    6    7
```

Structure Variables

MATLAB allows you to create structured variables, i.e., variables that contain data structures beneath them stored as separate *fields*.

For example, in the *Programming in* MATLAB chapter, we created separate arrays for the planets' names, periods, and orbits. For Mercury, the entries were:

```
      Name(1, *)  = 'Mercury';
      Period(1)   = 0.24;
      Orbit(1, *) = [0.39, 0.206, 0, 7.0];
```

We can create a structure variable called planet1 with three fields: .Name, .Period, and Orbit.

```
      planet1.Name   = 'Mercury';
      planet1.Period = 0.24;
      planet1.Orbit  = [0.39, 0.206, 0, 7.0];
```

Let's display the value of planet1.

```
>> planet1

planet1 =
       Name: 'Mercury'
     Period: 0.2400
      Orbit: [0.3900 0.2060 0 7]
```

Notice that each field can contain a different type of data (including other structure variables).

We can also create a "cell" array (called planets) whose elements are each structure variables, one for each planet.

```
planets = cell(9, 1);

planet{1}.Name   = 'Mercury';
planet{1}.Period = 0.24;
planet{1}.Orbit  = [0.39, 0.206, 0, 7.0];

...

planet{9}.Name   = 'Pluto';
planet{9}.Period = 247.7;
planet{9}.Orbit  = [39.46, 0.249, 0, 17.1];
```

An alternate way of defining a structure variable is to use the struct function. For example, we could define the first planet using:

```
planet{1} = struct('Name',   'Mercury', ...
                   'Period', 0.24,      ...
                   'Orbit',  [0.39, 0.206, 0, 7.0]);
```

In the following chapter, *Object-Oriented Programming*, this structure becomes the heart of a class of objects called planet.[92]

Command Listing

M_{out} = **cat**(int, M_1, M_2)

Concatenates matrices M_1 and M_2 along the int^{th} dimension.
Output: A matrix is assigned to M_{out}.
Argument options: M_{out} = cat(int, M_1, ..., M_n) to concatenate n matrices along the int^{th} dimension.
Additional information: The input matrices must be of appropriate dimensions.
✦ Values for int that are greater than 2 can be used to create multi-dimensional arrays from standard arrays.
See also: num2cell

M_c = **cell**(int_1, int_2)

Creates a $int_1 \times int_2$ cell array of empty matrices.

---

[92] In addition to having these properties, we define a set of functions (called *methods*) for the planet class, which have exclusive access to these properties.

Output: A cell matrix is assigned to M_c.
Argument options: M_c = cell(int) to create an int × int cell array of empty matrices.
* M_c = cell(int$_1$, ..., int$_n$) to create an n-dimensional cell array of empty matrices of the given dimensions.
See also: ones, zeros, cell2struct, celldisp, cellplot

M_s = cell2struct(M_c, V_{str}, int)

Converts cell array M_c into a structure array by changing the intth dimension into fields with names specified by V_{str}.
Output: A structure array is assigned to M_s.
Additional information: The number of strings in V_{str} must equal the size of intth dimension of M_c.
See also: struct2cell, cell, struct

celldisp(M_c)

Displays the contents of cell array M_c.
Output: Not applicable.
Additional information: Layers of values are displayed recursively.
See also: cell, cellplot

cellplot(M_c)

Displays the structure of cell array M_c graphically.
Output: Not applicable.
Argument options: V_{hndl} = cellplot(M_c) to assign the plot structures required to the vector of handles V_{hndl}.
Additional information: Nested, colored boxes are used to display the structure.
See also: cell, celldisp

V_{str} = fieldnames(M_s)

Returns the field names associated with structure array M_s.
Output: A cell array of strings is assigned to V_{str}.
See also: getfield, setfield

M_{out} = flipdim(M, int)

Flips matrix M along the intth dimension.
Output: A matrix is assigned to M_{out}.
Additional information: The resulting matrix is of the same dimensions as the original matrix.
See also: fliplr, flipud, rot90, permute, shiftdim

M = getfield(M_s, str)

Returns the value stored in the str field of 1 × 1 structure array M_s.
Output: A matrix is assigned to M.
Argument options: M = getfield(M_s, {int$_i$, int$_j$}, str, {int$_k$}) to return a value stored in str field of a two-dimensional structure array.
See also: struct, fieldnames, setfield

Data Structures

[int$_1$, ..., int$_n$] = ind2sub(V, int)
Computes the subscript values necessary in a matrix of dimensions defined by V to equal the linear (one-dimensional) index of int.
Output: Numeric values are assigned to int$_1$ through int$_n$.
Additional information: V must have n elements, each which will be greater than or equal to its respective int$_i$ value.
See also: sub2ind

M$_{out}$ = ipermute(M, V)
Performs an inverse permutation on the dimensions of matrix M, according to dimension-vector V.
Output: A matrix is assigned to M$_{out}$.
Additional information: The elements of V must be a permutation of the integers 1 through n, where n is the dimension of M. ✤ ipermute(permute(M, V), V) always returns M.
See also: permute

int = iscell(M)
Determines if matrix M is a cell array.
Output: If M is a cell array a value of 1 is assigned to int. Otherwise, a value of 0 is assigned to int.
See also: cell, isstruct

int = isstruct(M)
Determines if matrix M is a structure array.
Output: If M is a structure array a value of 1 is assigned to int. Otherwise, a value of 0 is assigned to int.
See also: struct, iscell

[M$_1$, ..., M$_n$] = ndgrid(V$_1$, ..., V$_n$)
Creates n matrices for n-dimensional functions from n vectors.
Output: Matrices are assigned to M$_1$ through M$_n$.
Additional information: For problems that are *spatially* based, use the meshgrid command.
See also: meshgrid, interpn

int = ndims(M)
Determines the number of dimensions of matrix M.
Output: A numeric value is assigned to int.
See also: size, length

M$_c$ = num2cell(M$_{num}$)
Converts each element of numeric array M$_{num}$ into a cell of a cell array.
Output: A cell array is assigned to M$_c$.
Argument options: M$_c$ = (M$_{num}$, [int$_1$, ..., int$_n$]) to place only the elements of M$_{num}$ corresponding to dimensions int$_1$ through int$_n$ into separate cells.

Additional information: in the two parameter version, M_{num} must have $n + 1$ dimensions. ✦ The cat command can be used to reverse this conversion.
See also: cell, cat

M_{out} = permute(M, V)
Permutes the dimensions of matrix M, according to dimension-vector V.
Output: A matrix is assigned to M_{out}.
Additional information: The elements of V must be a permutation of the integers 1 through n, where n is the dimension of M.
See also: ipermute, transpose

M_{out} = repmat(M_b, [num_r, num_c])
Creates an $num_r \times num_c$ matrix, which is tiled with copies of input matrix M_b.
Output: A matrix is assigned to num.
Argument options: M_{out} = repmat(M_b, [num_1, ..., num_n]) to create an n-dimensional array. ✦ M_{out} = repmat(num, [num_1, ..., num_n]) to fill the array with a single value, num.

$M_{s,out}$ = rmfield($M_{s,in}$, str)
Removes the str field of structure array $M_{s,in}$.
Output: An updated structure array is assigned to $M_{s,out}$.
See also: struct, fieldnames, setfield, getfield

$M_{s,out}$ = setfield($M_{s,in}$, str, M)
Sets the value stored in the str field of 1×1 structure array $M_{s,in}$ to M.
Output: An updated structure array is assigned to $M_{s,out}$.
Argument options: $M_{s,out}$ = setfield(M_s, {int_i, int_j}, str, {int_k}, M) to set the value in the str field of a two-dimensional structure array.
See also: struct, fieldnames, getfield

M_{out} = shiftdim(M, int)
Shifts the dimensions of matrix M int dimensions to the left.
Output: A matrix is assigned to M_{out}.
Argument options: [M_{out}, int] = shiftdim(M) to strip leading singleton dimensions from M. The number of singleton dimensions is returned in int.
Additional information: If int is negative, a shift to the right is performed. ✦ Singleton dimensions are removed, where applicable.
See also: flipdim, reshape, squeeze

M_{out} = squeeze(M)
Removes *all* singleton dimensions from matrix M.
Output: A matrix is assigned to M_{out}.
Additional information: Two-dimensional matrices are unaffected by squeeze.
See also: shiftdim

M_s = **struct**(str_1, M_1, ..., str_n, M_n)

Creates a structure array with values for fields str_1 through str_n defined by matrices M_1 through M_n, respectively.
Output: A structure array is assigned to M_s.
Additional information: The matrices must all be of equal dimensions. ✶ Instead of matrices, cell arrays or single numeric values can be used. The resulting structure array is size appropriately.
See also: cell, fieldnames, setfield, getfield

M_c = **struct2cell**(M_s)

Converts structure array M_s into a cell array.
Output: A cell array is assigned to M_c.
Additional information: If M_s in n-dimensional, then M_c is $(n+1)$-dimensional, with the extra dimension being equal in size to the number of fields in M_s.
See also: cell2struct, cell, struct

int = **sub2ind**(V, int_1, ..., int_n)

Computes the linear (one-dimensional) index of an element in the $int_1{}^{th}$ through $int_n{}^{th}$ position of a matrix of dimensions defined by V.
Output: A numeric value is assigned to int.
Additional information: V must have n elements, each of which is greater than or equal to its respective int_i value.
See also: ind2sub

Object-Oriented Programming

MATLAB allows you to write programs in **both** the traditional, procedural manner and in the newer, object-oriented style. The object-oriented syntax in MATLAB is not dissimilar to languages such as C++, but naturally it is much easier to use and is well integrated into the rest of the MATLAB language.

In this chapter we will use a modified version of our solar system program to illustrate the following main features and functions of object-oriented programming in MATLAB:

- class - defining classes
- objects
- methods
- operator overloading

Creating objects

In our solar system example we can consider the planets to be one *class* of objects where:

- each planet object has similar *properties* (e.g., period), and
- we perform similar *actions* on each planet object (e.g., compute position).

Let's write a script to define nine planet objects using the planet object *constructor* function:

```
c{1} = {'Mercury',   0.24, [0.39, 0.206], [], [], ''};
c{2} = {'Venus',     0.62, [0.72, 0.007], [], [], ''};
c{3} = {'Earth',     1.00, [1.00, 0.017], [], [], ''};
c{4} = {'Mars',      1.88, [1.52, 0.093], [], [], ''};
c{5} = {'Jupiter',  11.86, [5.20, 0.048], [], [], ''};
c{6} = {'Saturn',   29.46, [9.54, 0.056], [], [], ''};
c{7} = {'Uranus',   84.0,  [19.19, 0.047], [], [], ''};
c{8} = {'Neptune', 164.8,  [30.07, 0.009], [], [], ''};
c{9} = {'Pluto',   247.7,  [39.46, 0.249], [], [], ''};

%----------------------------------------------------------
```

```
names = {'Name', 'Period', 'Orbit', ...
         'Start_t', 'Start_th', 'Moons'};

for i=1:9
    s{i} = cell2struct(c{i}, names, 2);
end

%---------------------------------------------------

for i=1:9
    o{i} = planet( s{i} );
end
```

where `planet.m` contains the following:

```
function o = planet( s )

if nargin < 1
    s.Name     = '';
    s.Period   = [];
    s.Orbit    = [];
    s.Start_t  = [];
    s.Start_th = [];
    s.Moons    = [];
end

o = class(s, 'planet');
```

The planet object constructor function:

1. takes as its *argument* property fields in a **data structure** (e.g., `.Period`), and

2. passes the *data structure* to MATLAB's class function to create and return an **object** of the class planet.

Notes

`planet.m` must be stored in `<...>/solar/classes/@planet/planet.m`, and the `<...>/solar/classes` directory must be added to the MATLAB path before you can create planet objects. In the section *Where are Methods Stored?*, we go into these naming conventions in detail.

Once we run the script to define the nine planets, we can look at the first planet.

```
>> o{1}
```

displays

```
ans =

planet object: 1-by-1
```

We are not allowed to look at or change **individual** attributes of an object. For example:

```
>> o{1}.Orbit
```

results in:

```
??? Access to an object's fields is
    only permitted within its methods.
```

Defining Methods for Classes of Objects

Methods are the only functions which are allowed to directly access or change the data inside a class of objects.

Let's create a function called get to allow us to access the individual properties of a planet:

```
function v = get(    o,  varargin)

v = getfield( struct(o), varargin{:} );
```

get uses struct(o) to extract the structure from the planet object. Then it uses getfield to extract one field from the structure (if a field is specified), or extract all fields (if no field is specified).[93]

For example,

```
get(o{1}, 'Name')
```

returns

```
ans =

Mercury
```

while

```
get(o{1})
```

returns the entire data structure from the planet object:

```
ans =

      Name: 'Mercury'
    Period: 0.2400
     Orbit: [0.3900 0.2060]
   Start_t: []
  Start_th: []
     Moons: ''
```

Now let's create a function called set to allow us to set properties of a planet, especially:

- starting time (.Start_t),
- starting angular position (.Start_th), and

---

[93] See the *Programming in* MATLAB chapter for more information on the varargin function.

- moons (.Moons).

```
function o2 = set(            o1, field, value)
o2 = class(                                            ...
            setfield( struct(o1), field, value),       ...
                                                       ...
            class(            o1)                      ...
        );
```

set uses struct(o1) to extract the structure from the planet object. Then it uses setfield to redefine the specified field in the structure (with the specified value). Lastly, it reconstructs the object from the new structure using the class function.

We can then invoke the set method as follows for each planet:

```
t = 1995.0 + 3/365;

th = [345.7083
      138.8194
      102.1532
      119.2770
      242.7067
      348.7888
      291.6482
      293.7496
      231.6706] * pi/180;

for i=1:9
    o{i} = set( o{i}, 'Start_t' , t      );
    o{i} = set( o{i}, 'Start_th', th(i) );
end
```

And now,

```
get(o{1})
```

returns

```
ans =
         Name: 'Mercury'
       Period: 0.2400
        Orbit: [0.3900 0.2060]
      Start_t: 1.9950e+003
     Start_th: 6.0337
        Moons: ''
```

Notes

get.m must be stored in < ... >/classes/@planet/get.m, i.e., in the same directory as the class it serves. Being stored in this location is what makes the get function be a "method". (Similarly for set.)

When we pass set an object such as a planet, MATLAB first checks to see if there is a set method associated with that class of objects. If not, then MATLAB invokes its own set function (for graphics handles).

The Rest of the Program

The rest of this program consists of three parts:

1. First, we prompt for the final time (Final_t) and compute the final angle (Final_th) for each planet:

    ```
    Final_t = input('Enter Final time: ');

    Final_th = zeros(9,1);

    for id=1:9
        Final_th(id) = final( o{id}, Final_t );
    end

    Final_th
    ```

    Here we use the final method which invokes MATLAB's fzero function:

    ```
    function th = final(                         object, Final_t )

                 th = fzero('kepler', pi, [], [], object, Final_t);
    ```

    fzero calls our (modified) kepler method which in turn calls MATLAB's quad8 function:

    ```
    function diff = kepler(Final_th, ...
                           object, Final_t)

    % ----------------------------------------------------------------
    % Trying to find "Final_th" such that:
    %
    %                       A          t
    %         diff   =   --------  -  ---   =   0
    %                     Pi a b       T
    %
    % where:
    %
    %     A is the area swept between Start_th and Final_th
    %                                          ^^^^^^^^^
    %     t is the time taken between Start_t  and Final_t
    % ----------------------------------------------------------------

    a = object.Orbit(1);             %- semi-major axis
    e = object.Orbit(2);             %- eccentricity

    b = a * sqrt(1-e^2);             %- semi-minor axis

    tol   = 1e-6;
    trace = [];
    A = (1/2) * quad8('radius_2', object.Start_th, ...
                                  Final_th, ...
                                  tol, trace,         ...
                                  object);

    % ----------------------------------------------------------------

    t = Final_t - object.Start_t;

    T =             object.Period;
    ```

```
        %   -------------------------------------------------
        diff = A / (pi * a * b)    -  t / T;
```

quad8 calls our (modified) radius_2 method:

```
        function r_2 = radius_2(th,      ...
                                object)

        a     = object.Orbit(1);
        e     = object.Orbit(2);
        angle = 0;

        r = a*(1 - e^2) ./                ...
               (1 - e*cos(th-angle));

        r_2 = r.^2;
```

2. Then we plot the results:

```
        for ifig=1:2

             figure(ifig)

             if ifig==1,   from=1;  to=4;   end
             if ifig==2,   from=4;  to=9;   end

             %  - - - - - - - - - - - - - - - - - - - - - -

             for id=from:to
                   h(id) = plotit( o{id}, Final_th(id) );

                   set( h(id).text, 'Color', 'r' )

                   hold on
             end

             hold off

             %  - - - - - - - - - - - - - - - - - - - - - -

             axis equal
             axis('square')

             fname =  ['solar', num2str(ifig), '.eps'];
             %

             print('-deps', fname)

        end
```

Here we use the plotit method (which uses the ellipse function from the *Programming in* MATLAB chapter):

```
        function h = plotit( object, Final_th )

        %  - - - - - - - - - - - - - - - - - - - - - - - - -

                        a      = object.Orbit(1);
                        e      = object.Orbit(2);
                               angle = 0;
```

```
              h.ellipse = ellipse(a, e, 0, 2*pi, angle);

       % - - - - - - - - - - - - - - - - - - - - - - -

              h.arc = ellipse(a, e, object.Start_th, ...
                                      Final_th, ...
                              angle);

              set(h.arc, 'LineWidth', 4.0)

       % - - - - - - - - - - - - - - - - - - - - - - -

              th = 3 * pi / 2;

              r  = a*(1 - e^2) ./               ...
                   (1 - e*cos(th-angle));

                          [x, y] = pol2cart(th, r);
              h.text = text(x, y, object.Name);
```

Notice that we made plotit return the handles to the three graphics objects it created. Rather than storing the handles in an array, (h(1), h(2), h(3)), we stored them in a structure (h.ellipse, h.arc, h.text) to make the code easier to read.

3. And finally we prompt the user as to whether they wish to repeat the process:

```
       another = input(['Do you wish to trace ', ...
                        'another orbit? [y/n] '], 's');

       while strcmp(another, 'y')

              cont = input(['Do you wish to start from ', ...
                            'where you left off? [y/n] '], 's');

              if strcmp(cont, 'y')
                   Start_t  = Final_t;
                   Start_th = Final_th;
              end

              finals
              plots

              another = input(['Do you wish to trace ', ...
                               'another orbit? [y/n] '], 's');

       end
```

Where are Methods Stored?

As mentioned earlier, planet.m must be stored in < ... >/solar/classes/@planet/planet.m, where:

- < ... >/solar is simply the complete directory path where our program is stored,[94]

---

[94] For example, on our Windows 95 system, < ... >/solar is C:/Handbook/Object/solar.

Data Structures

- $< \ldots >$/solar/classes is the name of the directory where we will be storing our classes of objects (planets and, later, comets),

- $< \ldots >$/solar/classes/@planet is the name of the directory containing all the methods associated with just the planet class of objects,

- $< \ldots >$/solar/classes/@planet/planet.m is named after the planet class, and is used to create new planet objects.

The other methods we will store in this directory are get, set, final, and plotit as shown in this diagram:

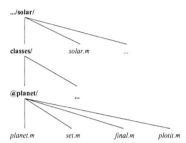

In order for MATLAB to be able to find our methods, we must add the $< \ldots >$/solar/classes directory to our MATLAB path:

```
>> path(path, '<...>/solar/classes')
```

Notes

The $< \ldots >$/solar/classes/@planet directory itself should **not** be added to the path.

Nested Object Classes

Objects can be defined *within* other objects. For example, we can fill in the .Moons data field of each planet with an array of **objects** representing that planet's moons.

Rather than create a whole new class for moons, we will simply **re-use** the planet class. That is, we will store our moons as an array of planet objects inside of our nine existing planet objects.

To add the moons to the .Moons data field of the planets, we use our multi-purpose set method.

```
%           ___  ___  ___
km = 1 / 149600000;      % fraction of an Astronomical Unit
dy = 1 / 365.24;         % fraction of a year

c{1} = {'The Moon', 27.32*dy, [384000*km, 0.05], [], [], ''};
c{2} = {'Phobos',    0.32*dy, [  9000*km, 0.02], [], [], ''};
c{3} = {'Deimos',    1.26*dy, [ 23000*km, 0.00], [], [], ''};
c{4} = {'Charon',    6.39*dy, [ 20000*km, 0.00], [], [], ''};

%-------------------------------------------------------
```

```
names = {'Name', 'Period', 'Orbit', ...
         'Start_t', 'Start_th', 'Moons'};

for i=1:4
    s{i} = cell2struct(c{i}, names, 2);
end

%------------------------------------------------------------

for i=1:4
    m{i} = planet( s{i} );
end
%- - - - - - - - - - - - - - - - - - - - - - - - - - - - - -
t  = 1995.0 + 3/365;
th = [0.0000
      0.0000
      0.0000
      0.0000] * pi/180;        % This is not real data!

for i=1:4
    m{i} = set( m{i}, 'Start_t',  t     );
    m{i} = set( m{i}, 'Start_th', th(i) );
end

%------------------------------------------------------------

o{3} = set( o{3}, 'Moons',  m{1}         );    % Earth

o{4} = set( o{4}, 'Moons', [m{2}, m{3}] );    % Mars

o{9} = set( o{9}, 'Moons',  m{4}         );    % Pluto

m = [];
```

Now let's display the periods of the Martian moons. If we simply type in

```
>> o{4}.Moons(1).Period      % Error!
>> o{4}.Moons(2).Period      % Error!
```

we get an error, because the periods are *inside* of objects (Phobos and Deimos) that are *inside* of another object (Mars).

Instead, we must first use the get method to get the moons from planet Mars, and then use get again for the period of each of the moons:

```
>> m = get( o{4}, 'Moons' );

>> get( m(1), 'Period' )
>> get( m(2), 'Period' )
```

and the results are:

```
ans =

   8.7614e-004

ans =

   0.0034
```

Creating a Class Based on Another Class

It is also possible to create objects *based on other objects*. These new objects *inherit* the fields and methods of the parent object(s), plus add their own fields and methods. For example, we could create a new class called comet, but base the new class heavily on the planet class. (Comets can essentially be treated as planets with the addition of *tails*.)

Space does not permit us to go further into this topic. However, you can see the comet class constructor function and associated methods, together with a program that uses them, at our web site (http://www.pracapp.com/matlab/)

Private Functions

Sometimes, methods call other functions to do a specialized task. If you wish to prevent these specialized functions from being called by anything other than the related methods, you can place them in the private subdirectory of the class.

For example, we would store *private* functions for the planet methods in the directory: < ... >/solar/classes/@planet/private.

Operator Overloading

Operator overloading consists of expanding the definitions of existing MATLAB functions and operators, so that they can handle new datatypes that you've created, i.e., objects.

We've already overloaded MATLAB's get and set functions to handle planet objects. Now, let's overload the minus operator (-) to compute the *differences* between two planets (e.g., Earth and Mars), so that the following command will work:

```
diffs = o{4} - o{3}
```

which returns

```
diffs =

    planet object: 1-by-1
```

Examining diffs, we get:

```
>> get(diffs)

ans =

         Name: 'Mars minus Earth'
       Period: 0.8800
        Orbit: [0.5200 0.0760]
      Start_t: 0
     Start_th: -5.5062
        Moons: {}
```

We achieve this by creating a planet method called minus defined as follows:

```
function o3 = minus(o1, o2)

o3 = o1;

s1 = struct(o1);
s2 = struct(o2);
s3 = struct(o3);

names = fieldnames(s3);
for i=1:length(names)
      name = char( names(i) );

      v1 = getfield( s1, name );
      v2 = getfield( s2, name );

      switch class( v1 )
      case {'double', 'sparse'}
            s3 = setfield(s3, name, v1 - v2);
      case 'char'
            s3 = setfield(s3, name, [v1, ' minus ', v2]);
      otherwise
            s3 = setfield(s3, name, {});
      end
end

o3 = class(s3, class(o3));
```

We can invoke this either of two ways:

```
diffs = minus(o{4}, o{3})
```

or

```
diffs = o{4} - o{3}
```

Similarly, we can overload other MATLAB operators, whose special "names" are:

minus	- a-b
plus	- a+b.
times	- a.*b
mtimes	- a*b.
mldivide	- a\b.
mrdivide	- a/b.
rdivide	- a./b.
ldivide	- a.\b.
power	- a.^b.
mpower	- a^b.
uminus	- -a.
uplus	- +a.
horzcat	- [a b].
vertcat	- [a;b].
le	- a<=b.
lt	- a<b.
gt	- a>b.
ge	- a>=b.
eq	- a==b.
ne	- a~=b.
not	- ~a.
and	- a&b.
or	- a|b.
subsasgn	- a(i)=b, a{i}=b, and a.field=b.

```
subsref     - a(i), a{i}, and a.field.
colon       - a:b.
transpose   - a.'
ctranspose  - a'
subsindex   - x(a)
```

Overloading operators further helps us to write highly compact and readable object-oriented code.

References

1. Data taken from "The Nine Planets", Appendix A, on the World Wide Web at: http://seds.lpl.arizona.edu/nineplanets/nineplanets/data.html.

Miscellaneous

Introduction

This chapter of *The* MATLAB *Handbook* is devoted to all the functions that did not find a home in previous chapters. This is not to say the material herein is unimportant. Some of the functions are obscure; many others are used by everyone, and are included here simply because they do not fit comfortably into any larger group.

The following list points out several of the themes found in this chapter and provides examples of the related command names:

- **statistics**: mean, median, std, corrcoef, cov

- **conversion between number types**: base2dec, bin2dec, dec2base, dec2bin, dec2hex, hex2dec, hex2num

- **bit-level integer operations**: bitand, bitcmp, bitget, bitmax, bitor, bitset, bitshift, bitxor

- **string manipulation**: blanks, cellstr, char, deblank, disp, findstr, int2str, iscellstr, ischar, isletter, isspace, isstr, lower, mat2str, num2str, numeric, strcat, str2num, strcmp, string, strjust, strmatch, strncmp, strrep, strtok, strvcat, upper

- **Fourier transforms**: fft, fft2, fftn, fftshift, ifft, ifft2, ifftn

- **polynomials**: inpolygon, poly, polyarea, polyfit, polyval, polyvalm, residue

- **set manipulation**: intersect, ismember, setdiff, setxor, union, unique

- **sorting**: sort, sortrows

Command Listing

int = base2dec(str, int_b)
Converts the numeric value in base int_b represented by string str to its decimal equivalent.
Output: An integer value is assigned to int.

Additional information: int_b must be between 2 and 36.
See also: dec2base, hex2dec, bin2dec, format

int = **bin2dec**(str)
Converts the binary value represented by string str into the equivalent value in base 10.
Output: An integer value is assigned to int.
See also: dec2bin, base2dec, format

int = **bitand**(int_1, int_2)
Performs a bit-wise *AND* on integers int_1 and int_2.
Output: An integer value is assigned to int.
Argument options: M = bitand(M_1, M_2) to perform operations on two matrices of integers, element by element.
Additional information: Both int_1 and int_2 must be non-negative integers less than the value bitmax.
See also: bitor, bitxor, bitshift, bitcmp, bitmax, bitget, bitset

int = **bitcmp**(int_1, int_c)
Computes the bit-wise *complement* of integer int_1.
Output: An int_c-bit non-negative integer value is assigned to int.
Argument options: M = bitcmp(M_1, int_c) to perform operations on a matrix of integers, element by element.
Additional information: int_1 must be a non-negative integer less than the value bitmax.
✢ If int_s is negative, the shift is made to the right.
See also: bitand, bitor, bitxor, bitshift, bitmax, bitget, bitset

int = **bitget**(int_1, int_b)
Returns the bit value in the int_b position of integer int_1.
Output: Either 1 or 0 is assigned to int.
Argument options: M = bitget(M_1, int_b) to perform operations on a matrix of integers, element by element.
Additional information: int_1 must be a non-negative integer less than the value bitmax.
See also: bitand, bitor, bitxor, bitshift, bitcmp, bitmax, bitset

int = **bitmax**
Returns the maximum integer value for your system.
Output: An integer value is assigned to int.
Additional information: On IEEE systems, bitmax is $2^{53} - 1$.
See also: bitand, bitor, bitxor, bitcmp, bitshift, bitget, bitset

int = **bitor**(int_1, int_2)
Performs a bit-wise *OR* on integers int_1 and int_2.
Output: An integer value is assigned to int.
Argument options: M = bitor(M_1, M_2) to perform operations on two matrices of integers, element by element.

Additional information: Both int_1 and int_2 must be non-negative integers less than the value bitmax.
See also: bitand, bitxor, bitshift, bitcmp, bitmax, bitget, bitset

int = bitset(int_1, int_b, int_v)
Sets the bit value in the int_b position of integer int_1 to int_v.
Output: An integer value is assigned to int.
Argument options: M = bitset(M_1, int_b) to perform operations on a matrix of integers, element by element.
Additional information: int_1 must be a non-negative integer less than the value bitmax.
* int_v must be either 0 or 1 (default = 1).
See also: bitand, bitor, bitxor, bitshift, bitcmp, bitmax, bitget

int = bitshift(int_1, int_s)
Performs a bit-wise *shift* of integer int_1, int_s places to the left.
Output: An integer value is assigned to int.
Argument options: M = bitshift(M_1, int_s) to perform operations on a matrix of integers, element by element.
Additional information: int_1 must be a non-negative integer less than the value bitmax.
* If int_s is negative, the shift is made to the right.
See also: bitand, bitor, bitxor, bitcmp, bitmax, bitget, bitset

int = bitxor(int_1, int_2)
Performs a bit-wise *exclusive OR* on integers int_1 and int_2.
Output: An integer value is assigned to int.
Argument options: M = bitxor(M_1, M_2) to perform operations on two matrices of integers, element by element.
Additional information: Both int_1 and int_2 must be non-negative integers less than the value bitmax.
See also: bitand, bitor, bitshift, bitcmp, bitmax, bitget, bitset

str = blanks(n)
Creates a string containing just n blank characters.
Output: A string is assigned to str.
Additional information: Use this command in conjunction with disp to scroll the cursor down n lines.
See also: deblank, disp

M_{str} = char(str_1, ..., str_n)
Creates a string matrix where each row is a given string.
Output: A string matrix is assigned to M_{str}.
Argument options: M_s = strcat($M_{s,1}$, ..., $M_{s,n}$) to create an arbitrarily large string matrix.
See also: Strvcat, strmatch

V_c = conv(V_1, V_2)
Computes the convolution of vectors V_1 and V_2.
Output: A vector whose length equals length(V_1) + length(V_2) - 1 is assigned to V_c.

Additional information: This operation can also be thought of as multiplying two polynomials whose coefficients are represented by vectors V_1 and V_2. ✤ For more information on the definition of convolution, see the on-line help file.
See also: deconv, residue

M_c = corrcoef(M)
Computes the correlation coefficients of matrix M.
Output: A square matrix is assigned to M_c.
Additional information: The rows of M correspond to observations and the columns of M correspond to variables. ✤ The correlation coefficients are strongly related to the covariance function, cov. ✤ For more information on the definition of correlation coefficients, see the on-line help file.
See also: cov, mean, std

M_c = cov(M)
Computes the covariance matrix of matrix M.
Output: A square matrix is assigned to M_c.
Argument options: num = cov(V) to computer the scalar variance of vector V and assign it to num. ✤ M_c = cov(V_1, V_2), where V_1 and V_2 are vectors of equal length to compute the covariance of the matrix [x y].
Additional information: The rows of M correspond to observations and the columns of M correspond to variables. ✤ For more information on the definition of covariance matrix, see the on-line help file.
See also: corrcoef, mean, std

V_{new} = cplxpair(V)
Sorts the elements of vector V so that the complex conjugates appear together.
Output: A vector is assigned to V_{new}.
Argument options: V_{new} = cplxpair(V, num) to set the tolerance for pairing conjugates to num. The default tolerance is 100*eps.
Additional information: V can be either a row vector or a column vector. ✤ If there are an odd number of complex numbers in V or if the complex values cannot all be paired up, an error message is printed. ✤ The pairs are sorted according to increasing values of their real parts. Within each pair the complex value with the negative imaginary part comes first. ✤ If any purely real values are found in V, they are sorted to the end of V_{new} in increasing order.
See also: conj, eps

str_{new} = deblank(str)
Removes all the null characters and trailing blanks from string str.
Output: A new string is assigned to str_{new}.
Additional information: A null character has an absolute value of 0.
See also: blanks

str = dec2hex(int)
Converts the decimal integer int to its hexadecimal representation.
Output: A string is assigned to str.

Additional information: Hexadecimal values are represented by strings containing the characters '0' through '9' and 'a' through 'f'.
See also: hex2dec, hex2num, base2dec, format

str = dec2base(int, int$_b$)
Converts the numeric value int (in base 10) into the equivalent value in base int$_b$.
Output: A string representing a value in base int$_b$ is assigned to str.
Additional information: int$_b$ must be between 2 and 36.
See also: base2dec, dec2bin, format

str = dec2bin(int)
Converts the numeric value int (in base 10) into the equivalent value in base 2.
Output: A string representing a value in base 2 is assigned to str.
Argument options: str = dec2bin(int, int$_n$) to ensure that str has at least int$_n$ bits.
See also: bin2dec, dec2base, format

[V$_q$, V$_r$] = deconv(V$_1$, V$_2$)
Computes the deconvolution of vector V$_1$ out of vector V$_2$.
Output: A vector representing the quotient of the division is assigned to V$_q$ and a vector representing the remainder of the division is assigned to V$_r$.
Additional information: This operation can also be thought of as division of polynomials whose coefficients are represented by vectors V$_1$ and V$_2$. ✦ The results are such that V$_b$ = conv(V$_q$, V$_1$) + V$_r$.
See also: conv, residue

M$_{new}$ = del2(M)
Computes the five-point discrete Laplacian of matrix M.
Output: A matrix of equal dimensions to M is assigned to M$_{new}$.
Additional information: Each element of M$_{new}$ is computed by taking the difference of that element with the average of all of its *direct* neighbors. ✦ Elements of a matrix can have four, three, or two neighbors, depending on their location.
See also: gradient, diff

disp(M)
Displays the elements of matrix M without the preface of M =.
Output: Not applicable.
Additional information: disp also works for string matrices.
See also: setstr, str2mat, blanks, sprintf

doc
Starts up the MATLAB hypertext help facility, if available.
Output: Not applicable.
Additional information: If you have access to this facility, we highly recommend you use it.
See also: help

var = eval(str)

Parses the string str as a MATLAB command and executes it if it is a statement.
Output: The result of the evaluated statement is assigned to var.
Argument options: eval(str, str$_{err}$) to execute the command represented by str$_{err}$ if there is an error encountered while evaluating str.
Additional information: If str parses to an expression (i.e., not an executable statement), the expression is assigned to var. ✦ disp is typical of a command to be invoked if there is a problem with the evaluation.
See also: disp, lasterr, feval

var = feval('fnc', expr$_1$, expr$_2$, ..., expr$_n$)

Evaluates the MATLAB function fnc with the parameters expr$_1$ through expr$_n$.
Output: The result of the evaluated function is assigned to var.
Argument options: [var$_1$, ..., var$_m$] = ('fnc', expr$_1$, expr$_2$, ..., expr$_n$) to evaluate a function that returns m values.
Additional information: The most common use of feval is within other functions that accept a function name as an input parameter.
See also: eval, fplot

V$_{new}$ = fft(V)

Computes the *fast Fourier transform* of V, a vector of complex values.
Output: A vector of complex values is assigned to V$_{new}$.
Argument options: V$_{new}$ = fft(V, n) to compute the transform of the first n values in V. If V has fewer than n values, it is padded with zeroes. ✦ M$_{new}$ = fft(M) to compute the transforms of each column of matrix M and return the results in the column of new matrix M$_{new}$.
Additional information: If the number of elements of V equals an integer power of 2, then a faster, more efficient algorithm is used. ✦ Fast fourier transforms are also known as *discrete Fourier transforms*. ✦ For more information on the definition, algorithm, and uses of fast Fourier transforms, see the on-line help file.
See also: ifft, fft2, fftshift

M$_{new}$ = fft2(M)

Computes the two-dimensional *fast Fourier transform* of M, a matrix of complex values.
Output: A matrix of complex values is assigned to M$_{new}$.
Argument options: M$_{new}$ = fft(M, m, n) to compute the transform of the first m rows and n columns in M. If M has fewer than m rows or n columns, it is padded with zeroes.
Additional information: Two-dimensional transforms are found by first taking the transform of the columns, then taking the transforms of the rows of the result. ✦ If the number of rows and/or columns of M equals an integer power of 2, then a faster, more efficient algorithm is used. ✦ Fast fourier transforms are also known as *discrete Fourier transforms*.
See also: ifft2, fft, fftshift

fftdemo

Runs a demonstration script showing how to work with Fast Fourier Transforms in MATLAB.
Output: Not applicable.
Additional information: Various commands are automatically entered for you. ✦ Occasionally, you will be prompted to strike any key to continue the demonstration. ✦ For a complete list of demonstrations available on your platform, see the on-line help file for demos.
See also: fft, quaddemo, odedemo, zerodemo, fplotdemo

M_{new} = fftshift(M)

Rearranges the results of the fft and fft2 commands.
Output: A matrix of complex values is assigned to M_{new}.
Argument options: V_{new} = fftshift(V) to to exchange the left and right halves of V, the vector result of fft.
Additional information: The first and third quadrants of matrix M are exchanged each other, as are the second and fourth quadrants. ✦ This command is useful in analyzing frequency components.
See also: fft, fft2

V_{pos} = findstr(str_1, str_2)

Finds the starting character positions of all the occurrences of the smaller of two strings in the larger.
Output: A vector of positive integers is returned.
Additional information: The two input strings can be given in any order.
See also: strcmp, strrep

int = hex2dec(str)

Converts the hexadecimal value represented by string str to a decimal integer.
Output: An integer value is assigned to int.
Additional information: Hexadecimal values are represented by strings containing the characters '0' through '9' and 'a' through 'f'.
See also: dec2hex, hex2num, base2dec, format

num = hex2num(str)

Converts the hexadecimal value represented by string str to a double precision floating-point number.
Output: A numeric value is assigned to num.
Additional information: Only valid IEEE numbers work with this function. ✦ Hexadecimal values are represented by strings containing the characters '0'–'9' and 'a'–'f'. ✦ The values of NaN, Inf, and denormalized numbers are also handled correctly.
See also: hex2dec, format

V_{new} = ifft(V)

Computes the inverse *f*ast *F*ourier *t*ransform of V, a vector of complex values.
Output: A vector of complex values is assigned to V_{new}.

Argument options: V_{new} = **ifft**(V, n) to compute the inverse transform of the first n values in V. If V has fewer than n values, it is padded with zeroes. ✦ M_{new} = ifft(M) to compute the inverse transforms of each column of matrix M and return the results in the column of new matrix M_{new}.
Additional information: If the number of elements of V equals an integer power of 2, then a faster, more efficient algorithm is used. ✦ Fast fourier transforms are also known as *discrete Fourier transforms*. ✦ For more information on the definition, algorithm, and uses of fast Fourier transforms, see the on-line help file for fft.
See also: fft, ifft2, fftshift

M_{new} = **ifft2**(M)

Computes the two-dimensional inverse *fast Fourier transform* of M, a matrix of complex values.
Output: A matrix of complex values is assigned to M_{new}.
Argument options: M_{new} = **ifft**(M, m, n) to compute the inverse transform of the first m rows and n columns in M. If M has fewer than m rows or n columns, it is padded with zeroes.
Additional information: If the number of rows and/or columns of M equals an integer power of 2, then a faster, more efficient algorithm is used. ✦ Fast fourier transforms are also known as *discrete Fourier transforms*. ✦ For more information on two-dimensional fast Fourier transforms, see the on-line help file for fft2.
See also: fft2, ifft, fftshift

num = **inpolygon**(num_x, num_y, V_x, V_y)

Determines whether the point [num_x, num_y] falls within the polygon defined by V_x and V_y.
Output: If the point lies within the polygon, a value of 1 is assigned to num. If the point lies on the polygon, a value of 0.5 is assigned to num. Otherwise, a value of 0 is assigned to num.
See also: polyval, poly, polyarea

str = **int2str**(int)

Converts integer int to a string.
Output: A string is assigned to str.
Additional information: This command is especially useful when concatenating an integer value into an already existing string.
See also: num2str, hex2num, setstr

num_y = **interp1**(V_x, V_y, num_x)

Calculates, using one-dimensional interpolation, the value of the function suggested by x and y vectors V_x and V_y at the x-value num_x.
Output: A numeric value is assigned to num_y.
Argument options: V_{new} = interp1(V_x, V_y, V) to interpolate the values at the elements of vector V. A vector of equal length is assigned to V_{new}. ✦ num_y = interp1(V_x, V_y, num_x, option) to control the method of interpolation used. Available options are *'linear'* for linear interpolation, *'cubic'* for cubic interpolation, and *'spline'* for cubic spline interpolation. The default is *'linear'*.

Additional information: It is always required that the elements of V_x be monotonic (i.e., either strictly increasing or decreasing). ✽ When the *cubic* method is specified, the elements of V_x must be equally spaced.
See also: interp2, interp3, interpn, interpft, griddata

num_z = **interp2**(M_x, M_y, M_z, num_x, num_y)

Calculates, using two-dimensional interpolation, the value of the function suggested by x, y, and z matrices M_x, M_y, and M_z at the x and y values num_x and num_y.
Output: A numeric value is assigned to num_z.
Argument options: num_z = interp2(V_x, V_y, M_z, num_x, num_y) to perform the same computation, where the values in M_z are related to the row and column vectors V_x and V_y, respectively. ✽ V_z = interp2(M_x, M_y, M_z, V_x, V_y) to interpolate the values at the elements of vectors V_x and V_y. A vector of equal length is assigned to V_z. ✽ num_z = interp2(M_x, M_y, M_z, num_x, num_y, option) to control the method of interpolation used. Available options are *'linear'* for linear interpolation and *'cubic'* for cubic interpolation. The default is *'linear'*.
Additional information: When performing interpolation, the elements of M_z represent the values corresponding to the appropriate elements of M_x and M_y. ✽ It is always required that the elements of V_x and V_y be monotonic (i.e., either strictly increasing or decreasing). ✽ When the *cubic* method is specified, the elements of V_x and V_y must be equally spaced.
See also: interp1, interp3, interpn, interpft, griddata

V_4 = **interp3**(M_1, M_2, M_3, M_4, V_1, V_2, V_3)

Calculates, using three-dimensional interpolation, the value of the function suggested by the four dimension matrices M_1 through M_4 at the values for the first three dimensions represented by V_1 through V_3.
Output: A vector is assigned to V_4.
Argument options: V_4 = interp3(M_1, M_2, M_3, M_4, V_1, V_2, V_3, option) to control the method of interpolation used. Available options are *'linear'* for linear interpolation, *'cubic'* for cubic interpolation, and *'nearest'* for nearest neighbor interpolation. The default is *'linear'*.
Additional information: For more information on interpolation, see the entries for interp1 and interp2 or the on-line help file.
See also: interp1, interp2, interpn, interpft, griddata

V_{n+1} = **interpn**(M_1, ..., M_{n+1}, V_1, ..., V_n)

Calculates, using n-dimensional interpolation, the value of the function suggested by the dimension matrices M_1 through M_{n+1} at the values for the first n dimensions represented by V_1 through V_n.
Output: A vector is assigned to V_{n+1}.
Argument options: V_{n+1} = interpn(M_1, ..., M_{n+1}, V_1, ..., V_n, option) to control the method of interpolation used. Available options are *'linear'* for linear interpolation, *'cubic'* for cubic interpolation, and *'nearest'* for nearest neighbor interpolation. The default is *'linear'*.
Additional information: For more information of interpolation, see the entries for interp1 and interp2 or the on-line help file.

See also: interp1, interp2, interpn, interpft, griddata

V_y = **interpft**(V_x, n)
Resamples the elements of vector V_x to n evenly spaced points, using the FFT method.
Output: A vector with n elements is assigned to V_y.
Additional information: The value n must be greater than or equal to max(size(V_x)).
See also: interp1, interp2

V = **intersect**(V_1, V_2)
Returns the unique elements common to vectors (sets) V_1 and V_2.
Output: A vector is assigned to V.
Argument options: [V, $V_{i,a}$, $V_{i,b}$] = intersect(V_1, V_2) to also return index vectors for the elements returned from M_a and M_b. ✦ M = intersect(M_1, M_2, 'rows') to return a matrix with rows common to both M_1 and M_2.
See also: union, setdiff, setxor, ismember, unique

V = **isletter**(str)
Determines whether each character in string str is a letter of the alphabet.
Output: A vector of 1s (is a letter) and 0s (not a letter) is assigned to V.
See also: isspace, isstr, findstr, lower, upper

int = **ismember**(num, V)
Determines if value num is an element of vector (set) V.
Output: If num is an element of S, then a 1 is assigned to int. Otherwise, a 0 is assigned to int.
Argument options: V = intersect(M_1, M_2, 'rows') to return a vector with 1s where the rows of M_1 are also rows of M_2.
See also: union, setdiff, setxor, setdiff, unique

V = **isspace**(str)
Determines whether each character in string str is a space, newline, carriage return, tab (horiz. or vert.), or formfeed.
Output: A vector of 1s (is a "space") and 0s (not a "space") is assigned to V.
See also: isletter, isstr, findstr, lower, upper

int = **isstr**(expr)
Determines if expr is of type string.
Output: If expr is a string, a value of 1 is assigned to int. Otherwise, a value of 0 is assigned to int.
Argument options: M_{new} = isstr(M) to determine if each element of matrix M is a string. An equally sized matrix of zeroes and ones is assigned to M_{new}.
Additional information: A string is stored internally as a vector with its text flag set.
See also: abs, setstr, strcmp, isletter, isspace, strings

M_K = **kron**(M_1, M_2)
Computes the Kronecker tensor product of X and Y.

Output: A matrix with a number of elements equal to the number of elements of M_1 times the number of elements of M_2 is assigned to M_K.
Additional information: The resulting matrix is computed by taking all possible products between elements of the two input matrices. ✦ For more information on Kronecker matrices, see the on-line help file.

num_{out} = lin2mu(num)
Converts linear audio signal amplitude num into a μ-law encoded value.
Output: An integer value between 0 and 255 is returned.
Argument options: V_{out} = lin2mu(V) or M_{out} = lin2mu(M) to convert a vector or matrix of values.
Additional information: The input values should be real values between -1 and 1.
See also: sound, mu2lin

str_{lc} = lower(str)
Converts all uppercase characters in str to lowercase.
Output: A completely lowercase string is assigned to str_{lc}.
Additional information: Characters already in lowercase are unaffected by lower. ✦ The resulting string is of equal length to the input string.
See also: upper, isstr, isletter, strcmp

M = magic(n)
Creates an n × n magic square.
Output: A square matrix is assigned to M.
Additional information: A magic square is a matrix such that the sums of all the rows and columns and the two main diagonals are equal.

str = mat2str(M)
Converts matrix M into a single string.
Output: A string is assigned to str.
Argument options: str = num2str(M, int) to use int digits of precision.
Additional information: This operation can be reversed with the eval function.
See also: num2str, int2str, eval, format

num = mean(V)
Determines the mean (average) value of the elements of vector V.
Output: A numeric value is assigned to num.
Argument options: V = mean(M) to assign a row vector containing the mean values in each column of matrix M to V.
Additional information: Using mean recursively determines the average of all values in a matrix. ✦ Remember that the mean of a set of values is not necessarily equal to any of the individual values.
See also: median, std, cov, corrcoef

num = median(V)
Determines the median (middle) value of the elements of vector V.
Output: A numeric value is assigned to num.

Argument options: V = median(M) to assign the median values in each column of matrix M to row vector V.
Additional information: Because the sort command is used, this command can cost a lot of time for large M.
See also: mean, sort, std

num$_{out}$ = **mu2lin(num)**
Converts μ-law encoded audio value num into a linear audio signal value.
Output: A real value between -1 and 1 is returned.
Argument options: V$_{out}$ = mu2lin(V) or M$_{out}$ = mu2lin(M) to convert a vector or matrix of values.
Additional information: The input values should be integer values between 0 and 255.
See also: sound, lin2mu

str = **num2str(num)**
Converts numeric value num to a string.
Output: A string is assigned to str.
Additional information: Approximately four digits of precision are used when converting. To change this default, the sprintf command in the M-file for num2str must be changed. ✽ This command is especially useful when concatenating a numeric value into an already existing string.
See also: int2str, hex2num, setstr, sprintf

M$_{int}$ = **numeric(M$_{str}$)**
Converts characters in matrix M$_{str}$ to their numeric counterparts.
Output: A numeric matrix is assigned to M$_{int}$.
See also: int2str, num2str, string, isnumeric

V$_r$ = **poly(M)**
Computes the characteristic polynomial of the square matrix M.
Output: A row vector containing the coefficients of the characteristic polynomial is assigned to V$_r$.
Argument options: V$_c$ = poly(V$_r$), where V$_r$ is a row vector containing the coefficients of a polynomial to assign the roots of that polynomial to column vector V$_c$. ✽ V$_r$ = poly(V$_c$), where V$_c$ is a column vector containing the roots of a polynomial to assign the coefficients of that polynomial to row vector V$_r$.
Additional information: Coefficients are ordered in descending powers. ✽ Computed roots are in no specific order. ✽ For the most part poly and roots are the inverse functions for each other. ✽ For more information about the algorithms used, see the on-line help file.
See also: eig, roots, polyval, conv, residue

num = **polyarea(V$_x$, V$_y$)**
Computes the area of polynomial defined by vectors V$_x$ and V$_y$.
Output: A numerical value is assigned to num.

Argument options: V_{num} = polyarea(M_x, M_y) to compute the area of polygons defined by the matching columns of M_x and M_y. ✤ V_{num} = polyarea(M_x, M_y, int) to compute the area of polygons defined along the dimension int.
Additional information: If a polygon's edges intersect, the difference between the subareas traced in different directions is computed.
See also: polyval, poly, inpoly

V_r = polyfit(V_x, V_y, n)

Computes a polynomial of degree n to fit the data in vector V_x versus vector V_y in a least-squares sense.
Output: A row vector containing n+1 coefficients of a polynomial is assigned to V_r.
Additional information: The polynomial computed is such that polyval(V_r, V_x) is as close to V_y as possible. ✤ Coefficients are ordered in descending powers. ✤ For more information about the algorithms used, see the on-line help file.
See also: spline, polyval, poly, vander

num_p = polyval(V_c, num)

Computes the value of the polynomial whose coefficients are represented by vector V_c at the value num.
Output: A numeric value is assigned to num_p.
Argument options: V_p = polyval(V_c, V) to evaluate the polynomial at each element of V and assign the results to V_p. ✤ M_p = polyval(V_c, M) to evaluate the polynomial at each element of M and assign the results to V_p.
Additional information: Coefficients are ordered in descending powers. ✤ If you want to compute a matrix polynomial, use polyvalm.
See also: polyvalm, poly, polyfit, roots, residue

M_p = polyvalm(V_r, M)

Computes the value of the matrix polynomial for V_r, a row vector representing a polynomial, and M, a square matrix.
Output: A square matrix is assigned to M_p.
Additional information: Coefficients of the polynomial are ordered in descending powers. ✤ If you want to compute the value of each element of M substituted into a particular polynomial, use polyval.
See also: polyval, poly

M = rand(m, n)

Creates an m × n matrix of uniformly distributed random values between 0 and 1.
Output: A matrix with numeric elements is assigned to M.
Argument options: rand to represent a scalar between 0 and 1 whose value changes each time it is accessed. ✤ M = rand(n) to create an n × n matrix of random values. ✤ M = rand(size(M_1)) to create a random matrix of equal dimensions to M_1. ✤ rand('seed', num) to set the random seed used to num. The seed is 0 when MATLAB starts. ✤ rand('seed') to display the current seed value.
Additional information: To create random numbers with a normal distribution, use randn. ✤ rand and randn maintain separate seeds. ✤ For more information on the random number generator for rand, see the on-line help file.

See also: randn, randperm, sprandn, sprandsym

M = randn(m, n)
Creates an m × n matrix of normally distributed random values between 0 and 1.
Output: A matrix with numeric elements is assigned to M.
Argument options: randn to represent a scalar between 0 and 1 whose value changes each time it is accessed. ✤ M = randn(n) to create an n × n matrix of random values. ✤ M = randn(size(M_1)) to create a random matrix of equal dimensions to M_1. ✤ randn('seed', num) to set the random seed used to num. The seed is 0 when MATLAB starts. ✤ randn('seed') to display the current seed value.
Additional information: To create random numbers with a uniform distribution, use rand. ✤ rand and randn maintain separate seeds. ✤ randn uses the same generator as rand but transforms the values before returning them.
See also: rand, randperm, sprandn, sprandsym

[V_r, V_p, V_{dt}] = residue(V_a, V_b)
Computes the partial fraction expansion of the ratio of two polynomials V_b / V_a.
Output: A column vector of residues is assigned to V_r. A column vector of poles is assigned to V_p. A row vector of direct terms is assigned to V_{dt}.
Additional information: Coefficients of the polynomials are ordered in descending powers. ✤ For details on the definitions of the various results and lengths of the accompanying vectors, see the on-line help file. ✤ Depending on the polynomials V_a and V_b, the algorithms used in residue can be inefficient.
See also: deconv, poly, roots

saxis([num_{min}, num_{max}])
Sets the minimum and maximum amplitude values for the sound command.
Output: Not applicable.
Argument options: saxis('auto') to cause sound to scale the input data itself. ✤ V = saxis to return the current minimum and maximum amplitude values in two-element vector V.
See also: caxis, axis, sound

V = setdiff(V_1, V_2)
Returns the unique elements from vector (set) V_1 that are not present in vector V_2.
Output: A vector is assigned to V.
Argument options: [V, $V_{i,a}$, $V_{i,b}$] = setdiff(V_1, V_2) to also return index vectors for the elements returned from M_a and M_b. ✤ M = setdiff(M_1, M_2, 'rows') to return a matrix with rows not in the intersection of M_1 or M_2.
See also: intersect, union, setdiff, ismember, unique

V = setxor(V_1, V_2)
Returns the unique elements which are not in the intersection of the vectors (sets) V_1 or V_2.
Output: A vector is assigned to V.

Argument options: [V, $V_{i,a}$, $V_{i,b}$] = setxor(V_1, V_2) to also return index vectors for the elements returned from M_a and M_b. ✦ M = setxor(M_1, M_2, 'rows') to return a matrix with rows not in the intersection of M_1 or M_2.
See also: intersect, union, setdiff, ismember, unique

sound(V)

Converts the signal represented by vector V into sound.
Output: Not applicable.
Additional information: V is automatically scaled to fit the ranges of your hardware. To override this scaling, use the saxis command.
See also: saxis

V_{sy} = spline(V_x, V_y, V_{sx})

Computes the corresponding y elements to the finely spaced x elements in V_{sx}, using cubic spline interpolation on the x and y values in V_x and V_y.
Output: A vector of equal dimension to V_x is assigned to V_{sy}.
Argument options: num_{sy} = spline(V_x, V_y, num_{sx}) to interpolate the single y elements num_{sy}.
Additional information: Typically, V_x and V_y contain more widely and erratically spaced values than V_{sx} and V_y. ✦ Use the plot function to plot the values in V_{sx} and V_{sy}.
See also: interp1, polyfit, interp3, interpn

num = std(V)

Calculates the standard deviation of the elements of vector V.
Output: A numeric value is assigned to num.
Argument options: V = mean(M) to assign a row vector containing the standard deviation of each column of matrix M to V.
Additional information: The *sample* standard deviation is computed.
See also: mean, median, cov, corrcoef

M_{str} = str2mat(str_1, ..., str_n)

Creates a string matrix out of the strings str_1 through str_n.
Output: A matrix, evaluated as strings, is assigned to M_{str}.
Argument options: M_{str} = str2mat($M_{str,1}$, ..., $M_{str,n}$) to create a larger string matrix from smaller string matrices $M_{str,1}$ through $M_{str,n}$.
Additional information: Each individual parameter is treated as a separate row of the string matrix. ✦ Each string is padded with trailing blanks to match the length of the longest string. ✦ Only 11 parameters can be used in str2mat.
See also: int2str, num2str, isstr, setstr

num = str2num(str)

Converts the string str to the appropriate numeric value.
Output: A numeric value is assigned to num.
Additional information: str may contain digits, a decimal point, a leading + or -, an e power factor, or the imaginary units i or j.
See also: num2str, hex2num, sscanf

str = **strcat**(str$_1$, ..., str$_n$)

Concatenates n strings into a single string.
Output: A string is assigned to str.
Argument options: M$_s$ = strcat(M$_{s,1}$, ..., M$_{s,n}$) to concatenate the respective strings in structure arrays M$_{s,1}$ through M$_{s,n}$.
See also: strvcat

int = **strcmp**(str$_1$, str$_2$)

Compares the string str$_1$ to str$_2$.
Output: If the two strings are identical, a value of 1 is assigned to int. Otherwise, a value of 0 is assigned to int.
Additional information: strcmp is case sensitive. Leading or trailing blanks are also compared. ✤ This command behaves differently from the conventional C language routine.
See also: strncmp, isstr, setstr, findstr, isletter, strrep, lower, upper

str = **string**(M$_{int}$)

Converts integer values in M$_{int}$ to a string.
Output: A string is assigned to str.
See also: int2str, num2str, numeric

V$_{ind}$ = **strmatch**(str, M$_{str}$)

Examines each row of string matrix M$_{str}$ to see if it begins with string str.
Output: A vector containing the matching row indices is assigned to V$_{ind}$.
Argument options: V$_{ind}$ = strmatch(str, M$_{str}$, 'exact') to return indices only if the match is exact.
Additional information: strmatch is case sensitive. Leading or trailing blanks are also compared.
See also: char, strncmp, findstr

int = **strncmp**(str$_1$, str$_2$, int)

Compares the first int characters of strings str$_1$ and str$_2$.
Output: If the two substrings strings are identical, a value of 1 is assigned to int. Otherwise, a value of 0 is assigned to int.
Additional information: strncmp is case sensitive. Leading or trailing blanks are also compared. routine.
See also: strcmp, isstr, setstr, findstr, isletter, strrep, lower, upper

str$_{new}$ = **strrep**(str$_1$, str$_2$, str$_3$)

Replaces all occurrences of string str$_2$ with str$_1$ in string str$_3$.
Output: A new string is assigned to str$_{new}$.
Additional information: If no occurrences of str$_2$ are found, the original string is returned.
See also: findstr, strcmp

str_{new} = **strtok(str, [char])**

Returns the portion of string str up to but not including the first occurrence of the character char.
Output: A new string is assigned to str_{new}.
Argument options: str_{new} = strtok(str) to check for "spaces". See the entry for isspace for the broader definition of spaces. ✦ str_{new} = strtok(str, [$char_1$, ..., $char_n$]) to check for the first occurrence of any of n characters.
Additional information: If no occurrences of char are found, the original string is returned. ✦ If strings containing more than one character are supplied in the second parameter, only the first character of each is used.
See also: findstr, strrep, isspace

M_{str} = **strvcat(str_1, ..., str_n)**

Vertically concatenates n strings into a single string matrix.
Output: A string matrix is assigned to M_{str}.
Argument options: M_{str} = strcat($M_{str,1}$, ..., $M_{str,n}$) to concatenate n string matrices, creating arbitrarily large string matrices.
Additional information: If the input strings are of differing lengths, padding is automatically performed.
See also: char, strcat

num_{int} = **trapz(V_x, V_y)**

Computes the definite integral of the function suggested by the values of vector V_y with respect to the values in the function V_x using trapezoidal numerical integration.
Output: A numerical value is assigned to num_{int}.
Argument options: num_{int} = trapz(V_y) to compute the integral assuming a starting point of 0 and a constant x value spacing of 1. ✦ V_{int} = trapz(V_x, M_y), where V_x is a column vector and M_y is a matrix with as many rows as V_x has elements, to compute the integrals of each column of M_y. The resulting row vector is assigned to V_{int}.
Additional information: Typically, a more accurate result is obtained from evenly spaced sample points. ✦ In the one parameter case, if the spacing is not consistently 1, the error can be overcome by multiplying the result by the true spacing.
See also: quad, quad8, sum

V = **union(V_1, V_2)**

Returns the unique elements which appear in either of the vectors (sets) V_1 or V_2.
Output: A vector is assigned to V.
Argument options: [V, $V_{i,a}$, $V_{i,b}$] = union(V_1, V_2) to also return index vectors for the elements returned from M_a and M_b. ✦ M = union(M_1, M_2, 'rows') to return a matrix with rows appearing in either M_1 or M_2.
See also: intersect, setdiff, setxor, ismember, unique

V_{out} = **unique(V)**

Removes duplicate elements from vector (set) V.
Output: A vector is assigned to V_{out}.

Argument options: [V_{out}, $V_{i,a}$, $V_{i,b}$] = unique(V) to also return index vectors for the elements returned from M_a and M_b. ✤ M_{out} = unique(M, 'rows') to return the unique rows of M_1.
See also: union, setdiff, setxor, setdiff, ismember

str_{uc} = **upper**(str)
Converts all lowercase characters in str to uppercase.
Output: A completely uppercase string is assigned to str_{uc}.
Additional information: Characters already in uppercase are unaffected by upper.
✤ The resulting string is of equal length to the input string.
See also: lower, isstr, isletter, strcmp

MATLAB Resources

This chapter presents a tour of various MATLAB resources. The first three stops are right in MATLAB itself, the others are available over the Internet.

- tour - tour of all MathWorks products.
- demo - MATLAB demonstrations.
- helpdesk - hypertext-based help system.
- **World Wide Web site:** for The MathWorks
- **FTP site** with user-contributed code, etc.
- **Newsletter:** MATLAB *News and Notes*.
- **Books:** A list of third-party books.
- **E-mail:** MATLAB Digest
- **Newsgroup:** comp.soft-sys.matlab

Resources within MATLAB

MATLAB comes with two excellent demonstration facilities which highlight the features of MATLAB, Simulink, and the various toolboxes - all using interesting examples.

To start the *MathWorks tour* type:

```
>> tour
```

To start the MATLAB *demo* type:

```
>> demo
```

MATLAB comes with a hypertext help system, which uses your Web-browser to display its information. To start the MATLAB *help desk* type:

```
>> helpdesk
```

You can now navigate about the help desk by clicking on the hypertext links or performing a keyword search of the content.

The MATLAB Web Site

The MathWorks can be reached on the World Wide Web. From within your Web browser (e.g., Internet Explorer, Netscape, etc.) *Open* the following "URL":

    http://www.mathworks.com

From here you can find information on the various toolboxes and blocksets available for MATLAB and Simulink, as well as user-contributed M-files and lists of third-party books.

Under *Solution Search*, you can search for answers to technical questions you have about MATLAB.

Under *News & Notes* you can view The MathWorks' newsletter and back issues on-line. (You can also order a free subscription of the paper form.)

Under MATLAB *Digest* you can view an electronic digest of recent MATLAB happenings, which you can have e-mailed to you whenever new material comes out.

The MATLAB Newsgroup

There is a Usenet newsgroup for MATLAB called comp.soft-sys.matlab, which is regularly contributed to by MATLAB users and developers. To access it from your Web browser, enter the following location:

    news:comp.soft-sys.matlab

Index

!, 391
<, 125
<=, 126
>, 126
>=, 126
\, 124
~, 127
~=, 127
', 125
*, 7, 124
+, 7, 123
-, 7, 123
.\, 124
.', 125
.*, 124
./, 125
.^, 125
/, 124
==, 126
%, 391
&, 127
^, 7, 125
|, 127

abs, 127
Accelerator, 359
acos, 128
acosh, 128
acot, 128
acoth, 128
acsch, 128
addpath, 391
airy, 129
align, 359
all, 391
AmbientLightColor, 290
AmbientStrength, 290
angle, 129

ans, 392
any, 392
area, 189
asec, 128, 129
asech, 129
asin, 129
asinh, 130
assignin, 392
atan, 130
atan2, 130
atanh, 130
AutomaticFileUpdates, 290
axes, 290
axis, 291

BackFaceLighting, 291
BackGroundColor, 359
BackingStore, 292
balance, 31
bar, 189
bar3, 222
bar3h, 223
barh, 189
base2dec, 458
bessel, 130
besseli, 130
besselj, 131
besselk, 131
bessely, 131
beta, 131
betainc, 131
betaln, 131
bicg, 68
bicgstab, 68
bin2dec, 459
bitand, 459
bitcmp, 459
bitget, 459

bitmax, 459
bitor, 459
bitset, 460
bitxor, 460
blanks, 460
Box, 292
break, 392
brighten, 292
BusyAction, 292
ButtonDownFnc, 292

calendar, 393
CallBack, 359
CallbackObject, 292
CameraPosition, 292
CameraPositionMode, 293
CameraTarget, 293
CameraTargetMode, 293
CameraUpVector, 293
CameraUpVectorMode, 293
CameraViewAngle, 293
CameraViewAngleMode, 293
cart2pol, 190
cart2sph, 223
cat, 441
caxis, 293
cbedit, 360
cd, 418
CData, 294
CDataMapping, 294
cdf2rdf, 31
ceil, 132, 136
cell, 441
cell2struct, 442
celldisp, 442
cellplot, 442
cgs, 68
char, 460
Checked, 360
Children, 294
choices, 360
chol, 32
cholinc, 32
cla, 294
clabel, 224
clc, 393
clear, 393
clf, 294

CLim, 294
CLimMode, 295
Clipping, 295
clock, 393
close, 295
CloseRequestFcn, 295
collect, 154
colmmd, 69
Color, 295
colorbar, 295
colordef, 296
Colormap, 296
colormap, 296
ColorOrder, 296
colperm, 69, 132
colspace, 155
comet, 190
comet3, 190
compan, 32
compass, 191
compose, 155
computer, 393
cond, 32
condeig, 33
condest, 33
conj, 132
constr, 98
contour, 224
contour3, 224
contourc, 225
contourf, 225
contrast, 297
conv, 460
convhull, 225
corrcoef, 461
cos, 132
cosh, 132
cosint, 155
cot, 132
coth, 133
cov, 461
cplxpair, 461
cputime, 393
CreateFcn, 297
cross, 133
csc, 133
csch, 133
ctlpanel, 360

cumprod, 133
cumsum, 133
cumtrapz, 103
CurrentAxes, 297
CurrentCharacter, 297
CurrentFigure, 297
CurrentObject, 297
CurrentPoint, 298
cylinder, 225

DataAspectRatio, 298
DataAspectRatioMode, 298
date, 394
datenum, 394
datestr, 394
datetick, 298
datevec, 394
dbclear, 434, 435
dbcont, 434
dbdown, 434
dblquad, 103
dbquit, 434
dbstack, 434
dbstatus, 435
dbstep, 435
dbtype, 435
dbup, 435
deblank, 461, 464
dec2base, 462
dec2bin, 462
dec2hex, 461
deconv, 462
del2, 462
delaunay, 226
delete, 418
DeleteFcn, 298
demo, 394
det, 33
diag, 72, 134
dialog, 360
Diary, 298
diary, 418
DiaryFile, 298
diff, 134, 155
diffuse, 226
DiffuseStrength, 299
digits, 156
dir, 418

disp, 394, 462
Dithermap, 299
DithermapMode, 299
dlmread, 418
dlmwrite, 419
dmperm, 69
doc, 462
dos, 395
dot, 134
dragrect, 360
DrawMode, 299
drawnow, 299
dsearch, 226
dsolve, 156

Echo, 299
echo, 395
EdgeColor, 299
EdgeLighting, 299
edit, 395
Editing, 300
editpath, 395
eig, 33
eigs, 69
ellipj, 134
ellipke, 135
Enable, 361
eomday, 395
eps, 395
EraseMode, 300
erf, 135
erfc, 135
erfcx, 135
erfinv, 135
error, 396
errorbar, 191
errordlg, 361
ErrorMessage, 300
etime, 396
eval, 462, 463
evalin, 396
exist, 419
exp, 136
expand, 156
expint, 136
expm, 136
Extent, 300
eye, 136

ezplot, 157

FaceColor, 300
FaceLighting, 300
Faces, 300
FaceVertexCData, 301
factor, 136, 157
fclose, 419
feather, 191
feof, 419
ferror, 419
fft, 463
fft2, 463
fftdemo, 463
fftshift, 464
fgetl, 419
fgets, 420
fieldnames, 442
figure, 301
filesep, 396
fill, 191
find, 396
finverse, 157
FixedColors, 301
flipdim, 442
fliplr, 137
flipud, 137
floor, 137
flops, 397
fmin, 98
fmins, 98
FontAngle, 301
FontName, 301
FontSize, 302
FontUnits, 302
FontWeight, 302
fopen, 420
for/end, 397
ForeGroundColor, 361
Format, 302
format, 420
FormatSpacing, 302
fplot, 192
fplotdemo, 192
fprintf, 421
fread, 421
frewind, 421
fscanf, 422

fseek, 422
fsolve, in *Optimization Toolbox*, 83
ftell, 422
full, 69
fullfile, 397
function, 397
funm, 137
funtool, 157
fwrite, 422
fzero, 84

gallery, 33
gallery(3), 34
gallery(5), 34
gamma, 137
gammainc, 137
gammaln, 138
gca, 302
gcd, 138
gcf, 302
gco, 303
get, 303
getenv, 397
getfield, 442
getframe, 245
ginput, 303
global, 398
gmres, 69
gplot, 192
gradient, 138
graymon, 303
grid, 292, 304
griddata, 226
GridLineStyle, 304
gtext, 304
guide, 361

hadamard, 34
HandleVisibility, 304
hankel, 34
help, 398
helpdesk, 398
helpdlg, 361
hess, 34
hex2dec, 464
hex2num, 464
hidden, 304
hilb, 34

hist, 193
hold, 304
home, 398
HorizontalAlignment, 305, 361
horner, 157
hsv, 305
hsv2rgb, 305, 314

i, 138
if/elseif/else/end, 398
ifft, 464
ifft2, 465
imag, 138
image, 305
imagesc, 306
ind2sub, 443
Inf, 139
inline, 399
inmem, 399
inpolygon, 465
input, 423
inputdlg, 362
inputname, 399
int, 157
int2str, 465
IntegerHandle, 306
interp1, 465
interp2, 466
interp3, 466
interpft, 467
interpn, 466
Interpreter, 306
Interruptible, 306
intersect, 467
inv, 35
InvertHardCopy, 306
invhilb, 35
ipermute, 443
iscell, 443
isempty, 399
isequal, 139
isfinite, 139
isglobal, 399
ishold, 306
isieee, 399
isinf, 139
isletter, 467
islogical, 399

ismember, 467
isnan, 139
isnumeric, 140
isprime, 139
isspace, 467
issparse, 70
isstr, 467
isstruct, 443

j, 140
jacobian, 158
jordan, 158

keyboard/return, 400
KeyPressFnc, 306
kron, 467

Label, 362
lambertw, 158
lasterr, 400
Layer, 307
lcm, 140
legend, 307
legendre, 140
length, 400
light, 307
lin2mu, 468
line, 307
LineStyle, 308
LineStyleOrder, 308
LineWidth, 308
linspace, 140, 141
listdlg, 362
load, 423
log, 141
log10, 141
log2, 141
logical, 400
loglog, 193
logm, 141
lookfor, 400
lower, 468
lp, 99
ls, 423
lscov, 35
lu, 35
luinc, 35

magic, 468
maple, 158
Marker, 308
MarkerEdgeColor, 308
MarkerFaceColor, 308
MarkerSize, 308
mat2str, 468
matlabrc, 400
matlabroot, 401
Max, 362
max, 141
mean, 468
median, 468
menu, 401
MenuBar, 309
menuedit, 362
mesh, 226
meshc, 227
meshgrid, 227
MeshStyle, 309
meshz, 227
mexext, 401
mfilename, 401
mfun, 159
mhelp, 159
Min, 363
min, 142
MinColormap, 309
mod, 142
more, 401
movie, 245
moviein, 245
msgbox, 363
mu2lin, 469

Name, 309
NaN, 142
nargchk, 401
nargin, 402
nargout, 402
nchoosek, 142
ndgrid, 443
ndims, 443
newplot, 309
NextPlot, 309
nextpow2, 143
nnls, 35
nnz, 70

nonzeros, 70
norm, 36
NormalMode, 309
normest, 36
now, 402
null, 36
num2cell, 443
num2str, 469
NumberTitle, 309
numden, 159
numeric, 469
nzmax, 70

ode113, 119
ode15s, 119
ode23, 120
ode23s, 120
ode45, 120
odedemo, 121
odeget, 121
odeset, 121
ones, 143, 148
orient, 309
orth, 36

pack, 402
PaperOrientation, 310
PaperPosition, 310
PaperPositionMode, 310
PaperSize, 310
PaperType, 310
PaperUnits, 310
Parent, 311
pascal, 36
patch, 311
path, 402
pathsep, 403
pause, 403
pcg, 70
pcode fnc), 403
pcolor, 227
perms, 143
permute, 444
pi, 143
pie, 193
pie3, 193
plot, 194
plot3, 194

PlotBoxAspectRatio, 311
PlotBoxAspectRatioMode, 311
plotyy, 194
Pointer, 311
PointerLocation, 312
PointerShapeCData, 312
PointerShapeHotSpot, 312
PointerWindow, 312
pol2cart, 195
polar, 195
poly, 469
poly2sym, 159
polyarea, 469
polyder, 103
polyeig, 37
polyfit, 470
polyval, 470
polyvalm, 470
Position, 312, 363
pow2, 143
pretty, 159
primes, 143
print, 312
printopt, 313
prism, 313
procread, 160
prod, 143
Profile, 313
profile, 436
ProfileFile, 313
ProfileInterval, 313
profsumm, 436
Projection, 314
propedit, 363
ps, 37

qmr, 71
qr, 37
qrdelete, 37
qrinsert, 37
quad, 104
quad8, 104
quaddemo, 104
questdlg, 364
quit, 403
quiver, 228
quiver3, 228
qz, 38

rand, 470
randn, 471
randperm, 144
rank, 38
rat, 144
rats, 144
rbbox, 364
rcond, 38
real, 144
realmax, 145
realmin, 145
refresh, 314
rem, 145
Renderer, 314
repmat, 444
reset, 314
reshape, 145
residue, 471
Resize, 314
ResizeFcn, 314
return, 403
ribbon, 228
rmfield, 444
rmpath(str), 403
root object, 315
roots, 84
rose, 195
rosser, 38
rotate, 315
Rotation, 315
round, 145
rref, 38
rrefmovie, 39
rsf2csf, 39
rsums, 160

save, 423
saxis, 471
schur, 39
ScreenDepth, 315
ScreenSize, 315
sec, 145
sech, 146
Selected, 315
SelectionHighlight, 316
SelectionType, 316
semilogx, 195
semilogy, 195

Separator, 364
set, 316
setdiff, 471
setfield, 444
setxor, 471
ShareColors, 316
shiftdim, 444
ShowHiddenHandles, 316
signum, 146
simple, 160
simplify, 160
sin, 146
sinh, 146
sinint, 160
size, 404
slice, 228
solve, 161
sort, 146
sortrows, 147
sound, 472
spalloc, 71
sparse, 71
spaugment, 71
spconvert, 72
specular, 229
SpecularColorReflectance, 316
SpecularExponent, 316
SpecularStrength, 317
speye, 72
spfun, 72
sph2cart, 229
sphere, 229
spinmap, 317
spline, 472
spones, 72
spparms, 73
sprand, 73
sprandn, 73
sprandsym, 73
sprank, 74
sprintf, 424
spy, 74
sqrt, 147
sqrtm, 147
squeeze, 444
sscanf, 424
stairs, 196
std, 472

stem, 229
stem3, 229
str2mat, 472
str2num, 472
strcat, 473
strcmp, 473
String, 317, 364
string, 473
strmatch, 473
strncmp, 473
strrep, 473
strtok, 473
struct, 444
struct2cell, 445
strvcat, 474
Style, 317, 364
sub2ind, 445
subplot, 317
subs, 161
subspace, 39
sum, 147
surf, 230
surface, 317
surfc, 230
surfl, 230
svd, 39
svds, 74
switch/case/otherwise/end, 404
sym, 161
sym2poly, 161
symmmd, 74
symrcm, 75
symsum, 162

Tag, 318
tan, 147
tanh, 148
taylor, 162
tempname, 424
terminal, 424
text, 318
tic, 404
TickDir, 318, 319
TickLength, 319
Title, 319
title, 319
toc, 404
toeplitz, 40

trace, 40
trapz, 474
tril, 40
trimesh, 231
trisurf, 231
triu, 40
tsearch, 231
Type, 319
type, 425

uicontrol, 365
uigetfile, 365
uimenu, 366
uiputfile, 366
uiresume, 366
uiwait, 366
union, 474
unique, 474
Units, 319
unix, 404
upper, 475
UserData, 319

Value, 367
vander, 40
varargin, 405
varargout, 405
ver, 405
VertexNormals, 319
VerticalAlignment, 320
Vertices, 320
view, 320
Visible, 320
voronoi, 231
vpa, 162

waitfor, 367
warndlg, 367
warning, 405
waterfall, 232
web, 406
weekday, 406
what, 425
whatsnew, 406
which, 406
while/end, 406
white, 320
whitebg, 321

who, 407
whos, 407
wilkinson, 41
WindowButtonDownFnc, 321
WindowButtonMotionFnc, 321
WindowButtonUpFnc, 321
WindowStyle, 321
wk1read, 425
wk1write, 425

XAxisLocation, 321
XColor, 322
XData, 322
XDir, 322
XGrid, 322
XLabel, 322
xlabel, 322
XLim, 322
XLimMode, 323
XScale, 323
XTick, 323
XTickLabel, 323
XTickLabelMode, 323
XTickMode, 323

YAxisLocation, 324
YColor, 324
YData, 324
YDir, 324
YGrid, 324
YLabel, 324
ylabel, 324
YLim, 325
YLimMode, 325
YScale, 325
YTick, 325
YTickLabel, 325
YTickLabelMode, 325
YTickMode, 325

ZColor, 326
ZData, 326
ZDir, 326
zerodemo, 84
zeta, 162
ZGrid, 326
ZLabel, 326
zlabel, 326

ZLim, 326
ZLimMode, 326
zoom, 327
ZScale, 327

ZTick, 327
ZTickLabel, 327
ZTickLabelMode, 327
ZTickMode, 327